D0914149

Stochastic Models In Biology

STOCHASTIC MODELS IN BIOLOGY

NARENDRA S. GOEL

Xerox Corporation
Rochester, New York

NIRA RICHTER-DYN

Department of Mathematical Sciences
Tel-Aviv University
Ramat-Aviv, Israel

ACADEMIC PRESS New York San Francisco London 1974
A Subsidiary of Harcourt Brace Jovanovich, Publishers

ACADEMIC PRESS, INC.
111 Fifth Avenue, New York, New York 10003

United Kingdom Edition published by
ACADEMIC PRESS, INC. (LONDON) LTD.
24/28 Oval Road, London NW1

Library of Congress Cataloging in Publication Data

Goel, N. S.
Stochastic models in biology.

Includes bibliographies.
1. Biology–Mathematical models. 2. Stochastic processes. I. Richter-Dyn, Nira, joint author. II. Title. [DNLM. 1. Biometry. 2. Models, Biological. 3. Probability. QT34 G595s 1974]
QH323.5.G6 574'01'82 73-18938
ISBN 0-12-287460-9

PRINTED IN THE UNITED STATES OF AMERICA

Contents

Preface

The interest in mathematical modeling of biological phenomena has grown considerably during the last two decades or so. This interdisciplinary field has attracted both biologists and physical scientists. Biologists are realizing that data by itself is not knowledge and are attempting to quantitatively describe and explain various biological phenomena and observations. Physical scientists are expanding their circle of inquisitiveness and challenge to include complex biological phenomena very much like they did in the past for astronomical phenomena. The basic approach taken by them is to treat a biological system as a physical state in which there is an interaction between energy and matter. This type of physical state is highly organized and complex, like a machine in which every part is more complex, more efficient, more precise, and much smaller than one is used to in physical systems. In addition to getting the satisfaction of providing insight into the biological phenomena, physical scientists are curious to know whether the present techniques of physical sciences are sufficient to explain biological phenomena. An added potential bonus is the emergence of new techniques which can potentially be powerful also for physical systems.

This monograph is an attempt to demonstrate the usefulness of the theory of stochastic processes in understanding biological phenomena at various levels of complexity—from the molecular to the ecological level. The modeling of biological systems via stochastic processes allows the incorporation of effects of secondary factors for which a detailed knowledge is missing.

Preface

In the first two chapters of this monograph we have presented the mathematical analysis used in the later chapters in modeling various biological systems. We have attempted to make the chapters self-contained and also to bring under one umbrella the various results derived by different authors using a variety of techniques and notations. Known mathematical results are compiled in the form of several tables for those who are not at ease with mathematical analysis. In other chapters, where models of various biological phenomena are discussed, introductory reviews of these phenomena are given for readers with a minimum biological background. For the readers who want to pursue the modeling further, additional references are given in each chapter.

We hope that this monograph will find audience among both professionals and amateurs. It should help an amateur to decide whether modeling of biological systems and the accompanying challenge are worth his effort and are enough of a stimulant for his creative thinking. It should allow a professional physical scientist to appreciate the problems of his colleague in biological sciences and a biologist to appreciate the methods used by his peer in physical sciences.

This monograph was, in most part, written while the authors were at the Institute for Fundamental Studies at the University of Rochester. We wish to thank its Director, Professor Elliott W. Montroll, for his hospitality and continuous encouragement. Both were essential for the completion of this work. Finally, we acknowledge the National Science Foundation for providing partial financial support for this work.

Stochastic Models In Biology

1

Introduction

Since the early years of this century and especially in the last two decades or so, a large number of investigators have attempted to understand biological phenomena using physical principles and both physical and mathematical techniques. The basic approach is to treat a biological system as a physical state in which there is an interaction between energy and matter. However, this type of physical state is highly organized and complex, like a machine in which every part is more complex, more efficient, more precise, and much smaller than one is used to in physics. Because of this complexity, ignorance, and lack of detailed knowledge, it is desirable to analyze biological systems via probabilistic models rather than deterministic ones. In this monograph we use two basically different approaches in making these probabilistic models. In both of these approaches, it is assumed that the behavior of the system is memoryless or Markovian; i.e., its future depends only on its present state and not on the past.

One of these approaches consists of taking time to change continuously and of assuming that the system can be described by a set of random variables which change either discretely or continuously in their state space. For the former case of discrete changes, the state of the variables is described by a set of discrete states. The system is characterized by a discrete probability distribution function which is the probability of the variables being in a certain state at a given time. The system is assigned a set of transition probabilities per unit time for the process to go from one state to another. The form of these transition probabilities depends on the process and reflects the nature of the interactions in the complex system. The equation describing the evolution in time of the probability distribution of the system's variables is a so-called master equation, which is a differential-difference equation, first order in time. Such equations have also been analyzed by physical scientists, especially those working in chemical kinetics and statistical mechanics, and by applied mathematicians and statisticians. In general, one requires many random variables to describe a system, but many systems are inherently describable or can be approximately described by a single random variable. Further, in case the interactions are such that transitions are allowed only between nearest neighboring states, then the random process describing such systems is called a univariate birth and death process. When the random variables defining the system change continuously instead of discretely, the system is characterized by a probability density function that satisfies a second-order partial differential equation, the so-called Fokker–Planck equation or diffusion equation. The form of the coefficients in the equation reflects the nature of the interactions in the complex system. This equation has been studied in connection with a variety of nonbiological phenomena, e.g., fluctuations in the stock market, Brownian motion, diffusion, heat and electric conduction, noise in electrical systems and lasers, and so on. This equation can be converted into an equation that is very similar to the Schrödinger equation of quantum mechanics.

In the second approach to the modeling of complex biological systems, one picks up one or two important components of a system whose evolution in time is describable by one or more known deterministic dynamical equations. The remaining components are assumed to behave like noise, e.g., white noise, and affect the system accordingly. This procedure converts the deterministic dynamical variables into random variables whose probability density satisfies the Fokker–Planck equation. This approach is quite general and can be applied to many complex systems. Of course, the resulting Fokker–Planck equation may not be amenable to an analytical solution in all cases.

There are other approaches that have been used in making probabilistic models in biology; these approaches are mentioned in Chapter 11. In this monograph, we present the two approaches mentioned above in a somewhat limited fashion. We limit ourselves to one random variable with time-independent transition probability rates in discrete state space, and time-independent coefficients of the diffusion equation in continuous state space. For some systems, this is not a limitation; for other systems, we hope our analysis will give an idea of the nature of the modeling required, and break the initial barrier one runs into in familiarizing a new topic.

In Chapter 2 we give a mathematical analysis of univariate birth and death processes. We have attempted to make the chapter self-contained and also to bring under one umbrella the various results derived by different authors using a variety of techniques and notations. We hope that a reader with a minimal background in probability theory and applied mathematics (e.g., a graduate student in physical or mathematical sciences) will be able to follow the detailed derivations easily. For those who are unable to do so, we have presented the results in the form of tables for quantities that provide insight into the process. The input to these tables is the forms of the transition probability rates between various states, the initial state of the systems, and the properties of special states, if any. In Chapter 11 we briefly discuss multivariate processes and provide references where the necessary mathematical analysis can be found. In Chapter 3 we give a detailed analysis of the Fokker–Planck equation for a univariate process in the same fashion and style as for the master equation for birth and death processes. In Chapter 3 we also describe and analyze the second approach for modeling a system, and give the procedure for obtaining the coefficients of the Fokker–Planck equation from a given deterministic equation when the noise is taken to be a white noise. We discuss the confusion and controversy that exist regarding the procedure, which depends on the type of calculus used. We also tabulate a set of dynamical equations for which the corresponding probabilistic equations are amenable to analytical solutions.

The remaining chapters, with the exception of Chapter 11, are devoted to illustrating the use of the above-mentioned approaches to a variety of biological systems. Chapters 4–6 deal with the growth and extinction of populations of one or more biological species. Chapter 4 deals with the population of a single species, and Chapters 5 and 6 with the populations of two and many species, respectively. The two-species systems discussed are of host populations of special type: epidemics and bacteriophage. We have chosen to illustrate these systems first, because we felt that these require minimal biological background, and thus a wider audience can appreciate the modeling approach and the insight into these systems gained thereby. Chapter 7 is devoted to population genetics where we do not regard the individuals constituting the population identical. Instead we consider the structure of the population at a somewhat microscopic scale by taking into account the differences among the individuals of the same species at the genetic level. Population genetics has been studied since Mendel's era and there exists an extensive literature on its various aspects. We have attempted to select the most important aspects which can be appreciated with a minimum background in genetics, which we have provided in order to reach a wider audience.

In Chapter 8 we discuss another complex system at a different level of complexity: the nervous system. For this system, the number of basic elements, the neurons, depends on the organism and on the specific part of the nervous system, but for a reasonably complex organism, it is in the hundreds of thousands and millions. We show that, by making reasonable assumptions about the effects of all the neurons on a single neuron, the behavior of a single neuron can be studied.

In Chapter 9 we model some biological phenomena at the molecular level. These phenomena are the conformational changes of biopolymers, biosynthesis, and enzyme kinetics. In Chapter 10 we investigate one of the basic processes in photosynthesis, namely, the motion of the photoexcitation (exciton) originating from the absorption of a photon anywhere in the network of chlorophyll molecules, through this network. A detailed investigation of biological phenomena at the molecular level and of photosynthesis cannot be carried out by using the mathematical analyses of Chapters 2 and 3, since these systems cannot be described by a single random variable. Therefore, the corresponding two chapters are less comprehensive than the preceding ones. In the Epilogue we take another look into the various systems studied in this monograph and also point out other biological systems that can be studied by using techniques which are similar to but different from the techniques of Chapters 2 and 3. In each chapter, we provide a list of suggested additional references for further study.

Finally, we hope that the experts in the field of applied statistics and theoretical physics will find some food for thought and new areas where their expertize can be applied usefully, and that the biologists will be motivated to learn some basic techniques of applied mathematics, physics, and statistics. We apologize to the experts in physical and mathematical sciences for our being somewhat pedagogic in Chapters 2 and 3, and to those in biological sciences for our introductions to various biological phenomena in the remaining chapters. We hope they will appreciate our motive of reaching a wider audience in doing so.

2

Random Processes Continuous in Time and Discrete in State Space

In this chapter we introduce and discuss the theory of random processes that change continuously in time and can be defined by a single random variable which is confined to a discrete set of states, with transitions allowed only between nearest neighboring states. One simple example is the population of a species; the population changes continuously in time, but its size changes in steps of one. We do not attempt to review all aspects of these processes, which are known as birth and death processes. At the end of this chapter we provide a list of review articles and books which serve this purpose. We restrict ourselves to a brief introduction of the formulation, discuss various general aspects and techniques with illustrations, and summarize the known relevant results. We have attempted to make the chapter self-contained and comprehensive, and also to bring together the results derived by various authors using a variety of techniques and notations.

2.0 Notation and Formulation

Let us designate the discrete states of the random variable by the integers $\ldots -2, -1, 0, 1, 2, \ldots$. Let us further assume that in the time interval $(t, t+\Delta t)$, the probability of transition from state n to $n+1$ is $\lambda_n \Delta t + O(\Delta t)$, and from state n to $n-1$ is $\mu_n \Delta t + O(\Delta t)$ [and hence the probability of staying in the state n is $1-(\lambda_n+\mu_n)\Delta t + O(\Delta t)$] where $\lim_{\Delta t \to 0} O(\Delta t)/\Delta t = 0$. Under these conditions, if the process was initially $(t=0)$ in the state m, then the probability $P_{n,m}(t)$ of the process being in the state n at time t satisfies two types of difference equations:

$$\begin{aligned} P_{n,m}(t+\Delta t) &= [\lambda_{n-1} \Delta t + O(\Delta t)] P_{n-1,m}(t) \\ &\quad + [1-(\mu_n+\lambda_n)\Delta t + O(\Delta t)] P_{n,m}(t) \\ &\quad + [\mu_{n+1} \Delta t + O(\Delta t)] P_{n+1,m}(t) \end{aligned} \tag{1}$$

$$\begin{aligned} P_{n,m}(t+\Delta t) &= [\mu_m \Delta t + O(\Delta t)] P_{n,m-1}(t) + [1-(\mu_m+\lambda_m)\Delta t + O(\Delta t)] P_{n,m}(t) \\ &\quad + [\lambda_m \Delta t + O(\Delta t)] P_{n,m+1}(t) \end{aligned} \tag{2}$$

Dividing both sides of the equation above by Δt and taking the limit $\Delta t \to 0$, we obtain

$$dP_{n,m}(t)/dt = \lambda_{n-1} P_{n-1,m}(t) - (\mu_n+\lambda_n) P_{n,m}(t) + \mu_{n+1} P_{n+1,m}(t) \tag{3}$$

$$dP_{n,m}(t)/dt = \mu_m P_{n,m-1}(t) - (\mu_m+\lambda_m) P_{n,m}(t) + \lambda_m P_{n,m+1}(t) \tag{4}$$

These equations are differential equations in the time variable. In addition, Eq. (3) is a difference equation in the state of the process at time t while Eq. (4) is in the initial state. Equations (3) and (4) belong to a general class of equations known as *master equations*. The general form of a master equation results when, in the time interval $(t, t+\Delta t)$, the transitions allowed are from state n to not only states $n-1$ and $n+1$ but also to any other state. Equation (3) is a so-called forward master equation, and Eq. (4) a backward master equation.

These equations, each of which is a set of difference-differential equations, are to be solved subject to an initial condition and some boundary conditions. It is sufficient to solve them for the initial condition

$$P_{n,m}(0) = \delta_{n,m} \tag{5}$$

i.e., for the case when the process is initially in a definite state m, because the solution for an initially distributed process is related to the solution for the definite initial state. If $\not{p}_m$ denotes the probability of the process initially being in the state m (with $\sum_m \not{p}_m = 1$), then the solution $P_n(t)$ of the master equation is given by

$$P_n(t) = \sum_{\substack{\text{state} \\ \text{space}}} P_{n,m}(t) \not{p}_m \tag{6}$$

The boundary conditions depend on the allowed set of states. Basically, the random processes may be of two types: one in which there are no restrictions on the allowed set of states (we call such a process unrestricted), and the other in which there are restrictions in the sense that some states have special properties (such a process is called restricted). The appropriate boundary conditions for these two types of processes are given in the next two sections, where the solutions of the master equations and further properties are discussed. As we will see the calculation of $P_{n,m}(t)$ is somewhat difficult, and if one is interested only in the first few moments of n, one has to solve a simpler set of equations. For example, the differential equation for the first moment,

$$\langle n \rangle = \sum nP_{n,m}(t) \tag{7}$$

where the summation is on the allowed states, is

$$d\langle n \rangle/dt = \langle \lambda_n \rangle - \langle \mu_n \rangle \qquad \langle \lambda_n \rangle \equiv \sum \lambda_n P_{n,m} \tag{8}$$

This equation results if we multiply Eq. (3) by n, and sum both sides over n to obtain

$$d\langle n \rangle/dt = \left[\sum n\lambda_{n-1} P_{n-1,m} - \sum n\lambda_n P_{n,m}\right] + \left[\sum n\mu_{n+1} P_{n+1,m} - \sum n\mu_n P_{n,m}\right] \tag{9}$$

and note that since $\sum n\lambda_{n-1} P_{n-1,m} = \sum (n+1)\lambda_n P_{n,m}$, the expression in the first bracket in Eq. (9) is $\sum \lambda_n P_{n,m}$, with an analogous expression for the second bracket.

In general, the kth moment

$$\langle n^k \rangle = \sum n^k P_{n,m}(t) \tag{10}$$

satisfies the differential equation

$$\begin{aligned} d\langle n^k \rangle/dt &= \langle \{(n+1)^k - n^k\} \lambda_n \rangle - \langle \{n^k - (n-1)^k\} \mu_n \rangle \\ &= \sum_{i=0}^{k-1} \binom{k}{i} \{\langle n^i \lambda_n \rangle + (-1)^{k-i} \langle n^i \mu_n \rangle\} \end{aligned} \tag{11}$$

where

$$\binom{k}{i} = \frac{k!}{i!(k-i)!} \tag{12}$$

In particular, $\operatorname{var}(n) \equiv \langle n^2 \rangle - \langle n \rangle^2$ satisfies the differential equation

$$\frac{d}{dt}\operatorname{var}(n) = 2\{\langle n\lambda_n \rangle - \langle n\mu_n \rangle\} + (1 - 2\langle n \rangle)\langle \lambda_n \rangle + (1 + 2\langle n \rangle)\langle \mu_n \rangle \tag{13}$$

The initial condition for $\langle n^k \rangle$ follows from Eq. (5), i.e.,

$$\langle n^k \rangle (t=0) = m^k \tag{14a}$$

$$\langle n \rangle (t=0) = m, \qquad \operatorname{var}(n)(t=0) = 0 \tag{14b}$$

When λ_n and μ_n are at most linear in n, the right-hand side of Eq. (8) is devoid of a higher than first moment of n and (8) can be solved for $\langle n \rangle$. Likewise, Eq. (11) can be solved for $\langle n^k \rangle$. On the other hand, if λ_n and μ_n have terms in n of degree higher than 1, in general Eqs. (8) and (11) cannot be solved. For this case, Eq. (8) for $\langle n \rangle$ involves $\langle n^2 \rangle$, the equation for which involves $\langle n^3 \rangle$, and so on. This hierarchy of equations can only be solved approximately by making any of a variety of approximations. In Appendix A, in addition to giving the details of these approximations, we also give the solutions of Eqs. (8) and (13) for the solvable linear case.

We now proceed directly to the discussion of unrestricted birth and death processes.

2.1 Unrestricted Processes

The probabilities $P_{n,m}(t)$ describe the unrestricted process uniquely and completely. Therefore we discuss first a method for calculating $P_{n,m}(t)$.

Since the process is unrestricted, the boundary conditions on $P_{n,m}$, based on the physical requirement that in finite time changes in the state space are finite, are $P_{n,m}(t) = 0$ for $n \to \pm\infty$. Let us define a so-called generating function G by

$$G_m(t, z) = \sum_{n=-\infty}^{\infty} P_{n,m}(t) z^n \tag{1}$$

Since by definition $P_{n,m}(t)$ is the coefficient of z^n in the expansion of G_m in powers of z, by knowing the generating function we can calculate $P_{n,m}(t)$. We can also obtain the moments of the random variable, describing the state of the process, by successive differentiation of G:

$$\langle n^k \rangle_m(t) \equiv \sum_{n=-\infty}^{\infty} n^k P_{n,m}(t) = \left. \frac{\partial^k G(t, z)}{\partial (\ln z)^k} \right|_{z=1} \tag{2}$$

Here $\langle n^k \rangle_m(t)$ stands for the kth moment and the final equation follows from Eq. (1).

The generating function G can be calculated by noting (see Appendix B) that, for a given set of transition probabilities λ_n $[\equiv f_1(n)]$ and μ_n $[\equiv f_{-1}(n)]$, G satisfies the differential equation

$$\frac{\partial G(t, z)}{\partial t} = (z-1) f_1\left(z \frac{\partial}{\partial z}\right) G(t, z) + \left(\frac{1}{z} - 1\right) f_{-1}\left(z \frac{\partial}{\partial z}\right) G(t, z) \tag{3}$$

The initial condition on G corresponding to the initial condition (2.0.5) is

$$G_m(0, z) = \sum_{n=-\infty}^{\infty} P_{n,m}(0) z^n = z^m \tag{4a}$$

and the boundary condition is

$$G_m(t,1) = \sum_{n=-\infty}^{\infty} P_{n,m}(t) = 1 \tag{4b}$$

To obtain the expression for G, one has to write Eq. (3) for a given set of λ_n and μ_n, and then solve the resulting equation subject to the conditions (4). For example, when the transition probabilities are linear in n, i.e.,

$$\lambda_n = \lambda n + \nu, \qquad \mu_n = \mu n \tag{5}$$

$$f_1(z\,\partial/\partial z) = \lambda z\,\partial/\partial z + \nu, \qquad f_{-1}(z\,\partial/\partial z) = \mu z\,\partial/\partial z \tag{6}$$

and Eq. (3) becomes

$$\partial G(t,z)/\partial t = (\lambda z + \mu z^{-1} - \lambda - \mu)\, z\, \partial G/\partial z + \nu(z-1)\, G \tag{7}$$

In Table 2.1 we give the solution of Eq. (7) for various values of the parameters λ, ν, and μ, together with the corresponding expressions for $P_{n,m}(t)$, $\langle n \rangle$, and $\mathrm{var}(n)$.

There are some other tractable quantities which give insight into the evolution of the process. These quantities are as follows:

(a) The steady-state probabilities $P_{n,m}(\infty)$ describing the process when it is in some dynamical equilibrium.

(b) The probability $R_{N,m}$ that a given state N is ever reached.

(c) The "first passage time" $T_{N,m}$, the time for the process to reach a given state N for the first time, its probability density function $F_{N,m}(t)$, and its arbitrary (ith) moment $\overset{i}{M}_{N,m}$, $i \geqslant 1$.

$F_{N,m}(t)$ is related to $P_{n,m}(t)$ through the relation

$$P_{n,m}(t) = \int_0^t F_{N,m}(t-\tau)\, P_{n,N}(\tau)\, d\tau, \qquad m < N < n \tag{8}$$

This relation reflects the fact that in order to reach a state n from state m, the process has to reach the intermediate state N first. This relation is general in the sense that it is true also for the restricted process, but in that case $P_{n,m}$ stands for the probability characterizing the restricted process in which the state N is an intermediate (not boundary) state. An equivalent and simpler form of Eq. (8) is in terms of Laplace transforms of the quantities occurring in the equation. Taking the Laplace transform of Eq. (8) and using the theorem on the Laplace transform of a convolution integral, we get

$$p_{n,m}(s) = f_{N,m}(s)\, p_{n,N}(s) \qquad \text{or} \qquad f_{N,m}(s) = p_{n,m}(s)/p_{n,N}(s) \tag{9}$$

(d) The probability $R_{N,m}(L)$ that a given state N is reached before another given state L is reached.

TABLE 2.1

Results for some unrestricted processes

Case	λ_n	μ_n	$G_m(*, z)$	$P_{n,m}(t)$	$\langle n \rangle$	$\mathrm{var}(n)$	Name of process, comments, and references
1	ν	0	$z^m \exp(\nu t(z-1))$	$\dfrac{e^{-\nu t}(\nu t)^{n-m}}{(n-m)!}$, $n \geqslant m$; 0, $n < m$	$m + \nu t$	νt	Poisson Process (Cox and Miller, 1965)
2	0	ρ	$z^m \exp(\rho t(z^{-1}-1))$	$\dfrac{e^{-\rho t}(\rho t)^{m-n}}{(m-n)!}$, $n \leqslant m$; 0, $n > m$	$m - \rho t$	ρt	Poisson process (Cox and Miller, 1965)
3	ν	ρ	$z^m \exp\{-(\nu+\rho)t + (\nu z + \rho z^{-1})t\}$	$\left(\dfrac{\nu}{\rho}\right)^{(n-m)/2} I_{n-m}(2t(\nu\rho)^{1/2})e^{-(\nu+\rho)t}$	$m + (\nu-\rho)t$	$(\nu+\rho)t$	$I_n(x)$—modified Bessel function of order n (Heathcote and Moyal, 1959)
4	λn	0	$[1 - e^{\lambda t}(1-z^{-1})]^{-m}$	$\dbinom{n}{m} e^{-m\lambda t}(1-e^{-\lambda t})^{n-m}$, $n \geqslant m$; 0, $n < m$	$me^{\lambda t}$	$me^{\lambda t}(e^{\lambda t}-1)$	Simple birth process (Bailey, 1964)
5	0	μn	$[1 - e^{-\mu t}(1-z)]^m$	$\dbinom{m}{n} e^{-n\mu t}(1-e^{-\mu t})^{m-n}$, $n \leqslant m$; 0, $n > m$	$me^{-\mu t}$	$me^{-\mu t}(1-e^{-\mu t})$	Simple death process (Bailey, 1964)
6	ν	μn	$[1 - e^{-\mu t}(1-z)]^m \times \exp\{\nu(z-1)(1-e^{-\mu t})/\mu\}$	$\exp\{-(\nu/\mu)(1-e^{-\mu t})\} \sum_{k=0}^{(m,n)} \dbinom{m}{k} \times \dfrac{e^{-\mu t k}(1-e^{-\mu t})^{m+n-2k}}{(n-k)!}\left(\dfrac{\nu}{\mu}\right)^{n-k}$	$me^{-\mu t} + \dfrac{\nu}{\mu}(1-e^{-\mu t})$	$\left(\dfrac{\nu}{\mu} + me^{-\mu t}\right)(1-e^{-\mu t})$	Simple death and Immigration process (Cox and Miller, 1965) $(m, n) = \min\{m, n\}$

7	λn	μn	$\left[\dfrac{(\sigma-1)+(\gamma-\sigma)z}{(\gamma\sigma-1)+\gamma(1-\sigma)z}\right]^m$	$\gamma^n \sum_{k=0}^{(m,n)} (-1)^k \binom{m+n-k-1}{n-k}\binom{m}{k} \times \left(\dfrac{1-\sigma}{1-\gamma\sigma}\right)^{m+n-k}\left(\dfrac{1-\sigma/\gamma}{1-\sigma}\right)^k,$ $\gamma = \lambda/\mu \neq 1,\ \sigma = e^{(\lambda-\mu)t}$	$m\sigma$	$\dfrac{m\sigma(\gamma+1)(\sigma-1)}{\gamma-1}$	Simple birth and death process (Bailey, 1964) $(m,n) = \min\{m,n\}$
	λn	λn	$\left[\dfrac{\lambda t+(1-\lambda t)z}{1+\lambda t-\lambda tz}\right]^m$	$\left(\dfrac{\lambda t}{1+\lambda t}\right)^{m+n} \sum_{k=0}^{(m,n)} \binom{m}{k}\binom{m+n-k-1}{n-k} \times \left(\dfrac{1-\lambda^2t^2}{\lambda^2t^2}\right)^k$	m	$2m\lambda t$	
8	$\lambda n+\nu$	μn	$\dfrac{(\gamma-1)^{\nu/\lambda}[\sigma-1+(\gamma-\sigma)z]^m}{[\gamma\sigma-1+\gamma(1-\sigma)z]^{m+\nu/\lambda}}$	$\left(\dfrac{1-\gamma}{1-\gamma\sigma}\right)^{\nu/\lambda} \gamma^n \sum_{k=0}^{(m,n)} \dfrac{(-1)^k(m+\nu/\lambda)_{n-k}}{(n-k)!} \times \binom{m}{k}\left(\dfrac{1-\sigma}{1-\gamma\sigma}\right)^{m+n-k}\left(\dfrac{1-\sigma/\gamma}{1-\sigma}\right)^k$	$m\sigma + \dfrac{\nu\gamma(\sigma-1)}{\lambda(\gamma-1)}$	$\dfrac{m\sigma(\gamma+1)(\sigma-1)}{\gamma-1} + \dfrac{\nu}{\lambda}\dfrac{\gamma(\sigma-1)(\gamma\sigma-1)}{(\gamma-1)^2}$	Simple birth and death process with immigration (Karlin and McGregor, 1958; Bailey, 1964)
	$\lambda n+\nu$	λn	$\dfrac{[\lambda t+(1-\lambda t)z]^m}{[1+\lambda t-\lambda tz]^{m+\nu/\lambda}}$	$\dfrac{(\lambda t)^{m+n}}{(1+\lambda t)^{m+n+(\nu/\lambda)}} \sum_{k=0}^{(m,n)} \binom{m}{k}\dfrac{(m+(\nu/\lambda))_{n-k}}{(n-k)!} \times \left(\dfrac{1-\lambda^2t^2}{\lambda^2t^2}\right)^k$	$m+\nu t$	$2m\lambda t + \nu t(\lambda t+1)$	$(m,n) = \min\{m,n\}$

(e) The time $T_{N,m}(L)$ for the process to reach a given state N for the first time before another given state L is reached, its probability density function $F_{N,m}(L,t) \equiv F_{N,m}(L)$, and its arbitrary ($i$th) moment $\overset{i}{M}_{N,m}(L)$, $i \geqslant 1$.

To derive $P_{n,m}(\infty) = P_n^*$, we put all the time derivatives in the forward equations (2.0.3) equal to zero and then sum up all the equations for $n \leqslant i-1$. The resulting equation is

$$\mu_i P_i^* = \lambda_{i-1} P_{i-1}^* \tag{10}$$

Using this relation recursively, we can express each P_i^* in terms of P_0^*:

$$P_i^* = \frac{\lambda_{i-1}\lambda_{i-2}\cdots\lambda_0}{\mu_i\mu_{i-1}\cdots\mu_1} P_0^*, \qquad i > 0 \tag{11a}$$

$$P_i^* = \frac{\mu_{i+1}\mu_{i+2}\cdots\mu_0}{\lambda_i\lambda_{i+1}\cdots\lambda_{-1}} P_0^*, \qquad i < 0 \tag{11b}$$

where, by the normalization condition $\sum_{-\infty}^{\infty} P_i^* = 1$, P_0^* is found to be

$$P_0^* = \left[1 + \sum_{i=1}^{\infty}\left(\frac{\lambda_0\lambda_1\cdots\lambda_{i-1}}{\mu_1\mu_2\cdots\mu_i} + \frac{\mu_0\mu_{-1}\cdots\mu_{-i+1}}{\lambda_{-1}\lambda_{-2}\cdots\lambda_{-i}}\right)\right]^{-1} \tag{12}$$

Equations (11) and (12) define P_i^*.

The quantities in (b) and (c) are independent of the behavior of the process once state N is reached. Therefore, one only has to consider a state space confined to $n \leqslant N$ when $m < N$ (and to $n \geqslant N$ when $m > N$). The expressions for these quantities are derived in Section 2.2D where we consider processes which are confined to such a state space. Similarly, the quantities in (d) and (e) are independent of the behavior of the process once state N or state L is reached, and their discussion is postponed to Section 2.2B where processes confined between two boundaries are discussed.

2.2 Restricted Processes

In the preceding section, we discussed those random processes for which there were no a priori restrictions on the allowed states of the random variable. There are many processes which are such that some of the states of the random variable are special, and as a consequence of this there are inherent restrictions on the allowed states. For example, for a growing population (simple birth and death process), once the number n of individuals is zero, the growth process stops. Thus the state $n = 0$ is a special state, and once the process reaches this state it is trapped forever. Such a trapping state is known as an *absorbing state*. Another possible special state is the so-called *reflecting state*, which is an impenetrable state. Once the process reaches this state, it must return to the previously occupied state. An example of such a state is the $n = 0$ state of a growing population in the presence of immigration. Once the population reaches the state $n = 0$, it can return to the state $n = 1$ because of immigration. In this section, we discuss and analyze the random processes that have either one or two special (absorbing or reflecting) states.

It should be pointed out that in the preceding section we did consider processes with some special states. For example, in Table 2.1 cases 6 and 7 define processes with reflecting and absorbing states, respectively. The analysis of Section 2.1 is valid only if the special states ($n = 0$ in the referred example) are built in the expressions for the transition probabilities per unit time, λ_n and μ_n. The analysis of this section is necessary when the values of λ_n and μ_n for certain n's have to be changed in order to incorporate the special properties of the states. For example, for a simple birth and death process with emigration, $\lambda_n = \lambda n$, $\mu_n = \mu n + \rho$, where λ and μ are the probabilistic rates per individual for birth and death and ρ is the probabilistic emigration rate, the state $n = 0$ is an absorbing state. But the expression for μ_n implies $\mu_0 = \rho \neq 0$; i.e., the process can go below the $n = 0$ state due to emigration, an absurd result. In this case we must demand $\mu_0 = 0$.

The most general case we consider here is when the process is confined to states $l \leqslant n \leqslant u$, where either both the boundary states l and u are special (absorbing or reflecting) states or only one of them is a special state while the other is at infinity ($l = -\infty$ or $u = +\infty$). The case when both $l = -\infty$ and $u = +\infty$ has been treated in Section 2.1 and therefore we do not consider it here. In the case where one of the two boundary states is infinite we limit our discussion to those "realistic" processes for which the probability of the process ever reaching the infinite boundary ($+\infty$ or $-\infty$) is zero.

When the process is confined between states l and u the probability density $P_{n,m}(t)$, for $l < n < u$, still satisfies the forward and backward master equations (2.0.3) and (2.0.4), but $P_{n,m}(t)$ for $n = l$ and $n = u$ (the boundary conditions) depends on the nature (absorbing or reflecting) of states l and u.

When state l is an *absorbing* state, by definition $\lambda_l = \mu_l = 0$, the $(l+1)$th forward equation becomes

$$dP_{l+1,m}/dt = -(\mu_{l+1} + \lambda_{l+1})\, P_{l+1,m} + \mu_{l+2}\, P_{l+2,m} \tag{1}$$

This is equivalent to taking

$$P_{l,m}(t) = 0 \tag{2}$$

as the boundary condition for the set of forward equations (2.0.3) for $l \leqslant n$. For the backward equations (2.0.4), the boundary condition is

$$P_{n,l}(t) = 0 \tag{3}$$

since any process which starts at state l is going to stay there forever.

On the other hand, if state l is a *reflecting* state, then $\lambda_l > 0$, $\mu_l = 0$, and from Eqs. (2.0.3) and (2.0.4) $P_{l,m}$ and $P_{n,l}$ satisfy the equations

$$dP_{l,m}/dt = \mu_{l+1} P_{l+1,m} - \lambda_l P_{l,m} \tag{4a}$$

$$dP_{n,l}/dt = \lambda_l P_{n,l+1} - \lambda_l P_{n,l} \tag{4b}$$

By a similar argument, it can be seen that when the upper state u is an *absorbing* state

$$P_{u,m}(t) = 0 \tag{5a}$$

$$P_{n,u}(t) = 0 \tag{5b}$$

and when it is a *reflecting* state

$$dP_{u,m}/dt = \lambda_{u-1} P_{u-1,m} - \mu_u P_{u,m} \tag{6a}$$

$$dP_{n,u}/dt = \mu_u P_{n,u-1} - \mu_u P_{n,u} \tag{6b}$$

When both states l and u are special states, the basic forward and backward master equations (2.0.3) and (2.0.4) are to be solved subject to the two appropriate boundary conditions. For example, if state l is reflecting and state u is absorbing, the boundary conditions for the forward master equation are (4a) and (5a) and for the backward master equation, (4b) and (5b). When only one of the two states l and u is a special state and the other is infinite, only one appropriate boundary condition is necessary. Of course, on physical grounds we require $P_{n,m}$ $(n \to \pm\infty)$ and $P_{n,m}$ $(m \to \pm\infty)$ to be zero.

It should be noted that when $l(u)$ is an absorbing state, $P_{l,m}(t)$ $[P_{u,m}(t)]$ does not represent the probability of the process being in state $l(u)$ at time t. This probability, a nonvanishing quantity, is equal to the probability of the process not being in any of the other states when $l(u)$ is the only absorbing state.

One method for calculating $P_{n,m}(t)$ for a restricted process is to introduce dummy variables d_n for the initial probabilities of the process being in the states $n < l$ and $n > u$, i.e.,

$$P_{n,m}(0) = d_n, \qquad n < l, \qquad n > u \tag{7}$$

and then solve for $P_{n,m}$ for an unrestricted process with the initial conditions (7) and

$$P_{n,m}(0) = \delta_{n,m}, \qquad l \leqslant n \leqslant u \tag{8}$$

by using the method of generating functions described in Section 2.1. The expressions for $P_{n,m}$ will involve d_n, which are then chosen to satisfy the appropriate boundary conditions on $P_{n,m}$.

In Appendix C we have illustrated the method for process 3 of Table 2.1

(immigration–emigration process). The results for this and other processes for which an analytical expression has been obtained are summarized in Table 2.2.

In the absence of an analytical expression for $P_{n,m}(t)$ (which if available describes the process uniquely and completely), it is possible to gain insight into the evolution of the process by deriving analytical expressions for the quantities described in Section 2.1. For convenience, we will divide the derivation of the expressions for the quantities (a), (b), and (c) according to the nature of the boundary states l and u. The quantities (d) and (e) are independent of the behavior of the process after either the state N or the state L is reached, and hence are independent of states outside the range (L, N), i.e., independent of l and u. For these quantities, $R_{N,m}(L)$, $F_{N,m}(L)$, and $\overset{j}{M}_{N,m}(L)$, the states N and L can be regarded as the absorbing boundaries of the process. We derive these quantities in Section 2.2B, where processes confined between two absorbing states are discussed.

We start with processes in which one of the boundary states is absorbing and the other is reflecting.

A. *Process Confined between an Absorbing State and a Reflecting State*

We shall take l to be a reflecting state and u an absorbing state. Since the case where state l is absorbing and u is reflecting can be treated likewise, we will only summarize the results for this at the end of this subsection.

As l is a reflecting state, starting from any state m, $l < m < u$, the process is bound to reach the state u eventually, i.e., eventually get absorbed. Therefore $P_{n,m}(\infty) = 0$ for all $l \leqslant n < u$.

Next, we consider $F_{u,m}(t)$ defined in (c) of Section 2.1, i.e., the probability density of $T_{u,m}$, the time spent by the process before absorption occurs. By arguments similar to those employed in deriving the backward equation (2.0.4) we find that $F_{u,m}(t)$ satisfies the backward equations:

$$dF_{u,m}(t)/dt = \lambda_m F_{u,m+1}(t) - (\lambda_m + \mu_m) F_{u,m}(t) + \mu_m F_{u,m-1}(t), \qquad l < m < u \tag{9}$$

Since $\mu_l = 0$ and the state $l-1$ is never reached, the lth backward equation becomes

$$dF_{u,l}(t)/dt = \lambda_l F_{u,l+1} - \lambda_l F_{u,l} \tag{10}$$

which is a boundary condition for the set of equations (9). To derive the second boundary condition we note that $F_{u,u}(t) = 0$, $t > 0$, and the $(u-1)$th equation, is of the form

$$dF_{u,u-1}/dt = -(\lambda_{u-1} + \mu_{u-1}) F_{u,u-1} + \mu_{u-1} F_{u,u-2}, \qquad t > 0 \tag{11}$$

TABLE 2.2[a]

Expressions for $P_{n,m}(t)$ for some restricted processes with λ_n and μ_n at most linear in n

λ_n	μ_n	Boundaries[b]	$P_{n,m}(t)$	Case
ν	ρ	u: abs l: $-\infty$	$\alpha[I_{n-m} - I_{2u-m-n}]$	0
		u: ∞ l: abs	$\alpha[I_{n-m} - I_{m+n-2l}]$	1
		u: refl l: $-\infty$	$\alpha\left[I_{n-m} + \left(\frac{\nu}{\rho}\right)^{1/2} I_{2u+1-m-n} + \left(1 - \frac{\rho}{\nu}\right)\sum_{j=2}^{\infty}\left(\frac{\nu}{\rho}\right)^{j/2} I_{2u-n-m+j}\right]$	2
		u: ∞ l: refl	$\alpha\left[I_{n-m} + \left(\frac{\rho}{\nu}\right)^{1/2} I_{m+n+1-2\rho} + \left(1 - \frac{\nu}{\rho}\right)\sum_{j=2}^{\infty}\left(\frac{\rho}{\nu}\right)^{j/2} I_{m+n-2l+j}\right]$	3
		u: abs l: abs	$\alpha\left\{\sum_{k=-\infty}^{\infty} I_{n-m+2k(u-l)} - \sum_{k=0}^{\infty}[I_{n+m-2l+2k(u-l)} + I_{2u-n-m+2k(u-l)}]\right\}$	4
$\lambda(n-l)+\lambda$	$\mu(n-l)$	u: abs l: refl	$\gamma^{l-m}\sum_{j=0}^{u-l-1} g_{m-l}g_{n-l}\sigma^{\xi_j}\left[\sum_{i=0}^{u-l-1} g_i\gamma^{-i}\right]^{-1}$	5
		u: refl l: refl	$\gamma^{l-m}\sum_{j=0}^{u-l} \hat{g}_{m-l}\hat{g}_{n-l}\sigma^{\hat{\xi}_j}\left[\sum_{i-0}^{u-l-1} \hat{g}_i\gamma^{-i}\right]^{-1}$	6
$\lambda(u-n)$	$\mu(u-n)+\mu$	u: refl l: abs	$\gamma^{u-m}\sum_{j=0}^{u-l-1} h_{u-m}h_{u-n}\sigma^{-\zeta_j}\left[\sum_{i=0}^{u-l-1} h_i\gamma^{i}\right]^{-1}$	7
		u: refl l: refl	$\gamma^{u-m}\sum_{j=0}^{u-l} \hat{h}_{u-m}\hat{h}_{u-n}\sigma^{-\hat{\xi}_j}\left[\sum_{i=0}^{u-l} \hat{h}_i\gamma^{i}\right]^{-1}$	8

$\lambda(u-l)$	$\mu(n-l)+\rho$	$u: \infty$ l: abs	$\gamma^{n'}\sigma^{1+\rho/\mu}\left(\frac{1-\gamma}{1-\sigma}\right)^{2+\rho/\mu}\sum_{i=0}^{(m',n')}(-1)^i\binom{n'}{i}\frac{(n'+2+\rho/\mu)_{m'-i'}}{(m'-i)!}\left(\frac{1-\sigma}{1-\sigma\gamma}\right)^{m'+n'-i}\left(\frac{1-\sigma\gamma^{-1}}{1-\sigma}\right)^{i}$	9
$\lambda(n-l)$	$\lambda(n-l)+\rho$	$u: \infty$ l: abs	$(\lambda t)^{-(2+\rho/\lambda)}\sum_{i=0}^{(m',n')}\binom{n'}{i}\frac{(n'+2+\rho/\lambda)_{m'-i}}{(m-i)!}\left(\frac{\lambda t}{1+\lambda t}\right)^{m'+n'-i}\left(\frac{1-\lambda t}{\lambda t}\right)^{i}$	10
$\lambda(u-n)+\nu$	$\mu(u-n)$	u: abs $l: -\infty$	$\gamma^{-n'}\sigma^{1+\nu/\lambda}\left(\frac{1-\gamma^{-1}}{1-\sigma}\right)^{2+\nu/\lambda}\sum_{i=0}^{(m',n')}(-1)^i\binom{n'}{i}\frac{(n'+2+\nu/\lambda)_{m'-i}}{(m'-i)!}\left(\frac{1-\sigma}{1-\sigma\gamma^{-1}}\right)^{m'+n'-i}\left(\frac{1-\sigma\gamma}{1-\sigma}\right)^{i}$	11
$\mu(u-n)+\nu$	$\mu(u-n)$	u: abs $l: -\infty$	$(\mu t)^{-(2+\nu/\mu)}\sum_{i=0}^{(m',n')}\binom{n'}{i}\frac{(n'+2+\nu/\mu)_{m'-i}}{(m'-i)!}\left(\frac{\mu t}{1+\mu t}\right)^{m'+n'-i}\left(\frac{1-\mu t}{\mu t}\right)^{i}$	12

[a] Auxiliary definitions and references:

0–4: $I_n = I_{-n} \equiv I_n(2(\rho\nu)^{1/2}t)$, where $I_n(x)$ is a modified Bessel function, $\alpha \equiv (\nu/\rho)^{(n-m)/2}e^{-t(\nu+\rho)}$ (Cox and Miller, 1965; Montroll, 1967).

5–6: $\gamma \equiv \lambda/\mu$, $\sigma \equiv e^{(\lambda-\mu)t}$; $g_n \equiv g_n(\xi_j,\gamma)$, $\hat{g}_n \equiv g_n(\hat{\xi}_j,\gamma)$, $g_n(x,\gamma) \equiv \gamma^n \sum_{i=0}^{n}(1-\gamma^{-1})^i\binom{n}{i}(x-i+1)_i/i! = \gamma^n F(-n,-x,1,1-\gamma^{-1})$; ξ_j, $\hat{\xi}_j$ roots of $g_{u-l}(\xi_j,\gamma)=0$, $j=0,\ldots,u-l-1$; $g_{u+1-l}(\hat{\xi}_j)=\gamma g_{u-l}(\hat{\xi}_j)$, $j=0,\ldots,u-l$, where g_n is a Gottlieb polynomial, F a hypergeometric function (Montroll and Shuler, 1958; Shuler *et al.*, 1962).

7–8: $h_n \equiv h_n(\zeta_j,\gamma)$, $\hat{h}_n \equiv h_n(\hat{\zeta}_j,\gamma)$, $h_n(x,\gamma) \equiv g_n(x,\gamma^{-1})$, $h_{u-l}(\zeta_j,\gamma)=0$, $j=0,\ldots,u-l-1$, $h_{u+l-1}(\hat{\zeta}_j,\gamma) = h_{u-l}(\hat{\zeta}_j,\gamma)/\gamma$, $j=0,\ldots,u-l$, where $g_n(x,\gamma)$, γ, and σ are as in 5–6.

9–10: $\sigma \equiv e^{(\lambda-\mu)t}$, $\gamma \equiv \lambda/\mu$, $m' \equiv m-l-1$, $n' \equiv n-l-1$, $(m',n') \equiv \min\{m',n'\}$ (Karlin and McGregor, 1958).

11–12: $\sigma \equiv e^{(\lambda-\mu)t}$, $\gamma \equiv \lambda/\mu$, $m' \equiv u-m-1$, $n' \equiv u-n-1$, $(m',n') \equiv \min\{m',n'\}$.

[b] Where abs is absorbing, and refl is reflecting.

This equation can be included in the general form (9) if we impose the boundary condition

$$F_{u,u}(t) = 0 \tag{12}$$

The initial condition necessary for the solution of the set (9) follows from the following simple argument. For the process to reach the state u for the first time at time $(t+\Delta t)$, it has to be in the state $u-1$ at time t and then go to state u in the remaining time Δt, i.e.,

$$F_{u,m}(t)\,\Delta t = P_{u-1,m}(t)(\lambda_{u-1}\,\Delta t) \tag{13}$$

Since $P_{n,m}(0) = \delta_{n,m}$ we get

$$F_{u,m}(0) = 0, \qquad l \leqslant m < u-1 \tag{14a}$$

$$F_{u,u-1}(0) = \lambda_{u-1} \tag{14b}$$

A procedure for solving Eq. (9) with the boundary conditions (10) and (12) and the initial condition (14) is to take the Laplace transform of these equations and solve the resulting equations for the Laplace transform $f_{u,m}(s)$ of $F_{u,m}(t)$, defined by

$$f_{u,m}(s) = \int_0^\infty e^{-st} F_{u,m}(t)\,dt \tag{15}$$

$F_{u,m}(t)$ can then be uniquely determined by using one of the many standard methods and extensive tables for inverting the Laplace transforms (Bellman *et al.*, 1966; Erdélyi, 1954). The detailed derivation of the expression for $f_{u,m}(s)$ is given in Appendix D where we show that

$$f_{u,m}(s) = Q_m(s)/Q_u(s) \tag{16}$$

where $Q_n(s)$ is a polynomial in s of degree $n-l$ and satisfies the recurrence relations

$$\lambda_i Q_{i+1}(s) = (\lambda_i+\mu_i+s)Q_i(s) - \mu_i Q_{i-1}(s), \qquad l+1 \leqslant i \leqslant u-1 \tag{17}$$

$$Q_l(s) = 1, \qquad Q_{l+1}(s) = (\lambda_l+s)/\lambda_l \tag{18}$$

The moments of $T_{u,m}$ can be derived directly from $f_{u,m}(s)$ without first inverting it to get $F_{u,m}(t)$ since

$$\overset{j}{M}_{u,m} = \int_0^\infty t^j F_{u,m}(t)\,dt = (-1)^j \frac{d^j}{ds^j} f_{u,m}(s)\bigg|_{s=0} \tag{19}$$

If one is interested only in the moments, one does not even have to calculate an explicit expression for $f_{u,m}(s)$. Instead, as shown in Appendix D, the moments can be derived from a set of backward equations satisfied by $f_{u,m}(s)$. The results are derived in Appendix D and summarized in Table 2.3.

TABLE 2.3

Expressions for various quantities of interest for a process confined between *a reflecting state* l and *an absorbing state* u, $l < u$

$P_{n,m}(\infty)$	$0 \quad (l < n < u)$	
	$l \leqslant m \leqslant N \leqslant u$	$l \leqslant N \leqslant m \leqslant u$
$R_{N,m}$	1	$\sum_{i=m}^{u-1} \Pi_{N+1,i} \Big/ \sum_{i=N}^{u-1} \Pi_{N+1,i}$
$M_{N,m}$	$\sum_{i=m}^{N-1} \sum_{n=l}^{i} \lambda_n^{-1} \Pi_{n+1,i} = \sum_{n=l}^{m} \lambda_n^{-1} \sum_{i=m}^{N-1} \Pi_{n+1,i} + \sum_{n=m+1}^{N-1} \lambda_n^{-1} \sum_{i=n}^{N-1} \Pi_{n+1,i}$	$\sum_{i=m}^{u-1} \sum_{n=N+1}^{i} \lambda_n^{-1} \Pi_{n+1,i} R_{N,n} - R_{N,n} \sum_{i=N+1}^{u-1} \sum_{n=N+1}^{i} \lambda_n^{-1} \Pi_{n+1,i} R_{N,n}$
$\overset{j}{M}_{N,m}$	$\sum_{i=m}^{N-1} \sum_{n=l}^{i} \lambda_n^{-1} \Pi_{n+1,i} \overset{j-1}{M}_{N,n}$	$\sum_{i=m}^{u-1} \sum_{n=N+1}^{i} \lambda_n^{-1} \Pi_{n+1,i} \overset{j-1}{M}_{N,n} - R_{N,m} \sum_{i=N+1}^{u-1} \sum_{n=N+1}^{i} \lambda_n^{-1} \Pi_{n+1,i} \overset{j-1}{M}_{N,n}$
$V_{N,m}$	$\sum_{i=m}^{N-1} \left[2 \sum_{n=l}^{i-1} \Pi_{n+1,i} (M_{n+1,n})^2 + (M_{i+1,i})^2 \right]$	$\overset{2}{M}_{N,m} - (M_{N,m})^2$

$$\Pi_{i,j} = \frac{\mu_i \mu_{i+1} \cdots \mu_j}{\lambda_i \lambda_{i+1} \cdots \lambda_j}, \qquad i \leqslant j, \qquad \Pi_{i,i-1} = 1.$$

TABLE 2.4

Expressions for various quantities of interest for a process confined between an *absorbing state* l and a *reflecting state* u, $l < u$

$P_{n,m}(\infty)$	$0 \quad (l < n < u)$	
	$l \leqslant m \leqslant N \leqslant u$	$l \leqslant N \leqslant m \leqslant u$
$R_{N,m}$	$\sum_{i=l}^{m-1} \Pi_{l+1,i} \Big/ \sum_{i=l}^{N-1} \Pi_{l+1,i}$	1
$M_{N,m}$	$R_{N,m} \sum_{i=l+1}^{N-1} \sum_{n=l+1}^{i} \lambda_n^{-1} \Pi_{n+1,i} R_{N,n} - \sum_{i=l+1}^{m-1} \sum_{n=l+1}^{i} \lambda_n^{-1} \Pi_{n+1,i} R_{N,n}$	$\sum_{i=N+1}^{m} \sum_{n=i}^{u} (\mu_n \Pi_{i,n-1})^{-1}$
$\overset{j}{M}_{N,m}$	$R_{N,m} \sum_{i=l+1}^{N-1} \sum_{n=l+1}^{i} \lambda_n^{-1} \Pi_{n+1,i} \overset{j-1}{M}_{N,n} - \sum_{i=l+1}^{m-1} \sum_{n=l+1}^{i} \lambda_n^{-1} \Pi_{n+1,i} \overset{j-1}{M}_{N,n}$	$\sum_{i=N+1}^{m} \sum_{n=i}^{u} (\mu_n \Pi_{i,n-1})^{-1} \overset{j-1}{M}_{N,n}$
$V_{N,m}$	$\overset{2}{M}_{N,m} - (M_{N,m})^2$	$\sum_{i=N+1}^{m} \left[2 \sum_{n=i+1}^{u} \Pi_{i,n-1}^{-1} (M_{n-1,n})^2 - (M_{i-1,i})^2 \right]$

$$\Pi_{i,j} = \frac{\mu_i \mu_{i+1} \cdots \mu_j}{\lambda_i \lambda_{i+1} \cdots \lambda_j}, \qquad i \leqslant j, \qquad \Pi_{i,i-1} = 1.$$

The above formulas, with u replaced by N, can be used to determine the first passage time $T_{N,m}$ (its probability density function and moments) to an arbitrary state N, $m \leqslant N < u$. This is so since $T_{N,m}$ is independent of the behavior of the process after state N is reached, and therefore the state N, for the purpose of these calculations, can be regarded as an absorbing state. In case $N < m < u$, by the same reasoning, the process to be considered is the one confined between two absorbing states N and u. One then uses the formulas given in Section 2.2B.

For the calculation of $R_{N,m}$, defined in (b) of Section 2.1, we note that if $m < N < u$, the process is bound to go through state N, since absorption is a certain event, and so $R_{N,m} = 1$, while if $l < N < m$, absorption at u can happen before state N is ever reached. Thus the probability $R_{N,m}$ of the process ever reaching state N is the same as $R_{N,m}(u)$, the expression for which is derived in Section 2.2B together with the expressions for $F_{N,m}(u)$ and $\overset{j}{M}_{N,m}(u)$.

When instead of state u, the state l is an absorbing state, and instead of state l, the state u is a reflecting state, $P_{n,m}(\infty)$ is still zero for all states other than the absorbing state, i.e., for $l < n \leqslant u$. The Laplace transform $f_{u,m}(s)$ of $F_{u,m}(t)$ is the ratio of two polynomials, i.e.,

$$f_{u,m}(s) = Q_m^*(s)/Q_l^*(s) \tag{20}$$

where $Q_i^*(s)$ is a polynomial of degree $u-i$ in s. It satisfies the recurrence relation

$$-\lambda_i Q_{i+1}^*(s) + (\lambda_i+\mu_i+s)\, Q_i^*(s) = \mu_i Q_{i-1}^*(s), \qquad l+1 \leqslant i \leqslant u-1 \tag{21}$$

with

$$Q_u^*(s) = 1, \qquad Q_{u-1}^*(s) = (\mu_u+s)/\mu_u \tag{22}$$

The expressions for moments of $T_{N,m}$ are summarized in Table 2.4 together with the expressions for other quantities of interest.

B. Process Confined between Two Absorbing States

Since both states l and u are absorbing states, the absorption is bound to occur eventually. Therefore $P_{n,m}(\infty) = 0$ for all $l < n < u$. The absorption can occur either at state l or at state u, and therefore it is of interest to know the probability $R_{l,m}$ ($R_{u,m}$) of absorption occurring at state l (u) and not at state u (l). Other quantities that give insight into the process of absorption are the (first passage) time $T_{u,m}$ ($T_{l,m}$) for absorption at the state u (l), the corresponding probability density $F_{u,m}(t)$ [$F_{l,m}(t)$], and the moments $\overset{j}{M}_{u,m}$ ($\overset{j}{M}_{l,m}$). Knowing these quantities one can derive the probability density and the moments of the conditional time for absorption at u (l), when absorption at

u (l) is a priori known to occur. These quantities are given by

$$F_{u,m}^{*}(t) = F_{u,m}(t)/R_{u,m}, \qquad \overset{j}{M}{}_{u,m}^{*} = \overset{j}{M}_{u,m}/R_{u,m} \tag{23a}$$

$$F_{l,m}^{*}(t) = F_{l,m}(t)/R_{l,m}, \qquad \overset{j}{M}{}_{l,m}^{*} = \overset{j}{M}_{l,m}/R_{l,m} \tag{23b}$$

One can also calculate the lifetime of the process, T_m, i.e., the time before absorption occurs (either at state u or at state l). This random variable, its probability density $F_m(t)$, and its moments $\overset{j}{M}_m$ are related to the above-mentioned quantities by

$$T_m = \min(T_{l,m}, T_{u,m}) \tag{24}$$

$$F_m(t) = F_{u,m}(t) + F_{l,m}(t) \tag{25a}$$

$$\overset{j}{M}_m = \int_0^\infty t^j F_m(t)\, dt = \overset{j}{M}_{u,m} + \overset{j}{M}_{l,m} \tag{25b}$$

We discuss the quantities related to absorption at state u in detail, and only summarize the results for the corresponding quantities related to absorption at l.

As in Section 2.2A, $F_{u,m}(t)$ for $l+1 < m < u-1$ satisfies the backward equation (9). Since both the lth and uth states are absorbing, the boundary condition (12) for an absorbing state (uth state) will be true for both of these absorbing states. In other words, the boundary conditions are

$$F_{u,l}(t) = 0, \qquad F_{u,u}(t) = 0 \tag{26}$$

The initial conditions follow from the same argument as the initial conditions (14) in Section 2.2A. The resulting initial conditions are

$$F_{u,m}(0) = 0, \qquad l+1 \leqslant m < u-1 \tag{27a}$$

$$F_{u,u-1}(0) = \lambda_{u-1} \tag{27b}$$

In Appendix E, we take the Laplace transform of the backward equation (9) satisfied by $F_{u,m}(t)$, and of the equations above describing the boundary conditions and solve the resulting equations for the Laplace transform $f_{u,m}(s)$ of $F_{u,m}(t)$. We show that

$$f_{u,m}(s) = q_m(s)/q_u(s), \qquad l+1 \leqslant m \leqslant u-1 \tag{28}$$

where $q_i(s)$ is a polynomial in s of degree $i-l-1$ and satisfies a recurrence relation (for $l+1 < m \leqslant u-1$) similar to Eq. (17) but with the boundary conditions

$$q_{l+1}(s) = 1, \qquad q_{l+2}(s) = (\lambda_{l+1} + \mu_{l+1} + s)/\lambda_{l+1} \tag{29}$$

In Appendix F we solve these recurrence relations for q_n in detail for the

process defined by

$$\lambda_n = \lambda n, \qquad \mu_n = \mu n$$

with $n = 0$ and $n = u$ as the absorbing states. We also give the expression for $f_{u,m}(s)$ explicitly, by using Eq. (28), which is used in Chapter 4 in connection with the problem of extinction of colonizing species. By a similar argument

$$f_{l,m}(s) = q_m^*(s)/q_l^*(s), \qquad l+1 \leqslant m \leqslant u-1 \tag{30}$$

where $q_n^*(s)$, $l \leqslant n \leqslant u-1$, satisfies the recurrence relation (21) for $l+1 \leqslant m < u-1$ with

$$q_{u-1}^*(s) = 1, \qquad q_{u-2}^*(s) = (s+\lambda_{u-1}+\mu_{u-1})/\mu_{u-1} \tag{31}$$

Here q_n^* is a polynomial in s of degree $u-1-n$.

The recurrence relations for $f_{u,m}$ can be used to derive the expression for the probability $R_{u,m}$ of absorption occurring at state u and not at state l. This probability is given by

$$R_{u,m} = R_{u,m}(l) = \int_0^\infty F_{u,m}(t)\,dt = f_{u,m}(0) \tag{32}$$

The details of the calculations are given in Appendix E. The final result is

$$R_{u,m} = R_{u,m}(l) = \sum_{n=l}^{m-1} \Pi_{l+1,n} \Big/ \sum_{n=l}^{u-1} \Pi_{l+1,n} \tag{33}$$

where

$$\Pi_{i,j} = \frac{\mu_i \mu_{i+1} \cdots \mu_j}{\lambda_i \lambda_{i+1} \cdots \lambda_j}, \qquad i \leqslant j, \qquad \Pi_{i,i-1} = 1 \tag{34}$$

Since absorption is a certain event, the probability of absorption at state l and not at state u is given by

$$R_{l,m} = R_{l,m}(u) = 1 - R_{u,m} = \sum_{n=m+1}^{u-1} \Pi_{l+1,n} \Big/ \sum_{n=l}^{u-1} \Pi_{l+1,n} \tag{35}$$

The expressions for the moments $\overset{j}{M}_{u,m}$ (and $\overset{j}{M}_{l,m}$) can also be derived directly by using the recurrence relations satisfied by $f_{u,m}$. The details are given in Appendix E and the results are summarized in Table 2.5. The expressions for $F_m(t)$ and $\overset{j}{M}_m$ which describe the lifetime of the process before absorption occurs (either at state u or at state l) can be obtained by using Eqs. (25). In particular,

$$M_m = M_{u,m} + M_{l,m} = \sum_{i=m}^{u-1} \sum_{n=l+1}^{i} \lambda_n^{-1} \Pi_{n+1,i} - R_{l,m} \sum_{i=l+1}^{u-1} \sum_{n=l+1}^{i} \lambda_n^{-1} \Pi_{n+1,i} \tag{36}$$

This expression has the physical meaning that the average time for absorption (either at state l or at state u) is equal to the average time for absorption at u

TABLE 2.5

Expressions for various quantities of interest for a process confined between *two absorbing states l and u*, $l < u$

$P_{n,m}(\infty)$	0 $\quad (l < n < u)$	
	$l \leqslant N \leqslant m \leqslant u$	$l \leqslant m \leqslant N \leqslant u$
$R_{N,m}$ $M_{N,m}$ $\overset{j}{M}_{N,m}$ $V_{N,m}$	See column 2, Table 2.3	See column 1. Table 2.4
R_m	1	
M_m	$\sum_{i=m}^{u-1} \sum_{n=l+1}^{i} \lambda_n^{-1} \Pi_{n+1,i} - R_{l,m} \sum_{i=l+1}^{u-1} \sum_{n=l+1}^{i} \lambda_n^{-1} \Pi_{n+1,i}$	
$\overset{j}{M}_m$	$\sum_{i=m}^{u-1} \sum_{n=l+1}^{i} \lambda_n^{-1} \Pi_{n+1,i} \overset{j-1}{M}_n - R_{l,m} \sum_{i=l+1}^{u-1} \sum_{n=l+1}^{i} \lambda_n^{-1} \Pi_{n+1,i} \overset{j-1}{M}_n$	
V_m	$\overset{2}{M}_m - (M_m)^2$	

$$\Pi_{i,j} = \frac{\mu_i \mu_{i+1} \cdots \mu_j}{\lambda_i \lambda_{i+1} \cdots \lambda_j}, \qquad i < j, \quad \Pi_{i,i-1} = 1.$$

when state $l+1$ is reflecting [compare the first term in Eq. (36) with Eq. (D.25)] minus the probability that absorption takes place at state l times the average time for absorption at state u for a process starting at the reflecting state $l+1$.

Finally we recall the important observation of Section 2.1 that the calculation of the quantities (d) and (e) described in that section involves the procedure described above for a process confined between two absorbing states. Thus, for example, $R_{N,m}(L)$, the probability of a process reaching state N before state L is the same as the probability of absorption occurring at state N in a process which is confined between the two absorbing states N and L. This observation is independent of the nature of the boundaries l and u of the original process for which $R_{N,m}(L)$ is to be calculated.

The expressions for $R_{N,m}(L)$ and $\overset{j}{M}_{N,m}(L)$, derived using this observation, are summarized in Table 2.6.

C. *Process Confined between Two Reflecting States*

Since both l and u are reflecting states, as time goes on the process is bound to reach each of the two reflecting states, and therefore $P_{n,m}(\infty)$ will be independent of the initial state m. For convenience in notation, let us denote

TABLE 2.6

Expressions for various quantities of interest which are *independent* of the nature of the boundary states

	$l \leqslant L \leqslant m \leqslant N \leqslant u$	$l \leqslant N \leqslant m \leqslant L \leqslant u$
$R_{N,m}(L)$	$\sum_{i=L}^{m-1} \Pi_{L+1,i} \Big/ \sum_{i=L}^{N-1} \Pi_{L+1,i}$	$\sum_{i=m}^{L-1} \Pi_{N+1,i} \Big/ \sum_{i=N}^{L-1} \Pi_{N+1,i}$
$M_{N,m}(L)$	$R_{N,m}(L) \sum_{i=L+1}^{N-1} \sum_{n=L+1}^{i} \lambda_n^{-1} \Pi_{n+1,i} R_{N,n}(L)$ $- \sum_{i=L+1}^{m-1} \sum_{n=L+1}^{i} \lambda_n^{-1} \Pi_{n+1,i} R_{N,n}(L)$	$\sum_{i=m}^{L-1} \sum_{n=N+1}^{i} \lambda_n^{-1} \Pi_{n+1,i} R_{N,n}(L)$ $- R_{N,m}(L) \sum_{i=N+1}^{L-1} \sum_{n=N+1}^{i} \lambda_n^{-1} \Pi_{n+1,i} R_{N,n}(L)$
$\overset{j}{M}_{N,m}(L)$	$R_{N,m}(L) \sum_{i=L+1}^{N-1} \sum_{n=L+1}^{i} \lambda_n^{-1} \Pi_{n+1,i} \overset{j-1}{M}_{N,n}(L)$ $- \sum_{i=L+1}^{m-1} \sum_{n=L+1}^{i} \lambda_n^{-1} \Pi_{n+1,i} \overset{j-1}{M}_{N,n}(L)$	$\sum_{i=m}^{L-1} \sum_{n=N+1}^{i} \lambda_n^{-1} \Pi_{n+1,i} \overset{j-1}{M}_{N,n}(L)$ $- R_{N,m}(L) \sum_{i=N+1}^{L-1} \sum_{n=N+1}^{i} \lambda_n^{-1} \Pi_{n+1,i} \overset{j-1}{M}_{N,n}(L)$

$$\Pi_{i,j} = \frac{\mu_i \mu_{i+1} \cdots \mu_j}{\lambda_i \lambda_{i+1} \cdots \lambda_j}, \qquad i \leqslant j, \qquad \Pi_{i,i-1} = 1.$$

the steady-state probability $P_{n,m}(\infty)$ by P_n^*. To calculate P_n^*, we equate all the time derivatives to zero in the forward equations (2.0.3) and in the boundary conditions (4a) and (6a), to get

$$0 = -\lambda_l P_l^* + \mu_{l+1} P_{l+1}^* \tag{37a}$$

$$0 = \lambda_{n-1} P_{n-1}^* - (\mu_n + \lambda_n) P_n^* + \mu_{n+1} P_{n+1}^*, \qquad l < n < u \tag{37b}$$

$$0 = \lambda_{u-1} P_{u-1}^* - \mu_u P_u^* \tag{37c}$$

Adding all the equations between the lth and nth equation, we get

$$-\lambda_n P_n^* + \mu_{n+1} P_{n+1}^* = 0, \qquad l \leqslant n \leqslant u-1 \tag{38}$$

Therefore,

$$P_n^* = \frac{\lambda_{n-1}}{\mu_n} P_{n-1}^* = \eta_{n,l} P_l^*, \qquad l \leqslant n \leqslant u-1 \tag{39}$$

where

$$\eta_{n,l} = \frac{\lambda_l \lambda_{l+1} \cdots \lambda_{n-1}}{\mu_{l+1} \mu_{l+2} \cdots \mu_n}, \qquad \eta_{l,l} = 1 \tag{40}$$

$\eta_{n,l}$ is the ratio between the probabilistic transition rates from state l to state n and from state n to state l. Since

$$\sum_{n=l}^{u} P_n^* = 1 \tag{41}$$

from Eq. (39) we get

$$P_l^* = \left(\sum_{n=l}^{u} \eta_{n,l} \right)^{-1} \tag{42}$$

Substituting Eq. (42) into (39), we finally get

$$P_n^* = \eta_{n,l} \Big/ \left(\sum_{i=l}^{u} \eta_{i,l} \right) \tag{43}$$

This expression for P_n^* can be used to calculate the steady-state values of the moments of n. Defining the function $Z(x)$ by

$$Z(x) = \sum_{n=l}^{u} \eta_{n,l} x^n = \sum_{n=l}^{u} \eta_{n,l} e^{n\theta}, \qquad \theta = \ln x \tag{44}$$

the jth moment is given by

$$\langle n^j \rangle = \sum_{n=l}^{u} n^j P_n^* = \sum_{n=l}^{u} n^j \eta_{n,l} \Big/ \sum_{n=l}^{u} \eta_{n,l}$$
$$= \frac{d^j}{d(\ln x)^j} \ln Z(x) \bigg|_{x=1} \tag{45}$$

In particular, the first moment and variance are given by

$$\langle n \rangle = \frac{d}{dx} \ln Z(x) \bigg|_{x=1} \tag{46a}$$

$$\mathrm{var}(n) = \langle n^2 \rangle - \langle n \rangle^2 = \frac{d^2}{dx^2} \ln Z(x) \bigg|_{x=1} \tag{46b}$$

To calculate $F_{N,m}(t)$, $\overset{j}{M}_{N,m}$, and $R_{N,m}$ we note that if $m < N$, the time to reach state N for the first time is the same as the time for absorption when the process is confined between the reflecting state l and the absorbing state N, and if $m > N$, a similar statement is true except the reflecting state is u instead of l. Therefore, the results for $F_{N,m}(t)$, $\overset{j}{M}_{N,m}$, and $R_{N,m}$ can be derived from the results of Section 2.2A. These are summarized in Table 2.7.

TABLE 2.7

Expressions for various quantities of interest for a process confined between *two reflecting states l and u, $l < u$*

$P_{n,m}(\infty)$	$\dfrac{\lambda_l \lambda_{l+1} \cdots \lambda_{n-1}}{\mu_{l+1}\mu_{l+2} \cdots \mu_n} \Bigg/ \sum_{i=l}^{u} \dfrac{\lambda_l \lambda_{l+1} \cdots \lambda_{i-1}}{\mu_{l+1}\mu_{l+2} \cdots \mu_i}$	
	$l \leqslant N \leqslant m \leqslant u$	$l \leqslant m \leqslant N \leqslant u$
$R_{N,m}$ $M_{N,m}$ $\overset{j}{M}_{N,m}$ $V_{N,m}$	See column 2, Table 2.4	See column 1, Table 2.3

D. *Process with Only One Special State*

If instead of having two special states l and u, the process has only one special state (reflecting or absorbing), the state space is a semiinfinite space. In case l (u) is the special state, the state space is $l \leqslant n < \infty$ ($-\infty < n \leqslant u$). The processes of biological interest are those in which the probability to drift to infinity is zero. For such processes the expressions for the various quantities of interest can be obtained by taking a suitable limit ($u \to +\infty$ when l is the special state, $l \to -\infty$ when u is the special state) of the expressions derived in one of the preceding subsections for a process in which the boundary corresponding to the present infinite boundary is reflecting. Thus when l is a reflecting state, we take the limit $u \to \infty$ of the results of Table 2.7, and of Table 2.4 when l is absorbing. Similarly when u is the special state, we take the limit $l \to -\infty$ of the results of Table 2.7 when u is reflecting, and of Table 2.3 when u is absorbing. Before taking the limit $l \to -\infty$ of $P_{n,m}(\infty)$ of Table 2.7, we note that this quantity can also be written as

$$P_{n,m}(\infty) = \frac{\mu_u \mu_{u-1} \cdots \mu_{n+1}}{\lambda_{u-1}\lambda_{u-2}\cdots\lambda_n} \Bigg/ \sum_{i=l}^{u} \frac{\mu_u \mu_{u-1} \cdots \mu_{i+1}}{\lambda_{u-1}\lambda_{u-2}\cdots\lambda_i}$$

The results obtained by taking the limits are summarized in Tables 2.8 and 2.9.

TABLE 2.8

Expressions for various quantities of interest for a process restricted below by *a special state* l (reflecting or absorbing), $l \leqslant n < \infty$

	l Absorbing	l Reflecting
$P_{n,m}(\infty)$	$0 \quad (n > l)$	$\dfrac{\lambda_l \lambda_{l+1} \cdots \lambda_{n-1}}{\mu_{l+1}\mu_{l+2}\cdots\mu_n} \Bigg/ \sum_{i=l}^{\infty} \dfrac{\lambda_l \lambda_{l+1} \cdots \lambda_{i-1}}{\mu_{l+1}\mu_{l+2}\cdots\mu_i}$
	$l \leqslant m \leqslant N$	
$R_{N,m}$ $M_{N,m}$ $\overset{j}{M}_{N,m}$ $V_{N,m}$	See column 1, Table 2.4	See column 1, Table 2.3
	$l \leqslant N \leqslant m \quad$ (l may also be $-\infty$)	
$R_{N,m}$	$\sum_{i=m}^{\infty} \Pi_{N+1,i} \Big/ \sum_{i=N}^{\infty} \Pi_{N+1,i}$	
$M_{N,m}$	$\sum_{i=N+1}^{m} M_{i-1,i}$	[a]
$\overset{j}{M}_{N,m}$	$\sum_{i=N+1}^{m} \sum_{n=0}^{\infty} (\mu_{i+n}\Pi_{i,i+n-1})^{-1} \overset{j-1}{M}_{N,i+n}$	[a]
$V_{N,m}$	$\sum_{i=N+1}^{m} \left[2 \sum_{n=0}^{\infty} \Pi_{i,i+n}^{-1} (M_{i+n,i+n+1})^2 - (M_{i-1,i})^2 \right]$	[a]

[a] The expressions for $l \leqslant N \leqslant m$ are valid only when

$$\sum_{n=l+1}^{\infty} \Pi_{l+1,n} = \sum_{n=l+1}^{\infty} \frac{\mu_{l+1}\mu_{l+2}\cdots\mu_n}{\lambda_{l+1}\lambda_{l+2}\cdots\lambda_n} = \infty$$

$M_{i-1,i}$ is defined by

$$M_{i-1,i} = \frac{1}{\mu_i} + \frac{\lambda_i}{\mu_i \mu_{i+1}} + \frac{\lambda_i \lambda_{i+1}}{\mu_i \mu_{i+1} \mu_{i+2}} + \cdots = \frac{\lambda_i}{\mu_i} M_{i,i+1} + \frac{1}{\mu_i}, \qquad i \geqslant N+1$$

TABLE 2.9

Expressions for various quantities of interest for a process restricted above by *a special state u* (reflecting or absorbing), $-\infty < n \leqslant u$

	u Absorbing	u Reflecting	
$P_{n,m}(\infty)$	$0 \quad (n < u)$	$\dfrac{\mu_u \mu_{u-1} \cdots \mu_{n+1}}{\lambda_{u-1}\lambda_{u-2} \cdots \lambda_n} \Big/ \sum_{i=0}^{\infty} \dfrac{\mu_u \mu_{u-1} \cdots \mu_{u-i}}{\lambda_{u-1}\lambda_{u-2} \cdots \lambda_{u-i-1}}$	
		$N \leqslant m \leqslant u$	
$R_{N,m}$, $M_{N,m}$, $\overset{j}{M}_{N,m}$, $V_{N,m}$	See column 2, Table 2.3	See column 2, Table 2.4	
	$m \leqslant N < u \quad$ (u may also be $+\infty$)		
$R_{N,m}$	$\sum_{i=0}^{\infty} \Pi_{m-i,N-1}^{-1} \Big/ \sum_{i=0}^{\infty} \Pi_{N-i,N-1}^{-1}$		
$M_{N,m}$	$\sum_{i=m}^{N-1} M_{i+1,i}$		[a]
$\overset{j}{M}_{N,m}$	$\sum_{i=m}^{N-1} \sum_{n=0}^{\infty} \lambda_{i-n}^{-1} \Pi_{i+1-n,i} \overset{j-1}{M}_{N,i-n}$		[a]
$V_{N,m}$	$\sum_{i=m}^{N-1} \left[2 \sum_{n=0}^{\infty} \Pi_{i-n,i} (M_{i-n,i-n-1})^2 + (M_{i+1,i})^2 \right]$		[a]

[a] The expressions for $m \leqslant N \leqslant u$ are valid only when

$$\sum_{n=1}^{\infty} (\Pi_{u-n,u-1})^{-1} = \sum_{n=1}^{\infty} \frac{\lambda_{u-1}\lambda_{u-2} \cdots \lambda_{u-n}}{\mu_{u-1}\mu_{u-2} \cdots \mu_{u-n}} = \infty$$

$M_{i+1,i}$ is defined by

$$M_{i+1,i} = \frac{1}{\lambda_i} + \frac{\mu_i}{\lambda_i \lambda_{i-1}} + \frac{\mu_i \mu_{i-1}}{\lambda_i \lambda_{i-1} \lambda_{i-2}} + \cdots = \frac{\mu_i}{\lambda_i} M_{i,i-1} + \frac{1}{\lambda_i}, \qquad i \leqslant N-1$$

A characterization of a process that does not drift to infinity can be determined by considering the expression for the probability $R_{i,m}$ of the process to reach a state i between the initial state m and the special state (l or u). If the process does not drift to infinity, this probability should be equal to 1. When the state l is the special state, this probability is given by Eq. (35) with $u \to \infty$, i.e.,

$$R_{i,m} = \lim_{u \to \infty} R_{i,m}(u) = \sum_{n=m+1}^{\infty} \Pi_{i+1,n} \Big/ \sum_{n=i}^{\infty} \Pi_{i+1,n}, \qquad i < m \tag{47}$$

This probability equals 1 if

$$\sum_{n=i}^{\infty} \Pi_{i+1,n} = 1 + \sum_{n=i+1}^{\infty} \frac{\mu_{i+1}\mu_{i+2}\cdots\mu_n}{\lambda_{i+1}\lambda_{i+2}\cdots\lambda_n} = \infty \tag{48}$$

Equation (48) is the required condition for the process to remain finite. It is satisfied if for all n above some n_0, $\mu_n/\lambda_n \geqslant 1$. When the state u is a special state, the corresponding condition is

$$\sum_{n=i-1}^{-\infty} \frac{\lambda_{i-1}\lambda_{i-2}\cdots\lambda_n}{\mu_{i-1}\mu_{i-2}\cdots\mu_n} \to \infty \tag{49}$$

which is satisfied if for all n below some n_1, $\lambda_n/\mu_n \geqslant 1$.

References

Bailey, N. T. J. (1964). "The Elements of Stochastic Processes with Applications to the Natural Sciences." Wiley, New York.

Bellman, R., Kalaba, R. R., and Lockett, J. (1966). "Numerical Inversion of the Laplace Transform." Amer. Elsevier, New York.

Cox, D. R., and Miller, H. D. (1965). "The Theory of Stochastic Processes." Wiley, New York.

Erdélyi, A. (1954). "Tables of Integral Transforms," 2 Vols. McGraw-Hill, New York.

Heathcote, C. R., and Moyal, J. E. (1959). The random walk (in continuous time) and its application to the theory of queues. *Biometrika* **46**, 400.

Karlin, S., and McGregor, J. L. (1958). Linear growth, birth and death processes. *J. Math. Mech.* **7**, 643.

Montroll, E. W. (1967). Stochastic processes and chemical kinetics, *in* "Energetics in Metallurgical Phenomenon" (W. M. Muller, ed.), Vol. 3, p. 123. Gordon & Breach, New York.

Montroll, E. W., and Shuler, K. E. (1958). The application of the theory of stochastic processes to chemical kinetics. *Advances in Chem. Phys.* **1**, 361.

Shuler, K. E., Weiss, G. H., and Anderson, K. (1962). Studies in nonequilibrium rate processes. V. The relaxation of moments derived from a master equation. *J. Mathematical Phys.* **3**, 550.

Additional References

Bharucha-Reid, A. T. (1960). "Elements of the Theory of Markov Processes and their Applications." McGraw-Hill, New York.

Doob, J. L. (1953). "Stochastic Processes." Wiley, New York.

Karlin, S. (1966). "A First Course in Stochastic Processes." Academic Press, New York.

Karlin, S., and McGregor, J. L. (1957). The classification of birth and death processes. *Trans. Amer. Math. Soc.* **86**, 366.

Keilson, J. (1962). The use of Green's functions in the study of bounded random walks, with application to queuing theory. *J. Math. and Phys.* **41**, 42.

Keilson, J. (1965). A review of transient behavior in regular diffusion and birth and death processes. *J. Appl. Probability* **1**, 247.
Keilson, J. (1965). A review of transient behavior in regular diffusion and birth and death processes. Part II. *J. Appl. Probability* **2**, 405.
Keilson, J. (1966). A technique for discussing the passage time distribution for stable systems. *J. Roy. Statist. Soc. Ser. B* **28**, 477.
Kemperman, J. H. G. (1962). An analytical approach to the differential equations of the birth-and-death processes. *Michigan Math. J.* **9**, 321.
Ledermann, W., and Reuter, G. E. H. (1954). Spectral theory for the differential equations of simple birth and death processes. *Philos. Trans. Roy. Soc. London Ser. A* **246**, 321.

3

Random Processes Continuous in Time and State Space

In the preceding chapter we discussed the theory of random processes that change continuously in time and can be defined by a single random variable for which the state space is discrete. The process had the additional characteristic that it is Markovian; i.e., its state at time $(t+\Delta t)$ depends on its state at time t, and in time Δt ($\Delta t \to 0$) the transition can occur between neighboring states only. For many processes the state variable, instead of changing discretely, changes continuously. For such processes the equations satisfied by the probabilities specifying the evolution of the process are partial differential equations. If the process satisfies certain conditions, derived later in this chapter, these equations are in the form of diffusion equations. This chapter is devoted to the study of such processes via solutions of the diffusion equations.

The diffusion equations are more amenable to analytical analysis than the differential-difference equations, which specify the evolution of the random processes discrete in state space. Therefore, by approximating the differential-difference equation by a partial differential equation in the form of a diffusion equation, a detailed, though approximate, knowledge of the behavior of a process with discrete state space can be obtained. The approximation improves as the ratio of distance between the allowed states and the value of the random variable describing the process decreases. A further incentive for analyzing the diffusion equations is that, as shown later in this chapter and alluded to in Chapter 1, these same equations arise when we model a dynamical system evolving in the presence of a so-called white noise. As noted in Chapter 1, this modeling approach is very powerful and general in analyzing and understanding complex physical and biological systems.

Since the reader is already familiar with processes that are discrete in space, we will introduce the processes described by the diffusion equation as a limit of processes described by the forward and backward master equations presented in Chapter 2.

3.0 Diffusion Equations—Derivations and Solutions

The limiting procedure for converting the differential-difference equation into a partial differential equation involves introducing a small parameter h and letting $x = nh$, $y = mh$, and $P_{n,m}(t) \equiv P(x \mid y, t)$. Thus $P(x \mid y, t)$ is the probability that the random variable has the value x at time t given that it had the value y at $t = 0$. Next we consider a sequence of birth and death processes corresponding to a sequence of values of h which tend to zero, with transition probability rates $\lambda_n(h)$ and $\mu_n(h)$ such that

$$h(\lambda_n(h) - \mu_n(h)) = a(nh) + O(h) \tag{1a}$$

$$h^2(\lambda_n(h) + \mu_n(h)) = b(nh) + O(h) \tag{1b}^\dagger$$

where $a(nh)$ is finite, $b(nh) > 0$ for all n, and $\lim_{h \to 0} O(h) = 0$. To obtain one of the partial differential equations for the process, we rewrite the forward equation (2.0.3) as

$$\begin{aligned} \partial P(x \mid y, t)/\partial t = {} & \tfrac{1}{2}[(\lambda_{n+1} + \mu_{n+1}) P(x+h \mid y, t) - 2(\lambda_n + \mu_n) P(x \mid y, t) \\ & + (\lambda_{n-1} + \mu_{n-1}) P(x-h \mid y, t)] \\ & - \tfrac{1}{2}[(\lambda_{n+1} - \mu_{n+1}) P(x+h \mid y, t) - (\lambda_n - \mu_n) P(x \mid y, t)] \\ & - \tfrac{1}{2}[(\lambda_n - \mu_n) P(x \mid y, t) - (\lambda_{n-1} - \mu_{n-1}) P(x-h \mid y, t)] \end{aligned}$$

Letting $h \to 0$ and using conditions (1), we arrive at the required equation

$$\frac{\partial}{\partial t} P(x \mid y, t) = -\frac{\partial}{\partial x}[a(x) P(x \mid y, t)] + \tfrac{1}{2} \frac{\partial^2}{\partial x^2}[b(x) P(x \mid y, t)] \tag{2}$$

† For the discrete birth and death processes, the probability for the process to stay in state n during the time Δt is $\exp(-(\lambda_n + \mu_n)\Delta t)$ which tends to 1 as $\Delta t \to 0$. On the other hand, in the continuous state space case, the transitions are infinitesimal but occur continuously. Thus the probability $\exp(-(\lambda_n + \mu_n)\Delta t)$ must tend to 0 as $h \to 0$ and $\Delta t \to 0$. This requirement is met by this condition, if Δt is of order h.

This equation is known as the forward Kolmogorov diffusion equation or the Fokker–Planck equation. By using a similar limiting procedure on the backward master equation (2.0.4), we get another partial differential equation, the so-called backward Kolmogorov diffusion equation,

$$\frac{\partial}{\partial t}P(x\,|\,y,t) = a(y)\frac{\partial P(x\,|\,y,t)}{\partial y} + \tfrac{1}{2}b(y)\frac{\partial^2 P(x\,|\,y,t)}{\partial y^2} \tag{3}$$

Equations (2) and (3) are to be solved with the initial condition

$$\lim_{t\to 0} P(x\,|\,y,t) = \delta(x-y) \tag{4}$$

which states that initially the random variable had the value y. As in Chapter 2, we note that if, instead of the random variable initially having a definite value y, its initial state is specified by a probability density $p(y)$ [with $\int_\Omega p(y)\,dy = 1$, where Ω denotes the state space], then the solution of the diffusion equations is

$$P(x\,|\,t) = \int_\Omega P(x\,|\,y,t)\,p(y)\,dy \tag{5}$$

The function $a(x)$ in Eqs. (2) and (3) is the rate of growth of the mean when the process is at x, i.e.,

$$a(x) = \lim_{\tau\to 0}\frac{1}{\tau}\int_\Omega (z-x)\,P(z\,|\,x,\tau)\,dz \tag{6a}$$

This can be seen by rewriting the right-hand side of this equation as

$$\int_\Omega (z-x)\frac{\partial P(z\,|\,x,0)}{\partial t}\,dz$$

using the differential equation (3) and the initial condition (4), and carrying out the integration by parts. Similarly, $b(x)$ in Eqs. (2) and (3) is the rate of growth of the variance when the process is at x, i.e.,

$$b(x) = \lim_{\tau\to 0}\frac{1}{\tau}\int_\Omega (z-x)^2 P(z\,|\,x,\tau)\,dz \tag{6b}$$

By similar arguments all the growth rates of the higher moments of the change in the state of the random variable vanish, i.e.,

$$\lim_{\tau\to 0}\frac{1}{\tau}\int_\Omega (z-x)^n P(z\,|\,x,\tau)\,dz = 0, \qquad n \geqslant 3 \tag{6c}$$

We now derive the partial differential equations satisfied by $P(x\,|\,y,t)$ for a Markovian process for which the state space is continuous, and derive the conditions under which these differential equations become diffusion equations [Eqs. (2) and (3)]. Our starting equation is the so-called Chapman–Kolmogorov equation:

$$P(x\,|\,y, t_1+t_2) = \int_{\Omega} P(z\,|\,y, t_1)\, P(x\,|\,z, t_2)\, dz \tag{7}$$

This equation is a mathematical manifestation of the Markovian nature of the process, according to which the behavior of the process in the time interval (t_1, t_1+t_2) depends on its state at time t_1, and not on its behavior in the previous time interval $(0, t_1)$. Therefore, if the process is in state y at time $t = 0$, the probability that it will be in state x at a later time t_1+t_2, must be equal to the probability that it will be in some state z at time t_1, times the probability that if it starts from state z at time t_1, it will be in state x at time t_1+t_2, summed over all the intermediate states z. In Eq. (7), we set $t_1 = t$, $t_2 = \tau$ $(\tau \to 0)$, $z = x-\mu$ to get

$$P(x\,|\,y, t+\tau) = \int_{\Omega'} P(x-\mu\,|\,y, t)\, P(x\,|\,x-\mu, \tau)\, d\mu \tag{8}$$

where $(-\Omega')$ is the state space translated by $-x$.

We limit the discussion to cases in which the probability distribution $P(x\,|\,y, t)$ does not change significantly in the short time τ. Therefore $P(x\,|\,y, t+\tau)$ will not differ much from $P(x\,|\,y, t)$, and we can expand it in a Taylor's series:

$$P(x\,|\,y, t+\tau) = P(x\,|\,y, t) + \tau \frac{\partial P}{\partial t}(x\,|\,y, t) + \cdots \tag{9}$$

Since from the initial condition (4), $P(x\,|\,y, 0) = \delta(x-y)$, for small τ $P(x\,|\,x-\mu, \tau)$ is sharply peaked around $\mu = 0$, and the integral in Eq. (8) need be integrated only in the neighborhood of $\mu = 0$. We can therefore expand the integrand of Eq. (8), regarded as a function of x, in a Taylor's series around the point $x+\mu$ to get

$$\begin{aligned} P(x-\mu\,|\,y, t)\, P(x\,|\,x-\mu, \tau) = {} & P(x\,|\,y,t)\, P(x+\mu\,|\,x, \tau) \\ & - \mu \frac{\partial}{\partial x}[P(x\,|\,y, t)\, P(x+\mu\,|\,x, \tau)] \\ & + \tfrac{1}{2}\mu^2 \frac{\partial^2}{\partial x^2}[P(x\,|\,y, t)\, P(x+\mu\,|\,x, \tau)] - \cdots \end{aligned} \tag{10}$$

Substituting this expansion into Eq. (8), carrying out the integration near $\mu = 0$, replacing the left-hand side of (8) by the expansion in (9), and noting that the integral over $P(x+\mu\,|\,x,\tau)$ equals 1, we obtain

$$\tau\left[\frac{\partial P(x\,|\,y,t)}{\partial t}+O(\tau)\right]=-\int_{\Omega^*}\frac{\partial}{\partial x}[P(x\,|\,y,t)\,\mu P(x+\mu\,|\,x,\tau)]\,d\mu$$
$$+\tfrac{1}{2}\int_{\Omega^*}\frac{\partial^2}{\partial x^2}[P(x\,|\,y,t)\,\mu^2 P(x+\mu\,|\,x,\tau)]\,d\mu-\cdots \tag{11}$$

where Ω^* is the neighborhood of $\mu=0$ in which $P(x\,|\,x-\mu,\tau)$ is concentrated. Interchanging the order of integration (with respect to μ) with differentiation (with respect to x), dividing by τ, and taking the limit $\tau\to 0$, we finally arrive at

$$\frac{\partial P(x\,|\,y,t)}{\partial t}=\sum_{n=1}^{\infty}\frac{(-1)^n}{n!}\frac{\partial^n}{\partial x^n}[M_n(x)\,P(x\,|\,y,t)] \tag{12}$$

where

$$M_n(x)=\lim_{\tau\to 0}\frac{1}{\tau}\int_{\Omega^*}\mu^n P(x+\mu\,|\,x,\tau)\,d\mu=\lim_{\tau\to 0}\int_{\Omega}(z-x)^n P(z\,|\,x,\tau)\,dz \tag{13}$$

Equation (12) is the required partial differential equation describing probabilistically the evolution of the process. It reduces to the forward diffusion equation (2) if $M_n(x)=0$ for all $n>2$, and these are the required conditions (in addition to the Markovian nature of the process) under which the forward diffusion equation describes the evolution of the process. By a similar procedure [which involves the expansion of $P(x\,|\,z,t_2)$ in Eq. (7) regarded as a function of z in a Taylor's series around the point y] one can show that under the above-stated conditions $P(x\,|\,y,t)$ also satisfies the backward diffusion equation (3).

We now introduce a wide class of processes for which $M_n(x)=0$ for all $n>2$ and relate the dynamical stochastic equation, obeyed by the processes, with the forward diffusion equation satisfied by the corresponding probability density $P(x\,|\,y,t)$.

Consider a process characterized by the stochastic dynamical equation satisfied by the process variable x,

$$dx/dt=h(x)+e(x)\,i(t) \tag{14}$$

where $i(t)$ is a stochastic memoryless input to the process, $e(x)$ describes the effect of this input, and $h(x)$ is the function describing the rate of change of the variable x in the absence of input. Both $e(x)$ and $h(x)$ are assumed to be differentiable functions; $i(t)$ is characterized by two parameters m and σ^2

defined by

$$\langle i(t)\rangle = m \tag{15a}$$

$$\langle [i(t)-m][i(t+t')-m]\rangle = \sigma^2\delta(t') \tag{15b}$$

with all the correlations of order greater than 3 of $i(t)$ assumed to be zero. The averages are taken over a suitable ensemble, e.g., a number of repetitive observations on the system. Such an $i(t)$ is said to be generated by a Gaussian random process, and σ^2 is known as an incremental variance or intensity of the input. We define a new quantity $F(t)$ by

$$F(t) = \frac{i(t)-m}{\sigma} \tag{16}$$

so that

$$\langle F(t)\rangle = 0 \tag{17a}$$

$$\langle F(t)F(t+t')\rangle = \delta(t') \tag{17b}$$

with vanishing correlations of order 3 and more of $F(t)$: $F(t)$ so defined is called a white noise. Substituting Eq. (16) into Eq. (14) we get

$$dx/dt = \alpha(x) + \beta(x)F(t) \tag{18}$$

where

$$\alpha(x) = h(x) + me(x) \tag{19a}$$

$$\beta(x) = \sigma e(x) \tag{19b}$$

We now show that for the process described by the stochastic dynamical equation (SDE) (14) or equivalently by SDE (18), $M_n(x) = 0$ for all $n \geqslant 3$,

$$M_1(x) = \alpha(x) + \frac{1}{4}\frac{\partial}{\partial x}\{\beta(x)\}^2 \equiv a(x) \tag{20a}$$

$$M_2(x) = \{\beta(x)\}^2 \equiv b(x) \tag{20b}$$

and the probability density $P(x\,|\,y,t)$ satisfies the forward diffusion equation (2).

Dividing Eq. (18) by $\beta(x)$, we obtain

$$dz/dt = \hat{a}(z) + F(t) \tag{21}$$

where

$$dz = dx/\beta(x) \tag{22a}$$

$$\hat{a}(z) = \alpha(x(z))/\beta(x(z)) \tag{22b}$$

To find $M_1(z)$, we integrate Eq. (21) over the short interval $(t, t+\tau)$ to get

$$\Delta z(t) \equiv z(t+\tau) - z(t) = \hat{a}(z)\tau + \int_t^{t+\tau} d\xi\, F(\xi) + O(\tau) \tag{23}$$

Therefore, $M_1(z)$, the growth rate of the mean value when the process is at z, is

$$M_1(z) = \lim_{\tau \to 0} \frac{\langle \Delta z \rangle}{\tau} = \hat{a}(z) + \lim_{\tau \to 0} \int_t^{t+\tau} d\xi \, \langle F(\xi) \rangle = \hat{a}(z) \tag{24}$$

where the last step follows from Eq. (17a). Further, from Eq. (23)

$$\langle (\Delta z)^2 \rangle = \int_t^{t+\tau} \int_t^{t+\tau} d\xi \, d\eta \, \langle F(\xi) F(\eta) \rangle + O(\tau)$$

and by Eq. (17b), the double integral in this equation is τ, so that

$$M_2(z) = \lim_{\tau \to 0} \frac{\langle (\Delta z)^2 \rangle}{\tau} = 1 \tag{25}$$

Since all correlations of $F(t)$ higher than second order vanish, following the same argument, we obtain

$$M_n(z) = 0, \qquad n \geqslant 3 \tag{26}$$

From Eqs. (24)–(26) and (12), $g(z \mid z, t)$, the probability density that the transformed variable defined by Eqs. (22) has the value z at time t when z_0 is its value at $t = 0$ (i.e., when $x = y$), satisfies the forward diffusion equation

$$\frac{\partial g}{\partial t} = -\frac{\partial}{\partial z}(\hat{a}(z) g) + \frac{1}{2} \frac{\partial^2 g}{\partial z^2} \tag{27}$$

Since by the transformation (22a)

$$\operatorname{prob}[x_1 \leqslant x \leqslant x_2] = \int_{x_1}^{x_2} P(x \mid y, t) \, dx = \int_{z_1}^{z_2} P(x(z) \mid y, t) \, \beta(x(z)) \, dz$$

$$= \operatorname{prob}[z_1 \leqslant z \leqslant z_2] = \int_{z_1}^{z_2} g(z \mid z_0, t) \, dz \tag{28a}$$

with $z_i = z(x_i)$, $i = 1, 2$, $g(z \mid z_0, t)$ is related to $P(x \mid y, t)$ by the relation

$$g(z \mid z_0, t) = P(x(z) \mid y, t) \, \beta(x(z)) \tag{28b}$$

Differentiating both sides with respect to z, and using Eqs. (22), we obtain

$$\frac{\partial g}{\partial z} = \frac{\partial(\beta P)}{\partial z} = \beta \frac{\partial(\beta P)}{\partial x} = \frac{\partial(\beta^2 P)}{\partial x} - \frac{P}{2} \frac{\partial \beta^2}{\partial x}, \qquad \frac{\partial^2 g}{\partial z^2} = \beta \frac{\partial}{\partial x}\left(\frac{\partial g}{\partial z}\right) \tag{28c}$$

so that Eq. (27), after the substitutions of Eqs. (28b), (28c), and (22b), becomes

$$\frac{\partial P}{\partial t} = -\frac{\partial}{\partial x}(a(x) P) + \frac{1}{2} \frac{\partial^2}{\partial x^2}(b(x) P) \tag{29}$$

where $a(x)$ and $b(x)$ are given by Eqs. (20a) and (20b), respectively. Equation (29) is the same forward diffusion equation as (2). For such an equation, we have already shown earlier in this section that $M_1(x) = a(x)$, $M_2(x) = b(x)$, and $M_n(x) = 0, n \geqslant 3$, which completes the proof of the assertions made above.

To summarize, we have shown that for a process described by the SDE (14), i.e.,

$$dx/dt = h(x) + e(x)\,i(t) \tag{30a}$$

with $i(t)$ a white noise with nonzero mean m and intensity σ^2, the forward diffusion equation or Fokker–Planck (FP) equation is

$$\frac{\partial P}{\partial t} = -\frac{\partial}{\partial x}\left[\{h(x) + me(x) + \frac{\sigma^2}{4}\frac{\partial}{\partial x}e^2(x)\}\,P\right] + \frac{\sigma^2}{2}\frac{\partial^2}{\partial x^2}[e^2(x)\,P] \tag{30b}$$

Similarly, the corresponding backward equation is

$$\frac{\partial P}{\partial t} = \left\{h(y) + me(y) + \frac{\sigma^2}{4}\frac{\partial}{\partial y}e^2(y)\right\}\frac{\partial P}{\partial y} + \frac{\sigma^2}{2}e^2(y)\frac{\partial^2 P}{\partial y^2} \tag{30c}$$

The results above are very useful in making primitive statistical models of complex biological systems. The deterministic behavior of one (or more) of the components of the complex system can be represented by the dynamical equation $dx/dt = h(x)$, and the remaining unknown fluctuating behavior, due to the presence of other components, can be approximated by $e(x)\,i(t)$ where $i(t)$ is a white noise. This white noise approximation is a reasonable approximation if the fluctuations occur extremely rapidly on the time scale defining the changes in x. This procedure converts the deterministic dynamical equation into a stochastic dynamical equation, and the probabilistic analysis of the process can be carried out by analyzing the diffusion equations (30b) and (30c).

It may be noted that although the FP equation for a given SDE is unique, the reverse is not true. In other words, the problem of finding an SDE for a random process satisfying a given FP equation does not have a unique solution. However, the solution will be unique if we restrict ourselves to equations of the type (18) containing a Gaussian delta-correlated random process $F(t)$ with zero mean and unit intensity. With this restriction, the SDE for the FP equation (29) is Eq. (18), or by a simple redifinition, the SDE for the FP equation

$$\frac{\partial P}{\partial t} = -\frac{\partial}{\partial x}(a(x)P) + \frac{1}{2}\frac{\partial^2}{\partial x^2}(b(x)\,P) \tag{31a}$$

is

$$\frac{dx}{dt} = a(x) - \frac{1}{4}\frac{\partial b(x)}{\partial x} + \{b(x)\}^{1/2}F(t) \tag{31b}$$

The formalism presented here for the derivation of an FP equation from a given SDE, follows the Stratonovich rules (Stratonovich, 1963). There is a controversy in this derivation when the coefficient of the white noise in the SDE [$e(x)$ in Eq. (14)] depends on the process variable x. This controversy arises from the pathological nature of the white noise (similar to the pathological nature of the δ function), which is well defined only in terms of its integral $\int_0^t F(\tau)\, d\tau$. We followed the Stratonovich approach since its rules are the same as ordinary calculus, and transformation of variables, as the one carried out in Eqs. (22), is valid. The other approach, taken by Doob (1953) and Ito (1944), uses the Ito calculus, where differentiation and integration rules differ from the ordinary ones. A discussion of this controversy from the point of view of modeling reality is given by Mortensen (1969). He concludes that, since a white noise is only an approximation to the stochastic behavior in nature, the stochastic process, derived by using any of the two approaches, should be checked against reality, and its capability to predict results which are an acceptable approximation to the actual behavior is the only criterion to be considered.

We now discuss the solution of the diffusion equations. Since $P(x\,|\,y,t)$ describes the evolution of the process completely, we need solve the forward diffusion equation—the FP equation. From the discussion given above, the FP equation of the type (29) can be transformed into an FP equation of the type (27) with the substitutions

$$dz = [b(x)]^{-1/2}\, dx, \qquad z = z(x) = \int^x [b(\xi)]^{-1/2}\, d\xi \tag{32a}$$

$$\hat{a}(z) = \left[a(x) - \frac{1}{4}\frac{db}{dx}\right][b(x)]^{-1/2} \tag{32b}$$

$$g(z\,|\,z_0,t) = [b(x)]^{1/2} P(x\,|\,y,t)|_{x=x(z)} \tag{32c}$$

and the initial condition

$$g(z\,|\,z_0,0) = \delta(z-z_0), \qquad z_0 = z(y) \tag{32d}$$

In Eq. (32b) the right-hand side is to be expressed in terms of z given by (32a). Therefore, it is sufficient for us to describe the method for solving Eq. (27) with the boundary conditions which depend on the allowed range of the variable z.

As in the case of discrete processes, the random processes may be basically of two types, one in which there are no forced restrictions on the allowed range of the random variable (we shall call such processes unrestricted), and the other type in which boundary conditions are imposed at one or two points of the state space (such processes will be called restricted). The appropriate boundary conditions imposed on the latter type of processes are given in Section 3.2, and the behavior of unrestricted processes near their "built-in" boundaries (finite or infinite) is discussed in Section 3.1. In the remaining part of this section we formally indicate a method for solving the FP equation, and in the next two sections we indicate how to apply this method for the various processes.

To solve Eq. (27), we use the standard method of separation of variables. Taking

$$g(z \mid z_0, t) = Q(z) e^{-Et/2} \tag{33}$$

as the trial solution, Eq. (27) becomes

$$\frac{d^2 Q}{dz^2} - \frac{d}{dz}(\hat{a}(z) Q) + EQ = 0 \tag{34}$$

This equation is to be solved subject to the boundary conditions on Q [implied by the boundary conditions on g through Eq. (33)], i.e., it is an eigenvalue problem. Depending on the form of the function $\hat{a}(z)$ and the boundary conditions, there will be a discrete and/or continuous set of E, with a corresponding set of Q. For a discrete set $\{E_n, Q_n\}$, the solution is

$$g(z \mid z_0, t) = \sum_n \alpha_n Q_n \exp(-E_n t/2) \tag{35a}$$

and for a continuous set

$$g(z \mid z_0, t) = \int \alpha(E) Q(E) e^{-Et/2} \, dE \tag{35b}$$

where α_n or $\alpha(E)$ is to be evaluated by using the initial condition (32d).

An equivalent and useful form of Eq. (34) is obtained by making a transformation which converts it into a differential equation free of a first-order partial derivative in z. Such a transformation is

$$Q = \psi(z)[\pi(z)]^{1/2} \tag{36a}$$

where

$$\pi(z) = \exp\left\{-2 \int^z \hat{a}(\xi) \, d\xi\right\} \tag{36b}$$

On carrying out this transformation, Eq. (34) becomes

$$d^2\psi/dz^2 + [E - U(z)]\psi(z) = 0 \tag{37}$$

where

$$U(z) = d\hat{a}/dz + \hat{a}^2 \tag{38}$$

The bonudary conditions are also transformed into the boundary conditions on ψ. Equation (37) and these boundary conditions once again constitute an eigenvalue problem.

In case z is confined between two finite boundaries and $U(z)$ is finite within these boundaries, the set of eigenvalues and eigenfunctions $\{E_n, \psi_n(z)\}$ for this eigenvalue problem is discrete, and the set of functions $\{\psi_n(z)\}$ is orthonormal [see, e.g., Titchmarsh (1962)], i.e.,

$$\int_\Omega \psi_n(z)\,\psi_m(z)\,dz = \delta_{mn} \tag{39}$$

In view of the transformation (36) the set of functions $\{Q_n(z)\}$ is an orthonormal set with respect to the weight function $\pi(z)$, i.e.,

$$\int_\Omega Q_n(z)\,Q_m(z)\,\pi(z)\,dz = \delta_{mn} \tag{40}$$

Using relation (40) and the initial condition (32d) the coefficients $\{\alpha_n\}$ in Eq. (35a) can be evaluated by putting $t = 0$ in this equation, multiplying both sides by $\pi(z)\,Q_n(z)$, and integrating over Ω. The resulting form of α_n is

$$\alpha_n = \int_\Omega \pi(z)\,Q_n(z)\,\delta(z-z_0)\,dz = Q_n(z_0)\,\pi(z_0)$$

Substituting α_n into Eq. (35a), we get the simple expression

$$g(z\,|\,z_0, t) = \pi(z_0) \sum_{n=0}^{\infty} Q_n(z)\,Q_n(z_0)\exp(-E_n t/2) \tag{41}$$

Relation (41) is valid also in case of an infinite state space, if $U(z)$ tends to infinity at the infinite boundaries. Otherwise, the set of eigenvalues $\{E_n\}$ is not discrete anymore, and there is a continuous interval of eigenvaiues.

There is an advantage in using the form (37) instead of (34), since this form is very similar to the time-independent Schrödinger equation of quantum physics

$$\frac{d^2\psi}{dz^2} + \frac{2m}{\hbar^2}[E - U(z)]\,\psi(z) = 0$$

which describes the motion, in one dimension, of a particle of mass m moving in a field with potential $U(z)$. E is the energy of the particle, $\hbar = 2\pi h$ is Planck's constant, and $\Psi(z,t) = \psi(z)\exp(-iEt/\hbar)$ is the wave function of the particle and has the physical significance that $|\Psi(z,t)|^2\,dz$ is the probability of finding the particle at a point between z and $z+dz$ at time t. Because Schrödinger's equation has been studied quite extensively in mathematical physics, together with many approximation methods for solving it, an extensive literature becomes immediately available for the solution of FP equations.

We now make an important observation by writing $\hat{a}(z)$ in the form

$$\hat{a}(z) = \phi'(z)/\phi(z) \tag{42}$$

Substituting Eq. (42) into (38) we get

$$\phi''(z) - U(z)\phi(z) = 0 \tag{43}$$

Comparing Eq. (37) with this equation, we note that $\phi(z)$ is a solution of (37) when E is taken to be equal to zero. Thus by choosing $U(z)$ such that (37) is analytically solvable for $U(z)$ and $U(z)+$constant, one can generate a set of $\hat{a}(z)$ [with $\hat{b}(z) = 1$] for which the diffusion equation (27) can be analytically solved. For these $\hat{a}(z)$ and $\hat{b}(z) = 1$, one can calculate $a(x)$ for a given $b(x)$ by using Eqs. (32) or calculate $a(x)$ and $b(x)$ from a given $\beta(x)$ from Eqs. (20) and (22).

One of the authors (Buff and Goel, 1969; Clay *et al.*, 1972) encountered a similar situation in connection with the electrostatic potential at an arbitrary point due to a point charge, in the presence of an interface between two dielectric media. The interface was taken to have an anisotropic and inhomogeneous dielectric tensor. A systematic search for the dielectric profiles for which the electrostatic potential was a quadrature was made by solving equations like (37) and (43).

On the basis of the literature on Schrödinger's equation on electrostatics and other areas of mathematical physics, we have compiled a list of $\hat{a}(z)$ for which one of the two second-order differential equations (34) and (37) has been solved. We give $\hat{a}(z)$ and the solutions in Table 3.1. In Table 3.2 we summarize a list of $a(x)$ and $b(x)$ for which the eigenfunctions of the FP equation (2) are known. For some of the more important cases we give the explicit expressions for $P(x|y,t)$ in Section 3.2 (Table 3.4) with specified boundary conditions.

When an analytical calculation of $P(x|y,t)$ is somewhat difficult, and if one is interested only in the first few moments of x, one has only to solve a set of differential equations which may be considerably simpler. The moments of x, if they are finite, satisfy ordinary differential equations, which can be derived from the forward diffusion equation (2) by multiplying it by x^n and then integrating over x from A to B, where the allowed range of the variable is $A < x < B$. The resulting equation is

$$\frac{d}{dt}\int_A^B x^n P(x|y,t)\,dx = \frac{d}{dt}\langle x^n\rangle = -\int_A^B x^n \frac{\partial}{\partial x}[a(x)P(x|y,t)]\,dx + \tfrac{1}{2}\int_A^B x^n \frac{\partial^2}{\partial x^2}[b(x)P(x|y,t)]\,dx$$

Integrating by parts we get

$$\frac{d}{dt}\langle x^n\rangle = n\int_A^B x^{n-1}a(x)P(x|y,t)\,dx + \frac{n(n-1)}{2}\int_A^B x^{n-2}b(x)P(x|y,t)\,dx - \left[x^n a(x)P - \frac{x^n}{2}\frac{\partial}{\partial x}[b(x)P] + \tfrac{1}{2}nx^{n-1}b(x)P\right]_A^B \tag{44}$$

TABLE 3.1

Solutions to the eigenvalue equation $\psi'' + (E - U)\psi = 0$, derived from an FP equation with $b(x) = 1$

$a(x)$	$U(x)$	$\psi(x) = h(x)\phi(x)$: $h(x)$	$\psi(x) = h(x)\phi(x)$: $\phi(x)$	References[a]
0 $1/x$ $\beta/(\alpha+\beta x)$	0	1	$\cos E^{1/2}x,\ \sin E^{1/2}x$	
α $\alpha \tanh \alpha x$ $\alpha \coth \alpha x$	α^2	1	$\exp[(\alpha^2-E)^{1/2}x],\ \exp[-(\alpha^2-E)^{1/2}x], \quad \alpha^2 > E$ $\cos(E-\alpha^2)^{1/2}x,\ \sin(E-\alpha^2)^{1/2}x, \quad \alpha^2 < E$	
$-\alpha \tan \alpha x$ $+\alpha \cot \alpha x$	$-\alpha^2$	1	$\cos(E+\alpha^2)^{1/2}x,\ \sin(E+\alpha^2)^{1/2}x$	
$\alpha \tan x$	$-\alpha^2 + \dfrac{\alpha(1-\alpha)}{\cos^2 x}$	$(\cos x)^\alpha$	Hypergeometric functions $F(-\nu, \nu+2\alpha; \alpha+\frac{1}{2}; \sin^2(\pi/4 - x/2))$ $\nu(\nu+2\alpha) = E$ ν = integer – Gegenbauer polynomials $T_\nu^{(\alpha)}(\sin x)$	NBS p. 781
$\alpha \cot x$	$-\alpha^2 - \dfrac{\alpha(1-\alpha)}{\sin^2 x}$	$(\sin x)^\alpha$	Hypergeometric functions $F(-\nu, \nu+2\alpha; \alpha+\frac{1}{2}; \sin^2 x/2)$ $\nu(\nu+2\alpha) = E$ ν = integer – $T_\nu^{(\alpha)}(\cos x)$	NBS p. 781
$\alpha x + \beta$	$\alpha + (\alpha x+\beta)^2$	$\exp[-(\alpha x+\beta)^2/2\alpha]$	Hermite functions with argument $(\alpha x+\beta)/\alpha^{1/2}$	NBS p. 781
$\alpha/x + \beta$	$\beta^2 + \dfrac{2\alpha\beta}{x} + \dfrac{\alpha(\alpha-1)}{x^2}$	$\exp[-(1-\beta^{-2}E)^{1/2}\beta x]\, x^{1-\alpha}$	Confluent hypergeometric functions $F(1-\alpha+\alpha(1+\beta^{-2}E)^{-1/2}; 2-2\alpha; (1-\beta^{-2}E)^{1/2}2\beta x)$	NBS p. 505
$\alpha x + \beta/x$	$\alpha^2 x^2 + \alpha(1+2\beta) - \dfrac{1-(2\beta-1)^2}{4x^2}$	$x^{\mu+1/2}\exp(-\|\alpha\| x^2/2)$ $\mu = \pm(\beta - \frac{1}{2})$	Confluent hypergeometric functions $F(-\lambda; \mu+1; \|\alpha\| x^2),\ 4\lambda = \dfrac{E}{\|\alpha\|} - \dfrac{\alpha}{\|\alpha\|}(1+2\beta) - 2\mu - 2$ $\lambda = n$, an integer – Laguerre polynomials $L_n^{(\mu)}(\|\alpha\| x^2)$	NBS p. 781

[a] NBS—Abramowitz and Stegun (1964).

TABLE 3.2

Solutions to the eigenvalue equation $\frac{\partial^2}{\partial x^2}[b(x)\,Q(x)] - 2\frac{\partial}{\partial x}[a(x)\,Q(x)] + EQ(x) = 0$

$a(x)$	$b(x)$	Eigenfunctions $Q(x)$	References[a]
$\alpha x + \beta$	$x(1-x)$	Hypergeometric functions $F(a, b; 2(1-\beta); x)$	MOS p. 42
		$ab = 2\alpha + 2 - E,\ a + b = 2\alpha + 3$	
$\alpha x + \beta$	$1 - x^2$	Hypergeometric functions $F\left(-\lambda, 2\alpha+\lambda+3; \alpha+\beta+2; \frac{1-x}{2}\right)$	
		$\lambda(\lambda+2\alpha+3) = E - 2(\alpha+1)$	
		$\alpha = -1; \beta = 0$ Legendre functions $P_\lambda(x)$, $Q_\lambda(x)$	MOS p. 151
		$\lambda = n$, an integer, Jacobi polynomials $P_n^{(\alpha+\beta+1,\,\alpha-\beta+1)}(x)$	MOS p. 209
		$\beta = 0$, Gegenbauer polynomials $T_n^{(\alpha+3/2)}(x)$	MOS p. 218
		$\alpha = -\frac{3}{2}$, Chebycheff polynomials of first kind $T_n(x)$	MOS p. 256
		$\alpha = -\frac{1}{2}$, Chebycheff polynomials of second kind $U_n(x)$	
		$\alpha = -1$, Legendre polynomials $P_n(x)$	MOS p. 227
$\alpha x + \beta$	x	Confluent hypergeometric functions (Kumer's functions)	MOS p. 268
		$F(1-E/2\alpha; 2(1-\beta); 2\alpha x)$	
		$E/2\alpha = n$, an integer, Laguerre polynomials $L_{n-1}^{(1-2\beta)}(2\alpha x)$	
$\alpha(1-x^2)$	$1 - x^2$	$e^{\alpha x}S(x)$: $S(x)$ oblate spheroidal functions	SMCLC p. 2
		$(1-x^2)\,S''(x) - 4xS'(x) + \{E-2-\alpha^2+\alpha^2x^2\}\,S(x) = 0$	
$\alpha x(1-x)$	$x(1-x)$	$e^{\alpha x}R(1-2x)$: $R(x)$ oblate spheroidal functions	SMCLC p. 2
		$(1-x^2)\,R''(x) - 4xR'(x) + \left\{E - 2 - \frac{\alpha^2}{4} + \frac{\alpha^2}{4}x^2\right\}R(x) = 0$	

[a] MOS—Magnus *et al.* (1966), SMCLC—Stratton *et al.* (1956).

If the term in square brackets vanishes at $x = A$ and B, (44) reduces to

$$\frac{d}{dt}\langle x^n\rangle = n\langle x^{n-1}a(x)\rangle + \frac{n(n-1)}{2}\langle x^{n-2}b(x)\rangle \tag{45}$$

Equation (45) has then to be solved subject to the initial conditions

$$\langle x^n\rangle(t=0) = y^n \tag{46}$$

For $a(x)$ and $b(x)$ polynomials in x, Eq. (45) is a set of coupled linear ordinary differential equations in $\langle x^n\rangle$, $n \geqslant 1$. When the polynomial $a(x)$ is of degree at most 1 and the polynomial $b(x)$ is of degree at most 2, the nth equation in the set (45) involves moments up to order n, and hence the equations can be solved successively. For example, if

$$a(x) = a_0 + a_1 x, \qquad b(x) = b_0 + b_1 x + b_2 x^2 \tag{47}$$

Eq. (45) for $n = 1$ and 2, becomes

$$d\langle x\rangle/dt = \langle a(x)\rangle = a_0 + a_1\langle x\rangle \tag{48}$$

$$d\langle x^2\rangle/dt = 2\langle xa(x)\rangle + \langle b(x)\rangle = b_0 + (b_1+2a_0)\langle x\rangle + (2a_1+b_2)\langle x^2\rangle \tag{49}$$

Equation (48) is linear in $\langle x\rangle$ and can be easily integrated to give

$$\langle x\rangle = \left(y + \frac{a_0}{a_1}\right)\exp(a_1 t) - \frac{a_0}{a_1} \tag{50}$$

This solution, when substituted in Eq. (49), gives a linear equation in $\langle x^2\rangle$ which can be easily integrated.

When either $a(x)$ is a polynomial of degree greater than 1 or $b(x)$ is of degree greater than 2, the coupled system of equations (45) can only be solved approximately† by using one of the methods presented in Appendix A in connection with birth and death processes.

In addition to moments of x, there are some other quantities which provide insight into the evolution of the process, and for which an analytical expression can be derived. These quantities are the continuous analogs of the quantities defined in Section 2.1 (p. 9). Here we list these quantities together with some general results which are independent of the type of process.

(a) The steady-state probability density $P(x\,|\,y,\infty)$, which describes the process when it is in some dynamical equilibrium. Such a dynamical equilibrium is reached since $a(x)$ and $b(x)$ do not depend on time.

† The solution of Eq. (45) for $n = 1$ is independent of $b(x)$ as long as $a(x)$ is at most linear, and $\langle x\rangle$ is given by Eq. (50) even when $b(x)$ is of degree greater than 2.

To calculate $P(x\,|\,y,\infty)$, we set $\partial P/\partial t = 0$ in Eq. (2) to get

$$J(x\,|\,y,\infty) = \text{constant} \equiv J(\infty) \tag{51}$$

where

$$J(x\,|\,y,t) \equiv a(x)P(x\,|\,y,t) - \frac{1}{2}\frac{\partial}{\partial x}[b(x)P(x\,|\,y,t)] \tag{52}$$

Since in terms of this function J, the FP equation (2) is

$$\partial P/\partial t + \partial J/\partial x = 0 \tag{53}$$

J can be interpreted as the probability current, and Eq. (53) as the equation of conservation of probability. The general solution of Eq. (52) for $t=\infty$ is obtained by writing

$$P(x\,|\,y,\infty) = v(x)/b(x) \tag{54}$$

so that Eq. (52) becomes

$$\frac{dv}{dx} - 2\frac{a(x)}{b(x)}v = -2J(\infty)$$

This is a linear equation in v, which admits the solution

$$v(x) = -2J(\infty)\int^{x}\exp\left\{2\int_{x'}^{x}\frac{a(\xi)}{b(\xi)}\,d\xi\right\}dx' + C\exp 2\int^{x}\frac{a(\xi)}{b(\xi)}\,d\xi \tag{55}$$

where C is an arbitrary constant of integration. Substituting this expression for $v(x)$ into Eq. (54) we get the expression for $P(x\,|\,y,\infty)$ in terms of two constants $J(\infty)$ and C. The constant C is determined by a normalization condition, and $J(\infty)$ depends on the nature of the process. In this monograph we limit the discussion to processes in which there is no flow of probability into the state space from the outside. Therefore, at the boundaries, $J(x\,|\,y,t) \geqslant 0$. In case $J(x\,|\,y,t)$ vanishes at both boundaries for all $t \geqslant 0$, $J(\infty) = 0$ by Eq. (51), and the steady-state probability density is of the form

$$P(x\,|\,y,\infty) = \frac{C}{b(x)}\exp\left\{2\int^{x}[a(\xi)/b(\xi)]\,d\xi\right\} \tag{56a}$$

where C is determined by the condition

$$\int_{\Omega} P(x\,|\,y,\infty)\,dx = 1 \tag{56b}$$

In case there is a positive flow of probability out from the state space, at least at one of the boundaries, as $t \to \infty$ all probability is bound to be outside the state space, and the steady-state density is the trivial solution of Eq. (51), i.e.,

$$P(x\,|\,y,\infty) = 0 \tag{56c}$$

(b) The probability $R(z|y)$ that the random variable ever takes the value z.

(c) The "first passage time" $T(z|y)$, the time for the process to take the value z for the first time, its probability density function $F(z|y,t)$, and its arbitrary (ith) moment $M_i(z|y,t)$, $i \geqslant 1$.

$F(z|y,t)$ is related to $P(x|y,t)$ through the relation

$$P(x|y,t) = \int_0^t F(z|y,t-\tau)\,P(x|z,\tau)\,d\tau, \qquad y \leqslant z \leqslant x \quad \text{or} \quad x \leqslant z \leqslant y \tag{57}$$

This relation is the continuous analog of relation (2.1.8) and reflects the fact that the random variable can take the value x at time t when initially it has the value y, only if it takes an intermediate value z at some time $t-\tau$ in the time interval $(0,t)$, and then in the remaining time τ its value changes from z to x. An equivalent and simpler form of Eq. (57) is obtained by taking the Laplace transform of Eq. (57). Using the theorem on the Laplace transform of a convolution integral, we get

$$f(z|y,s) = p(x|y,s)/p(x|z,s) \tag{58}$$

where $f(z|y,s)$ and $p(x|y,s)$ are the Laplace transforms of $F(z|y,t)$ and $P(x|y,t)$, respectively.

In the derivations of $F(z|y,t)$ and the moments of $T(z|y)$, the differential equation satisfied by $F(z|y,t)$ is the starting point, and we will now show that this equation is the backward diffusion equation. Inserting Eq. (57) into the backward diffusion equation (3), we get

$$-F(z|y,0)\,P(x|y,t) = \int_0^t \left[\frac{\partial}{\partial t} - a(y)\frac{\partial}{\partial y} - \tfrac{1}{2}b(y)\frac{\partial^2}{\partial y^2}\right] F(z|y,t-\tau)\,P(x|z,\tau)\,d\tau$$

Since this equation is valid for all t and since, by definition, $F(z|y,0) = 0$ for $y \neq z$, $F(z|y,t)$ has to satisfy the backward equation

$$\frac{\partial F(z|y,t)}{\partial t} = a(y)\frac{\partial F(z|y,t)}{\partial y} + \tfrac{1}{2}b(y)\frac{\partial^2 F(z|y,t)}{\partial y^2} \tag{59}$$

The initial condition follows from the initial condition (4) on $P(x|y,t)$ and from the observation that in Eq. (57), $x = y$ only if $x = y = z$. Therefore,

$$F(z|y,0) = \delta(z-y) \tag{60}$$

One of the two boundary conditions to be imposed on $F(z|y,t)$ follows directly from Eq. (57), i.e.,

$$F(z|z,t) = \delta(t) \tag{61}$$

while the second condition depends on the nature of the process, and is given in subsequent sections where the different types of processes are discussed.

(d) The probability $R(z \mid y, w)$ that the random variable takes the value z before taking the value w.

(e) The time $T(z \mid y, w)$ for the random variable to take the value z for the first time before it takes the value w, its probability density function $F(z \mid y, w)$, and its arbitrary (ith) moment $M_i(z \mid y, w)$, $i \geqslant 1$.

The quantities in (d) and (e) are independent of the behavior of the process once the variable takes the value z or w, and therefore their discussion is postponed to Section 3.2B, where a process confined between two absorbing boundaries is discussed.

With this general formulation of continuous processes in hand, we now proceed to discuss processes according to whether they are unrestricted or restricted.

3.1 Unrestricted Processes (Singular Processes)

In this section we discuss processes that are completely defined by the coefficients $a(x)$ and $b(x)$ of the FP equation with no additional boundary conditions. Such processes may have an unlimited state space if $b(x) > 0$ and $a(x)$ is finite for all x, or may be confined by some "built-in" boundaries (which are either inaccessible, absorbing, or reflecting) in case $b(x)$ vanishes at some point or $a(x)$ becomes infinite. These processes are the continuous counterpart of the birth and death processes discussed in Section 2.1, for which any limitation on the set of states is due to states where either λ_n or μ_n vanishes.

An extensive study of this type of process can be found in the work of Feller (1952) who used the semigroup methods. His characterization of the "built-in" boundaries of such processes is represented here in the simplified form given by Keilson (1965). The type of boundary is determined by the integrability of the following two functions

$$h_1(x) = \pi(x) \int_{x_0}^{x} [b(\xi)\pi(\xi)]^{-1}\, d\xi \tag{1a}$$

$$h_2(x) = [b(x)\pi(x)]^{-1} \int_{x_0}^{x} \pi(\xi)\, d\xi \tag{1b}$$

over the interval $I \equiv [x_0, r]$, where x_0 is any interior point of the state space of the process and r is the boundary (r might be infinite in case of no "built-in" finite boundary). The function $\pi(x)$ in Eqs. (1) is defined by

$$\pi(x) = \exp\left\{-2 \int^{x} [a(\xi)/b(\xi)]\, d\xi\right\} \tag{2}$$

in accordance with definition (3.0.36b) for a process with $b(\xi) = 1$. When both h_1 and h_2 are integrable over I, the boundary is called a *regular* boundary since there are two possible solutions: in one $P(r \mid y, t) < \infty$ and the boundary acts as an absorbing boundary, and in the other $\lim_{x \to r} P(x \mid y, t) = \infty$ but $\lim_{x \to r} J(x \mid y, t) = 0$, where $J(x \mid y, t)$ is defined by Eq. (3.0.52), and the boundary acts as a reflecting boundary. When h_1 is integrable over I but h_2 is not, the boundary is called an *exit* boundary and it acts as an absorbing boundary. In this case $P(r \mid y, t) < \infty$. When h_1 is not integrable over I but h_2 is, the boundary is called an *entrance* boundary. An entrance boundary is inaccessible from inside the open interval (x_0, r), but any probability assigned to it initially flows into the open interval. When both h_1 and h_2 are not integrable over I, the boundary is called a *natural* boundary. This boundary is inaccessible from inside the open interval, and any probability assigned to it initially is trapped there forever.

It is shown in the next section (see Tables 3.5 and 3.6) that

$$2 \int_{x_0}^{z} h_1(x)\, dx = M_1(z \mid x_0) = \begin{array}{l}\text{average time to reach } z \text{ starting from } x_0 \\ (x_0 \text{ a reflecting boundary})\end{array}$$

$$2 \int_{x_0}^{z} h_2(x)\, dx = M_1(x_0 \mid z) = \begin{array}{l}\text{average time to reach } x_0 \text{ starting from } z \\ (z \text{ a reflecting boundary})\end{array}$$

Therefore a boundary r is inaccessible (entrance or natural) when h_1 is not integrable, i.e., when the average time to reach r from the inside is infinite, $\lim_{z \to r} M_1(z \mid x) = \infty$. On the other hand, when h_2 is not integrable, the boundary acts as a trap since the average time to reach any interior point starting from the boundary is infinite, $\lim_{z \to r} M_1(x_0 \mid z) = \infty$. One can further show that in case of accessible boundaries, since h_1 is integrable, the function $\pi(x)$ is also integrable over I and the probability $R(r \mid y, x_0)$ (see Table 3.8) of reaching r before any other state x_0, when starting from y, $x_0 < y < r$, is nonzero. For an entrance boundary $\pi(x)$ is not integrable and $R(r \mid y, x_0) = 0$ while for a natural boundary this probability can be zero or positive, but in case it is positive the average time to get there is infinite. For clarity, the classification of boundaries just given is summarized in Table 3.3.

The steady-state density $P(x \mid y, \infty)$ of unrestricted processes is given by Eq. (3.0.56a) when no flow of probability from the state space occurs, i.e., when no boundary is an exit boundary or a regular absorbing boundary. By introducing the function $\pi(x)$, defined in Eq. (2), we can rewrite the steady-state density [Eq. (3.0.56a)] as

$$P(x \mid y, \infty) = [b(x)\pi(x)]^{-1} \Big/ \int_{\Omega} [b(x)\pi(x)]^{-1}\, dx \tag{3}$$

In case one of the boundaries is an exit or a regular absorbing boundary, $P(x \mid y, \infty) = 0$ for all the interior points of the process.

TABLE 3.3

Classification of singular boundaries

Type of boundary	$h_1(x)$	$h_2(x)$	$\pi(x)$	Properties
Regular	Integrable	Integrable	Integrable	Accessible, absorbing or reflecting
Exit	Integrable	Not integrable	Integrable	Accessible, acts as absorbing
Entrance	Not integrable	Integrable	Not integrable	Inaccessible, no flow to the outside
Natural	Not integrable	Not integrable	Integrable	Inaccessible in finite time, acts as absorbing
			or	
			Not integrable	Inaccessible, no flow to the outside

Among processes with a state space extending without limit in one or two directions, we will consider only those "physical" processes in which the infinite boundaries are inaccessible in a finite time. For such boundaries

$$\lim_{x \to r} P(x \mid y, t) = 0, \qquad |r| = \infty \tag{4}$$

As was indicated in the preceding section, the solution to the FP equation can be found by separation of variables and expansion of $P(x \mid y, t)$ in terms of eigenfunctions. For an FP equation with nonconstant $b(x)$, the transformation given in Eq. (3.0.32a) can transform a finite boundary to an infinite boundary, or vice versa, but it does not change the nature of the boundary, being a one-to-one transformation $[(b(x))^{1/2} > 0$ and $z(x)$ is monotone increasing]. To determine the eigenfunctions, in the absence of imposed boundary conditions, one has to choose those functions which behave near the boundary according to the nature of the boundary; e.g., in the case of an exit boundary the eigenfunctions are the solutions of the corresponding eigenvalue problems which are finite at the exit boundary. In Appendix G we have illustrated this procedure for the so-called Ornstein–Uhlenbeck (OU) process which is defined by $a(x) = -rx$, $b(x) = \sigma^2$. The resulting expression for $P(x \mid y, t)$ is given in Table 3.4. This process has been used in modeling many physical and biological systems and is discussed in the context of biological systems later in this monograph. In Table 3.4 we also give the expression for $P(x \mid y, t)$ corresponding to another widely used process, the so-called Wiener process, defined by $a(x) = r$, $b(x) = \sigma^2$.

TABLE 3.4

Expressions for $P(x \mid y, t)$ for some processes with various boundary conditions, where $b(x)$ is at most quadratic in x

$a(x)$	$b(x)$	A	B	$P(x \mid y, t)$	Auxiliary definitions and references
Wiener process r	σ^2	$-\infty$ Natural	∞ Natural	$(2\pi\sigma^2 t)^{-1/2} \exp\{-(x-y-rt)^2/2\sigma^2 t\}$	Cox and Miller (1965, p. 203)
		Absorbing	∞ Natural	$(2\pi\sigma^2 t)^{-1/2} [\exp\{-(x-y-rt)^2/2\sigma^2 t\}$ $- \exp\{2r(A-y)/\sigma^2 - (x+y-2A-rt)^2/2\sigma^2 t\}]$	Cox and Miller (1965, p. 221)
		Reflecting	∞ Natural	$(2\pi\sigma^2 t)^{-1/2} [\exp\{-(x-y-rt)^2/2\sigma^2 t\}$ $+ \exp\{-[4r(y-A)t + (x+y-2A-rt)^2]/2\sigma^2 t\}$ $+ (2r/\sigma^2) \exp\{-2r(x-A)/\sigma^2\}\{1-\Phi(\xi)\}]$; $\xi \equiv (x+y-2A+rt)/(\sigma^2 t)^{1/2}$	Cox and Miller (1965, p. 224) Φ standard normal integral
		$-\infty$ Natural	Absorbing	$(2\pi\sigma^2 t)^{-1/2} [\exp\{-(x-y-rt)^2/2\sigma^2 t\}$ $- \exp\{2r(B-y)/\sigma^2 - (x+y-2B-rt)^2/2\sigma^2 t\}]$	
		$-\infty$ Natural	Reflecting	$(2\pi\sigma^2 t)^{-1/2} [\exp\{-(x-y-rt)^2/2\sigma^2 t\}$ $+ \exp\{-[4ryt + (2B-x-y-rt)^2]/2\sigma^2 t\}$ $- (2r/\sigma^2) \exp\{-2r(B-x)/\sigma^2\} \Phi((2B-x-y+rt)/(\sigma^2 t)^{1/2})$	Φ standard normal integral
		Absorbing	Absorbing	$(2/L) \exp\{r(x-y-rt/2)/\sigma^2\} \sum_{n=1}^{\infty} \sin[n\pi(x-A)/L]$ $\times \sin[n\pi(y-A)/L] \exp\{-(\sigma n\pi/L)^2 t/2\}$,	Appendix G
		Reflecting	Absorbing	$(2/L) \exp\{r(x-y-rt/2)/\sigma^2\} \sum_{n=0}^{\infty} \cos[(2n+1)\pi(x-A)/2L]$ $\times \cos[(2n+1)\pi(y-A)/2L] \exp\{-[(2n+1)\sigma\pi/2L]^2 t/2\}$	Appendix G
		Absorbing	Reflecting	$(2/L) \exp\{r(x-y-rt/2)/\sigma^2\} \sum_{n=0}^{\infty} \cos[(2n+1)\pi(B-x)/2L]$ $\times \cos[(2n+1)\pi(B-y)/2L] \exp\{-[(2n+1)\sigma\pi/2L)^2 t/2\}$	Appendix G

Ornstein–Uhlenbeck process $-rx$, $r > 0$	σ^2	$-\infty$ Natural	∞ Natural	$[2\pi V^2(t)]^{-1/2} \exp\{-[\{x-m(t)\}/V(t)]^2/2\}$	$m(t) \equiv y \exp(-rt)$
		0 Reflecting	∞ Natural	$[8\pi V^2(t)]^{-1/2} [\exp\{-[\{(x-m(t)\}/V(t)]^2/2\} + \exp\{-[\{(x+m(t)\}/V(t)]^2/2\}]$	$V^2(k) \equiv \sigma^2(1-e^{-2rt})/2r$
		0 Absorbing	∞ Natural	$[8\pi V^2(t)]^{-1/2} [\exp\{-[\{(x-m(t)\}/V(t)]^2/2\} - \exp\{-[\{(x+m(t)\}/V(t)]^2/2\}]$	Appendix G
Rayleigh process $s/x - rx$	σ^2	0 Singular $\mu \leqslant -1$ exit; $-1 < \mu < 0$ reflecting; $0 \leqslant \mu$ entrance	∞ Natural	$[4r/\sigma^2\Gamma(\mu+1)](rxy/\sigma^2)^{2\mu+1} \exp\{-r(x^2+y^2)/\sigma^2\} \sum_{n=0}^{\infty} L_n^{(\mu)}(rx^2/\sigma^2) \times L_n^{(\mu)}(ry^2/\sigma^2) \exp(-2nrt)/n!\Gamma(\mu+n+1)$; $\mu > -1,\ \mu \equiv s/\sigma^2 - 1/2,$	L_n generalized Laguerre polynomials. Stratonovitch (1963, p. 75)
$-rx + \beta$	$\sigma^2 x$	0 Singular $\mu \leqslant 0$ exit; $0 < \mu < 1$ reflecting; $1 \leqslant \mu$ entrance	∞ Natural	$(2r/\sigma^2)(x/y)^{(\mu-1)/2} \dfrac{e^{\mu rt/2}}{e^{rt/2} - e^{-rt/2}} \exp\left\{-\dfrac{2r(x+ye^{-rt})}{\sigma^2(1-e^{-rt})}\right\} I_{\mu-1}\left[\dfrac{4r(xy)^{1/2}}{\sigma^2(e^{rt/2}-e^{-rt/2})}\right];$ $\mu \equiv 2\beta/\sigma^2$	
0	$\sigma^2 x(1-x)$	0 Exit	1 Exit	$\sum_{i=1}^{\infty} [(2i+1)(1-\eta^2)/i(i+1)]\, T_{i-1}^{(1)}(\eta)\, T_{i-1}^{(1)}(\xi) \exp\{-\sigma^2 i(i+1)\,t/2\};$ $\xi \equiv 1 - 2x,\ \eta \equiv 1 - 2y$	T_n Gegenbauer-polynomials Crow and Kimura (1970, p. 383)
$-rx + \beta$	$\sigma^2 x(1-x)$	0 Singular $\mu \leqslant 0$ exit; $0 < \mu < 1$ reflecting; $1 \leqslant \mu$ entrance	1 Singular $\nu - \mu \leqslant 0$; $0 < \nu - \mu < 1$; $\nu - \mu \geqslant 1$	$x^{\mu-1}(1-x)^{\nu-\mu-1} \sum_{i=0}^{\infty} \dfrac{(\nu+2i-1)\,\Gamma(\nu+i-1)\,\Gamma(\nu-\mu+i)}{i!\,\Gamma^2(\nu-\mu)\,\Gamma(\mu+i)} \times \tilde{F}_i(1-x)\,\tilde{F}_i(1-y) \exp(-\lambda_i t);$ $\mu \equiv 2\beta/\sigma^2,\ \nu \equiv 2r/\sigma^2,\ \lambda_i = i[2r + (i-1)\sigma^2]/2,$ $\tilde{F}_i(x) \equiv F(\nu+i-1, -i; \nu-\mu; x)$	Crow and Kimura (1970, p. 391)

We do not discuss here the quantities introduced at the end of Section 3.0, since for the quantities (b) and (c), which are independent of the behavior of the process once the variable takes the value z, the state space to be considered is $x \leqslant z$ when $y < z$ (and $x \geqslant z$ when $y > z$). The expressions for these quantities are therefore derived in Section 3.2D, where we consider processes restricted by one boundary. Similarly, expressions for the quantities (d) and (e) are derived in Section 3.2B.

3.2 Restricted Processes

When one or two of the states of the process are of some special nature not incorporated into the form of the coefficients of the FP equation, the continuous process, as its discrete counterpart, is restricted by one or two boundaries, where boundary conditions are imposed. The analysis of these processes is carried out in a way similar to the analysis presented in Section 2.2 for the restricted discrete processes.

In general we consider a process that is confined to the interval $[A, B]$, and is restricted by at least one boundary, which is either absorbing or reflecting. We do not discuss other types of boundaries nor the case when both A and B are "built-in" (finite or infinite) boundaries, which is treated in the preceding section.

A restricted process is described by a probability density function $P(x \mid y, t)$ which satisfies the forward and backward diffusion equations (3.0.2) and (3.0.3) for $A < x < B$, with one boundary condition at each imposed boundary. The form of the boundary condition depends on the type of the boundary (absorbing or reflecting).

One simple way to obtain the boundary conditions involves the passage to the limit in the corresponding boundary conditions for the discrete process by the procedure described in Section 3.0 for the derivation of the diffusion equations.

When boundary A is an *absorbing* boundary, the boundary conditions analogous to conditions (2.2.2) and (2.2.3) are

$$P(A \mid y, t) = 0 \tag{1a}$$

$$P(x \mid A, t) = 0 \tag{1b}$$

Similarly, if boundary B is an *absorbing* boundary

$$P(B \mid y, t) = 0 \tag{2a}$$

$$P(x \mid B, t) = 0 \tag{2b}$$

When boundary A is a *reflecting* boundary, we take the limit of the boundary condition (2.2.4a) for the corresponding discrete process. When state l is a reflecting state, the boundary condition (2.2.4a) together with the forward master equation (2.0.3) is

$$\lambda_{l-1} P_{l-1,m} - \mu_l P_{l,m} = 0 \tag{3}$$

Defining $A = lh$ and taking the limit of this equation, with the assumptions (3.0.1a) and (3.0.1b), we get

$$\begin{aligned}
\lim_{h\to 0} \tfrac{1}{2}[(\lambda_{l-1}+\mu_{l-1}) P(A-h \mid y,t) - (\lambda_l+\mu_l) P(A \mid y,t) \\
+ (\lambda_{l-1}-\mu_{l-1}) P(A-h \mid y,t) + (\lambda_l-\mu_l) P(A \mid y,t)] \\
= \lim_{h\to 0} \frac{1}{h}\left[\frac{1}{2h}\{b(A-h) P(A-h \mid y,t) - b(A) P(A \mid y,t)\}\right. \\
\left. + \tfrac{1}{2}\{a(A-h) P(A-h \mid y,t) + a(A) P(A \mid y,t)\} + 0(h)\right] \\
= \lim_{h\to 0} \frac{1}{h}\left\{\frac{1}{2}\frac{\partial}{\partial x} b(x) P(x \mid y,t) - a(x) P(x \mid y,t)\right\}_{x=A} = 0
\end{aligned}$$

i.e.,

$$J(A,t) = 0 \tag{4}$$

which is the required boundary condition. Similarly, when boundary B is a *reflecting* boundary, the boundary condition is

$$J(B,t) = 0 \tag{5}$$

Boundary condition (5) for the reflecting boundary can also be derived by noting that $J(A,t)$ and $J(B,t)$ are the loss or gain of the overall probabilities at boundaries A and B, respectively. Whatever enters the reflecting boundary A (B) is bound to be retracted back and hence the net flow $J(A,t)$ $[J(B,t)]$ is zero.

Using a similar limiting procedure on Eqs. (2.2.4b) and (2.2.6b), the boundary conditions for the backward diffusion equation, when A and B are reflecting boundaries, are

$$\left.\frac{\partial}{\partial y} P(x \mid y,t)\right|_{y=A} = 0 \tag{6a}$$

$$\left.\frac{\partial}{\partial y} P(x \mid y,t)\right|_{y=B} = 0 \tag{6b}$$

The boundary conditions on $g(z \mid z_0,t)$, $Q(z)$, and $\psi(z)$ required for the solution of the FP equation (see Section 3.0) can be obtained by using the above boundary conditions on $P(x \mid y,t)$ and the transformation equations (3.0.32),

(3.0.33), and (3.0.36). These conditions at boundary A are

for A absorbing,

$$g(z(A)\,|\,z_0,t) = 0, \qquad Q(z(A)) = 0, \qquad \psi(z(A)) = 0 \tag{7}$$

for A reflecting,

$$\left[\frac{1}{2}\frac{\partial g}{\partial z} - \hat{a}(z)g\right]_{z=z(A)} = 0, \qquad \left[\frac{1}{2}\frac{\partial Q}{\partial z} - \hat{a}(z)Q\right]_{z=z(A)} = 0, \qquad \frac{\partial \psi}{\partial z}(z(A)) = 0 \tag{8}$$

and similar conditions hold at boundary B.

When both boundaries are imposed boundaries, each eigenfunction Q of Eq. (3.0.34) [ψ of Eq. (3.0.37)] is constructed by a linear combination of the two linearly independent solutions of Eq. (3.0.34) [or Eq. (3.0.37)]. The eigenvalue E and the coefficient in the linear combination are determined by the two boundary conditions of the form (7) or (8) imposed on the eigenfunction. In case one boundary is a "built-in" boundary, only one boundary condition can be imposed. The eigenfunction is chosen according to the expected behavior near the "built-in" boundary, and according to the boundary condition at the other boundary.

In Appendix G we have solved the FP equation for the OU process when $x = 0$ is a reflecting boundary, and for the Wiener process when it is confined either between two absorbing boundaries or between one reflecting and one absorbing boundary. In Table 3.4 are listed explicit expressions for $P(x\,|\,y,t)$ for a variety of processes with different boundary conditions.

In Section 3.0 we gave one boundary condition on $F(z\,|\,y,t)$ at $y = z$. The second boundary condition on $F(z\,|\,y,t)$ required for solving Eq. (3.0.59) can be derived from the corresponding conditions on $P(x\,|\,y,t)$ [Eqs. (1b), (2b), and (6)]. In case $y < z$, the boundary of interest is the boundary A ($<y$). When A is an *absorbing* boundary from Eqs. (3.0.57) and (1b),

$$F(z\,|\,A,t) = 0 \tag{9a}$$

and when A is a *reflecting* boundary from Eqs. (3.0.57) and (6a)

$$\left.\frac{\partial}{\partial y}F(z\,|\,y,t)\right|_{y=A} = 0 \tag{9b}$$

In case $y > z$, the boundary of interest is the boundary B, and the boundary conditions on $F(z\,|\,y,t)$ are (9) with A replaced by B.

We wish to point out that there is an important class of restricted processes where one of the boundaries varies with time. The general theory for such processes is not yet developed; we do discuss them only in Chapter 8 in connection with the problem of firing of a neuron. There the process is restricted between $-\infty$ and an absorbing boundary $B(t)$.

The detailed derivation of the quantities (b) and (c) defined in Section 3.0 for various types of boundaries is carried out in the following subsections. We start first with processes confined between a reflecting boundary and an absorbing boundary.

A. Process Confined between an Absorbing Boundary and a Reflecting Boundary

For explicit analysis, we will take A to be the reflecting and B to be the absorbing boundary. The complementary case, of A the absorbing and B the reflecting boundary, can be treated likewise, and we only summarize the results at the end of this section.

Since A is finite and reflecting and B is an absorbing boundary, absorption is bound to take place and thus $P(x \mid y, \infty) = 0$ for all $A \leqslant x < B$.

We now proceed to derive explicit expressions for quantities related to the lifetime of the process $T(B \mid y)$, which is the time spent by the process before it is absorbed at boundary B. The probability density $F(B \mid y, t)$ satisfies the backward equation (3.0.59) with the initial and boundary conditions [see Eqs. (3.0.60), (3.0.61), and (9b)]

$$F(B \mid y, 0) = \delta(y - B) \tag{10}$$

$$F(B \mid B, t) = \delta(t) \tag{11}$$

$$\left.\frac{\partial}{\partial y} F(B \mid y, t)\right|_{y=A} = 0 \tag{12}$$

Taking Laplace transform of Eqs. (11), (12), and (3.0.59), we obtain the differential equation and boundary conditions for the Laplace transform

$$f(B \mid y, s) \equiv \int_0^\infty e^{-st} F(B \mid y, t)\, dt \tag{13}$$

of $F(B \mid y, t)$:

$$sf(B \mid y, s) = a(y) \frac{\partial}{\partial y} f(B \mid y, s) + \tfrac{1}{2} b(y) \frac{\partial^2}{\partial y^2} f(B \mid y, s), \qquad A < y < B \tag{14}$$

$$f(B \mid B, s) = 1, \qquad \left.\frac{\partial}{\partial y} f(B \mid y, s)\right|_{y=A} = 0 \tag{15}$$

This second-order ordinary linear differential equation, together with these boundary conditions, can be solved by standard methods. Having a solution to Eq. (14) which satisfies the boundary condition at A and is nonzero at B, $u(y,s)$, we can write $f(B\,|\,y,s)$ as the ratio

$$f(B\,|\,y,s) = u(y,s)/u(B,s) \tag{16}$$

This form of $f(B\,|\,y,s)$ is analogous to the form of $f_{N,m}(s)$ in the corresponding discrete process, where $f_{N,m}(s)$ is given as a ratio between two polynomials [Eq. (2.2.16)]. Equation (16) can be Laplace inverted by any of the standard methods to obtain $F(B\,|\,y,t)$. The moments of $T(B\,|\,y)$ can be derived directly from Eq. (14) by using the defining relation

$$M_j(B\,|\,y) = \int_0^\infty t^j F(B\,|\,y,t)\,dt = (-1)^j \frac{\partial^j}{\partial s^j} f(B\,|\,y,s)\bigg|_{s=0} \tag{17}$$

Differentiating Eq. (14) j times and putting $s = 0$, we get the differential equation

$$a(y)\frac{\partial M_j(B\,|\,y)}{\partial y} + \tfrac{1}{2}b(y)\frac{\partial^2 M_j(B\,|\,y)}{\partial y^2} + jM_{j-1}(B\,|\,y) = 0, \qquad j \geqslant 1 \tag{18}$$

satisfied by the jth moment. The zeroth moment, $M_0(B\,|\,y)$, can be simply calculated by recalling that absorption is a certain event, and therefore

$$M_0(B\,|\,y) = f(B\,|\,y,0) = \int_0^\infty F(B\,|\,y,t)\,dt = 1 \tag{19}$$

The boundary conditions to be satisfied by $M_j(B\,|\,y)$, easily derived from Eqs. (15) and (17), are

$$M_j(B\,|\,B) = 0, \qquad \frac{\partial}{\partial y} M_j(B\,|\,y)\bigg|_{y=A} = 0 \tag{20}$$

The solution of Eq. (18) which satisfies boundary conditions (20) is derived in Appendix H, and is found to be

$$M_j(B\,|\,y) = 2j\int_y^B d\eta\,\pi(\eta)\int_A^\eta \frac{M_{j-1}(B\,|\,\xi)}{b(\xi)\,\pi(\xi)}\,d\xi \tag{21}$$

where $\pi(\eta)$ is defined as in Section 3.1, i.e.,

$$\pi(\eta) = \exp\left\{-\int^\eta \frac{2a(u)}{b(u)}\,du\right\} \tag{22}$$

In particular, in view of Eq. (19), the first moment is given by

$$M_1(B\,|\,y) = 2\int_y^B d\eta\,\pi(\eta)\int_A^\eta [b(\xi)\,\pi(\xi)]^{-1}\,d\xi \tag{23}$$

Formula (21) can also be used, with B replaced by z, $z > y$, to derive the moments of $T(z\,|\,y)$—the first passage time to the value z of a process that starts with value y, since $T(z\,|\,y)$ is independent of the behavior of the process after the variable takes the value z, and the process to be considered is restricted to $[A, z]$. For $z < y < B$, by the same reasoning, the process to be considered is the one confined between two absorbing boundaries z and B; the results are discussed in the following subsection. These results will also apply for the quantities (d) and (e) defined in Section 3.0.

We can get a simple relation between $F(B\,|\,y,t)$ and $P(x\,|\,y,t)$ near $x = B$ by noting that for the process under consideration the probability of absorption at the boundary B at time t is due to the loss of probability density in the interval (A, B) at time t, i.e.,

$$F(B\,|\,y,t) = -\frac{d}{dt}\int_A^B P(x\,|\,y,t)\,dx = a(x)\,P(x\,|\,y,t) - \frac{1}{2}\frac{\partial}{\partial x}[b(x)\,P(x\,|\,y,t)]\bigg|_A^B$$

or

$$F(B\,|\,y,t) = -\frac{1}{2}\frac{\partial}{\partial x}[b(x)\,P(x\,|\,y,t)]_{x=B} \tag{24}$$

Here we have used the FP equation (3.0.2) and the boundary conditions on P at the reflecting boundary A [Eq. (3.2.4)] and at the absorbing boundary B [Eq. (3.2.2a)]. The resulting equation, (24), is a simple relation between $F(B\,|\,y,t)$ and $P(x\,|\,y,t)$. This relation is also valid in case the absorbing boundary $B = B(t)$ is a function of time, since in this case $P(B(t)\,|\,y,t) = 0$, and

$$F(B(t)\,|\,y,t) = -\frac{d}{dt}\int_A^{B(t)} P(x\,|\,y,t)\,dx = -B'(t)\,P(B(t)\,|\,y,t) + J(B(t),t)$$

$$= -\frac{1}{2}\frac{\partial}{\partial x}[bP]_{x=B(t)}$$

In case A is an absorbing boundary and B is a reflecting boundary, the probability density of the first passage time $T(A\,|\,y)$ satisfies Eq. (3.0.59), with initial and boundary conditions

$$F(A\,|\,y,0) = \delta(y-A) \tag{25}$$

$$F(A\,|\,A,t) = \delta(t) \tag{26}$$

$$\frac{\partial}{\partial y}F(A\,|\,y,t)\bigg|_{y=B} = 0 \tag{27}$$

while its Laplace transform $f(A\,|\,y,s)$ satisfies Eq. (14), with $f(B\,|\,y,s)$ replaced by $f(A\,|\,y,s)$. and the boundary conditions

$$f(A\,|\,A,s) = 1, \qquad \frac{\partial}{\partial y}f(A\,|\,y,s)\bigg|_{y=B} = 0 \tag{28}$$

$f(A|y,s)$ can be written in the form $f(A|y,s) = \tilde{u}(y,s)/\tilde{u}(A,s)$, where $\tilde{u}(y,s)$ is a solution of Eq. (14) which satisfies the boundary condition at B and is nonzero at A. The moments of $T(A|y)$ are given by an equation analogous to Eq. (21), i.e., by

$$M_j(A|y) = 2j \int_A^y d\eta \, \pi(\eta) \int_\eta^B \frac{M_{j-1}(A|\xi)}{b(\xi)\,\pi(\xi)} \, d\xi \tag{29}$$

with

$$M_0(A|y) = 1 \tag{30}$$

and $F(A|y,t)$ is related to $P(x|y,t)$ by a relation analogous to relation (24), i.e., by

$$F(A|y,t) = \frac{1}{2} \frac{\partial}{\partial x} [b(x)\, P(x|y,t)]_{x=A} \tag{31}$$

All the results derived in this subsection are summarized in Tables 3.5 and 3.6, together with the results derived in the next subsection which also apply to the present process.

B. *Process Confined between Two Absorbing Boundaries*

Any process confined between two absorbing boundaries is bound to be absorbed, and therefore $P(x|y,\infty) = 0$, $A < x < B$. The more interesting quantity is the probability $R(A|y)$ $[R(B|y)]$ of absorption occurring at boundary A (B) and not at the other boundary B (A). Other quantities which provide more insight into the process of absorption are quantities similar to those for the discrete process (Section 2.2B). These quantities are the first passage time $T(B|y)$ $[T(A|y)]$ for absorption at boundary B (A), the corresponding probability density $F(B|y,t)$ $[F(A|y,t)]$, and its moments $M_j(B|y,t)$ $[M_j(A|y,t)]$, and the probability density $F^*(B|y,t)$ $[F^*(A|y,t)]$ and its moments $M_j^*(B|y)[M_j^*(A|y)]$ of the conditional time for absorption at B (A) when absorption at boundary B (A) is a priori known to occur. These latter quantities are related to the unconditional quantities by

$$F^*(A|y,t) = \frac{F(A|y,t)}{R(A|y)}, \qquad M_j^*(A|y) = \frac{M_j(A|y)}{R(A|y)} \tag{32}$$

$$F^*(B|y,t) = \frac{F(B|y,t)}{R(B|y)}, \qquad M_j^*(B|y) = \frac{M_j(B|y)}{R(B|y)} \tag{33}$$

The lifetime $T(y)$ of the process before any absorption occurs, its probability density $F(y,t)$, and moments $M_j(y)$ are related to the above-mentioned

TABLE 3.5

Expressions for various quantities of interest for a process confined between the *reflecting boundary* A and the *absorbing boundary* B ($A < B$)

$P(x \mid y, \infty)$	$0 \quad (A \leqslant x < B)$	
	$A \leqslant y \leqslant z \leqslant B$	$A \leqslant z \leqslant y \leqslant B$
$R(z \mid y)$	1	$\int_y^B \pi(\eta)\, d\eta \Big/ \int_z^B \pi(\eta)\, d\eta$
$M_1(z \mid y)$	$2 \int_y^z d\eta\, \pi(\eta) \int_A^\eta [b(\xi)\pi(\xi)]^{-1}\, d\xi$	$2\left[\int_y^B d\eta\, \pi(\eta) \int_z^\eta \frac{R(z \mid \xi)}{b(\xi)\pi(\xi)}\, d\xi - R(z \mid y) \int_z^B d\eta\, \pi(\eta) \int_z^\eta \frac{R(z \mid \xi)}{b(\xi)\pi(\xi)}\, d\xi\right]$
$M_j(z \mid y)$	$2j \int_y^z d\eta\, \pi(\eta) \int_A^\eta \frac{M_{j-1}(z \mid \xi)}{b(\xi)\pi(\xi)}\, d\xi$	$2j\left[\int_y^B d\eta\, \pi(\eta) \int_z^\eta \frac{M_{j-1}(z \mid \xi)}{b(\xi)\pi(\xi)}\, d\xi - R(z \mid y) \int_z^B d\eta\, \pi(\eta) \int_z^\eta \frac{M_{j-1}(z \mid \xi)}{b(\xi)\pi(\xi)}\, d\xi\right]$

$$\pi(y) = \exp\left\{-\int^y \frac{2a(\eta)}{b(\eta)}\, d\eta\right\}$$

TABLE 3.6

Expressions for various quantities of interest for a process confined between the *absorbing boundary A* and the *reflecting boundary B* $(A < B)$

$P(x\mid y,\infty)$	$0 \quad (A < x \leqslant B)$	
	$A \leqslant y \leqslant z \leqslant B$	$A \leqslant z \leqslant y \leqslant B$
$R(z\mid y)$	$\int_A^y \pi(\eta)\,d\eta \Big/ \int_A^z \pi(\eta)\,d\eta$	1
$M_1(z\mid y)$	$2\left[R(z\mid y)\int_A^z d\eta\,\pi(\eta)\int_A^\eta \frac{R(z\mid\xi)}{b(\xi)\pi(\xi)}\,d\xi - \int_A^y d\eta\,\pi(\eta)\int_A^\eta \frac{R(z\mid\xi)}{b(\xi)\pi(\xi)}\,d\xi\right]$	$2\int_z^y d\eta\,\pi(\eta)\int_\eta^B [b(\xi)\pi(\xi)]^{-1}\,d\xi$
$M_j(z\mid y)$	$2j\left[R(z\mid y)\int_A^z d\eta\,\pi(\eta)\int_A^\eta \frac{M_{j-1}(z\mid\xi)}{b(\xi)\pi(\xi)}\,d\xi - \int_A^y d\eta\,\pi(\eta)\int_A^\eta \frac{M_{j-1}(z\mid\xi)}{b(\xi)\pi(\xi)}\,d\xi\right]$	$2j\int_z^y d\eta\,\pi(\eta)\int_\eta^B \frac{M_{j-1}(z\mid\xi)}{b(\xi)\pi(\xi)}\,d\xi$

$$\pi(y) = \exp\left\{-\int^y \frac{2a(\eta)}{b(\eta)}\,d\eta\right\}$$

quantities by

$$T(y) = \min[T(A|y), T(B|y)] \tag{34}$$

$$F(y,t) = F(A|y,t) + F(B|y,t) \tag{35}$$

$$M_j(y) = \int_0^\infty t^j F(y,t)\,dt = M_j(A|y) + M_j(B|y) \tag{36}$$

We will explicitly derive the expressions for the various quantities when absorption takes place at boundary B. The corresponding expressions for absorption at boundary A can be derived similarly and are summarized at the end of this section.

The basic equation to be discussed is the backward equation (3.0.59) satisfied by the probability density $F(B|y,t)$. The initial condition and the boundary condition at $y = B$ are the same as in Section 3.2A [Eqs. (10) and (11)], but the boundary condition (12) at $y = A$ is replaced by the corresponding boundary condition at an absorbing boundary, i.e., by [see Eq. (9a)]

$$F(B|A,t) = 0 \tag{37}$$

The Laplace transform $f(B|y,s)$ of $F(B|y,t)$ also satisfies Eq. (14), but with boundary conditions

$$f(B|B,s) = 1, \qquad f(B|A,s) = 0 \tag{38}$$

As in Section 3.2A, the solution of Eq. (14) subject to these boundary conditions is of the form

$$f(B|y,s) = v(y,s)/v(B,s) \tag{39a}$$

where $v(y,s)$ is any solution of Eq. (14) satisfying the conditions

$$v(B,s) \neq 0, \qquad v(A,s) = 0 \tag{39b}$$

Equation (39a) can be Laplace inverted by any of the standard methods to obtain $F(B|y,t)$.

To calculate the overall probability $R(B|y)$ of absorption at B (and not at A), we note that

$$R(B|y) = \int_0^\infty F(B|y,t)\,dt = f(B|y,0) \tag{40}$$

Therefore from Eq. (14), $R(B|y)$ satisfies the differential equation

$$a(y)\frac{\partial}{\partial y}R(B|y) + \tfrac{1}{2}b(y)\frac{\partial^2}{\partial y^2}R(B|y) = 0 \tag{41}$$

and from Eq. (38), the boundary conditions

$$R(B|B) = 1, \qquad R(B|A) = 0 \tag{42}$$

Equation (41) is of the form of Eq. (H. 4) and therefore the solution [see Eq. (H. 5a)] that satisfies conditions (42) is

$$R(B\,|\,y) = \int_A^y \pi(\eta)\, d\eta \Big/ \int_A^B \pi(\eta)\, d\eta \tag{43}$$

where $\pi(\eta)$ is defined by Eq. (22). Since absorption occurs at either A or B

$$R(A\,|\,y) = 1 - R(B\,|\,y) = \int_y^B \pi(\eta)\, d\eta \Big/ \int_A^B \pi(\eta)\, d\eta \tag{44}$$

To get expressions for the moments of $T(B\,|\,y)$, we note that these moments satisfy Eq. (18) with the boundary conditions [derived from relation (17) and conditions (38)]

$$M_j(B\,|\,B) = 0, \qquad M_j(B\,|\,A) = 0 \tag{45}$$

The explicit solution of Eq. (18) subject to these conditions is derived in Appendix H, and is given by

$$\begin{aligned} M_j(B\,|\,y) &= 2j\Bigg[\int_y^B d\eta\, \pi(\eta) \int_A^\eta \frac{M_{j-1}(B\,|\,\xi)}{b(\xi)\,\pi(\xi)}\, d\xi \\ &\qquad - R(A\,|\,y) \int_A^B d\eta\, \pi(\eta) \int_A^\eta \frac{M_{j-1}(B\,|\,\xi)}{b(\xi)\,\pi(\xi)}\, d\xi\Bigg] \\ &= 2j\Bigg[R(B\,|\,y) \int_A^B d\eta\, \pi(\eta) \int_A^\eta \frac{M_{j-1}(B\,|\,\xi)}{b(\xi)\,\pi(\xi)}\, d\xi \\ &\qquad - \int_A^y d\eta\, \pi(\eta) \int_A^\eta \frac{M_{j-1}(B\,|\,\xi)}{b(\xi)\,\pi(\xi)}\, d\xi\Bigg] \end{aligned} \tag{46}$$

where the lower limit in the integrations over ξ can be replaced by any point within the state space. Note that with the lower limit A, as in Eq. (46), this result reduces to the corresponding result when A is a reflecting boundary [Eq. (21)] by putting $R(B\,|\,y) = 1$. Since by definition

$$M_0(B\,|\,\xi) = \int_0^\infty F(B\,|\,y, t)\, dt = R(B\,|\,y) \tag{47}$$

from Eqs. (44), (46), and (47) the first moment is given by

$$\begin{aligned} M_1(B\,|\,y) &= 2\left[\int_A^B \pi(\eta)\, d\eta\right]^{-1} \Bigg[\int_y^B d\eta\, \pi(\eta) \int_A^\eta d\xi\, [b(\xi)\,\pi(\xi)]^{-1} \int_A^\xi \pi(\rho)\, d\rho \\ &\qquad - \frac{\int_y^B \pi(\eta)\, d\eta}{\int_A^B \pi(\eta)\, d\eta} \int_A^B d\eta\, \pi(\eta) \int_A^\eta d\xi\, [b(\xi)\,\pi(\xi)]^{-1} \int_A^\xi \pi(\rho)\, d\rho\Bigg] \end{aligned} \tag{48}$$

By following a procedure similar to the one described above for absorption at boundary B, the Laplace transform of the probability density $F(A\,|\,y, t)$ for absorption at boundary A is given by

$$f(A\,|\,y,s) = \tilde{v}(y,s)/\tilde{v}(A,s) \tag{49}$$

where $\tilde{v}(y,s)$ satisfies Eq. (14) with the boundary condition $\tilde{v}(B,s) = 0$. The moments of $T(A\,|\,y)$ satisfy the same equation [Eq. (18)] and the same boundary conditions [Eq. (45)] as the moments of $T(B\,|\,y)$ and are given by a recurrence relation similar to the relation (46), but with

$$M_0(A\,|\,y) = \int_0^\infty F(A\,|\,y,t)\,dt = R(A\,|\,y) \tag{50}$$

Explicit formulas for $M_j(A\,|\,y)$ and $M_1(A\,|\,y)$ are listed in Table 3.7.

TABLE 3.7

Expressions for various quantities of interest for a process confined between *two absorbing boundaries*, *A* and *B* (*A* < *B*)

$P(x\,|\,y,\infty)$	$0 \quad (A < x < B)$	
	$A \leqslant z \leqslant y \leqslant B$	$A \leqslant y \leqslant z \leqslant B$
$R(z\,|\,y)$ $M_1(z\,|\,y)$ $M_j(z\,|\,y)$	See column 2, Table 3.5	See column 1, Table 3.6
$R(y)$	1	
$M_1(y)$	$2\left[\int_y^B d\eta\,\pi(\eta)\int_A^\eta [b(\xi)\pi(\xi)]^{-1}\,d\xi - R(A\,|\,y)\int_A^B d\eta\,\pi(\eta)\int_A^\eta [b(\xi)\pi(\xi)]^{-1}\,d\xi\right]$	
$M_j(y)$	$2j\left[\int_y^B d\eta\,\pi(\eta)\int_A^\eta \frac{M_{j-1}(\xi)}{b(\xi)\pi(\xi)}\,d\xi - R(A\,|\,y)\int_A^B d\eta\,\pi(\eta)\int_A^\eta \frac{M_{j-1}(\xi)}{b(\xi)\pi(\xi)}\,d\xi\right]$	

$$\pi(y) = \exp\left\{-\int^y \frac{2a(\eta)}{b(\eta)}\,d\eta\right\}.$$

The expressions for $F(y, t)$ and $M_j(y)$, describing the lifetime of the process, can now be derived by using Eqs. (35) and (36). Thus from Eqs. (39a) and (49) we get

$$f(y,s) = \frac{v(y,s)}{v(B,s)} + \frac{\tilde{v}(y,s)}{\tilde{v}(A,s)} \tag{51}$$

and from Eq. (46) and the corresponding equation for $M_j(A\,|\,y)$, we get

$$M_j(y) = 2j\left[\int_y^B d\eta\,\pi(\eta)\int_A^\eta \frac{M_{j-1}(\xi)}{b(\xi)\pi(\xi)}\,d\xi - R(A\,|\,y)\int_A^B d\eta\,\pi(\eta)\int_A^\eta \frac{M_{j-1}(\xi)}{b(\xi)\pi(\xi)}\,d\xi\right] \tag{52}$$

where

$$M_0(y) = R(A|y) + R(B|y) = 1 \tag{53}$$

In particular the first moment is given by

$$M_1(y) = 2\left[\int_y^B d\eta\, \pi(\eta) \int_A^\eta \frac{d\xi}{b(\xi)\pi(\xi)} - \frac{\int_y^B \pi(\xi)\, d\xi}{\int_A^B \pi(\xi)\, d\xi} \int_A^B d\eta\, \pi(\eta) \int_A^\eta \frac{d\xi}{b(\xi)\pi(\xi)}\right] \tag{54}$$

Comparing this expression with the expression for $M_1(B|y)$ when A is a reflecting boundary [Eq. (23)], we conclude that the average time for absorption (either at A or at B) is equal to the average time for absorption at B when A is a reflecting boundary, minus the probability that absorption occurs at boundary A, times the average time for absorption at B when a process starts at the reflecting boundary A.

As noted in the preceding section, the first passage time densities $F(y,t)$, $F(A|y,t)$, and $F(B|y,t)$ are related to the probability densities of the process by "the conservation of probability" argument [whatever flows out of the interval (A, B) is absorbed either at A or at B]. Therefore,

$$F(B|y,t) + F(A|y,t) = F(y,t) = -\frac{\partial}{\partial t}\int_A^B P(x|y,t)\, dx = J(B,t) - J(A,t) \tag{55}$$

$$F(A|y,t) = \frac{1}{2}\frac{\partial}{\partial x}[b(x)P(x|y,t)]_{x=A} \tag{56}$$

$$F(B|y,t) = -\frac{1}{2}\frac{\partial}{\partial x}[b(x)P(x|y,t)]_{x=B} \tag{57}$$

The expressions for the probability densities and moments of $T(A|y,t)$ and $T(B|y,t)$ can be used to express the corresponding quantities for the first passage time $T(z|y,w)$ to any intermediate point z before another prescribed point w is ever reached, starting at a point y between z and w. These expressions are independent of the type of boundaries at A and B and therefore are given together in Table 3.8.

C. *Process Confined between Two Reflecting Boundaries*

A process that is confined between two reflecting boundaries is bound to hit either one of the two boundaries as time goes on, and from then on its evolution is independent of the initial state of the process. Therefore, the steady state probability density

$$P(x|y,\infty) = P^*(x) \tag{58}$$

TABLE 3.8

Expressions for various quantities of interest which are *independent* of the nature of the boundaries

	$A \leqslant w \leqslant y \leqslant z \leqslant B$	$A \leqslant z \leqslant y \leqslant w \leqslant B$
$R(z \mid y, w)$	$\int_w^y \pi(\eta)\, d\eta \Big/ \int_w^z \pi(\eta)\, d\eta$	$\int_y^w \pi(\eta)\, d\eta \Big/ \int_z^w \pi(\eta)\, d\eta$
$M_1(z \mid y, w)$	$2\left[R(z \mid y, w) \int_w^z d\eta\, \pi(\eta) \int_w^\eta \frac{R(z \mid \xi, w)}{b(\xi)\pi(\xi)}\, d\xi - \int_w^y d\eta\, \pi(\eta) \int_w^\eta \frac{R(z \mid \xi, w)}{b(\xi)\pi(\xi)}\, d\xi\right]$	$2\left[\int_y^w d\eta\, \pi(\eta) \int_z^\eta \frac{R(z \mid \xi, w)}{b(\xi)\pi(\xi)}\, d\xi - R(z \mid y, w) \int_z^w d\eta\, \pi(\eta) \int_z^\eta \frac{R(z \mid \xi, w)}{b(\xi)\pi(\xi)}\, d\xi\right]$
$M_j(z \mid y, w)$	$2j\left[R(z \mid y, w) \int_w^z d\eta\, \pi(\eta) \int_w^\eta \frac{M_{j-1}(z \mid \xi, w)}{b(\xi)\pi(\xi)}\, d\xi - \int_w^y d\eta\, \pi(\eta) \int_w^\eta \frac{M_{j-1}(z \mid \xi, w)}{b(\xi)\pi(\xi)}\, d\xi\right]$	$2j\left[\int_y^w d\eta\, \pi(\eta) \int_z^\eta \frac{M_{j-1}(z \mid \xi, w)}{b(\xi)\pi(\xi)}\, d\xi - R(z \mid y, w) \int_z^w d\eta\, \pi(\eta) \int_z^\eta \frac{M_{j-1}(z \mid \xi, w)}{b(\xi)\pi(\xi)}\, d\xi\right]$

$$\pi(y) = \exp\left\{-\int^y \frac{2a(\eta)}{b(\eta)}\, d\eta\right\}$$

is independent of the initial state of the process. To calculate $P^*(x)$, we note that from the boundary condition (4) for reflecting boundaries A and B, $J(A,t) = J(B,t) = 0$. Therefore, there is no flow of probability away from the interval $[A, B]$, and $P^*(x)$ is given by Eq. (3.0.56a), i.e.,

$$P^*(x) = \frac{c}{b(x)} \exp\left[2 \int^x d\xi \, a(\xi)/b(\xi)\right] \tag{59}$$

Using the normalization condition

$$\int_A^B P^*(x) \, dx = 1 \tag{60}$$

the constant c can be calculated; the resulting expression is

$$P^*(x) = \frac{[b(x)\pi(x)]^{-1}}{\int_A^B [b(x)\pi(x)]^{-1} \, dx} \tag{61}$$

where $\pi(x)$ is defined by Eq. (3.1.2).

To calculate $F(z|y,t)$, $M_j(z|y)$, and $R(z|y)$, as in the case of discrete processes (see p. 27), we note that if $y < z$, the time for the random variable to reach the value of z for the first time is the same as the time for absorption when the process is confined between reflecting boundary A and absorbing boundary z, and if $y > z$, similar statement is true but with the reflecting boundary B instead of A. Therefore, the expressions for these quantities can be derived by using the results of Section 3.2A. These results are summarized in Table 3.9.

TABLE 3.9

Expressions for various quantities of interest for a process confined between *two reflecting boundaries A and B* ($A < B$)

$P(x\|y,\infty)$	$\left\{b(x)\pi(x)\int_A^B [b(\eta)\pi(\eta)]^{-1} d\eta\right\}^{-1}$	
	$A \leqslant z \leqslant y \leqslant B$	$A \leqslant y \leqslant z \leqslant B$
$R(z\|y)$ $M_1(z\|y)$ $M_j(z\|y)$	See column 2, Table 3.6	See column 1, Table 3.5

$$\pi(y) = \exp\left\{-\int^y \frac{2a(\eta)}{b(\eta)} \, d\eta\right\}.$$

$P(x|y,\infty)$ is extremal at x satisfying the condition $a(x) = \frac{1}{2}\, db(x)/dx$.

D. *Process with Only One Imposed Boundary*

For a process with state space restricted by one imposed boundary at A (or at B), the results of the previous subsections can be used by taking a proper limit.

When the state space is confined between the imposed boundary A (B) and a finite "built-in" boundary r such that

$$\lim_{B \to r} \int_A^B \pi(\eta)\, d\eta < \infty \qquad \left[\lim_{A \to r} \int_A^B \pi(\eta)\, d\eta < \infty\right] \tag{62}$$

i.e., the boundary r is an exit or a natural boundary (Table 3.3), or if r is a regular-absorbing boundary, the limiting procedure $B \to r$ ($A \to r$) is applied to the expressions for a state space with the same boundary A (B) and an absorbing boundary B (A).

TABLE 3.10

Expressions for various quantities of interest for a process restricted *only by one imposed boundary* A (reflecting or absorbing), $A \leqslant x < \infty$

	A absorbing	A reflecting
$P(x \mid y, \infty)$	$0 \quad (A < x)$	$\left\{b(x)\pi(x)\int_A^\infty [b(\eta)\pi(\eta)]^{-1}\, d\eta\right\}^{-1}$
	$A \leqslant y \leqslant z$	
$R(z \mid y)$ $M_1(z \mid y)$ $M_j(z \mid y)$	See column 1, Table 3.6	See column 1, Table 3.5
	$A \leqslant z \leqslant y$ (independent of the nature of A)	
$R(z \mid y)$	$\int_y^\infty \pi(\eta)\, d\eta \Big/ \int_z^\infty \pi(\eta)\, d\eta$	
$M_1(z \mid y)$	$2\left[\int_z^y d\eta\, \pi(\eta)\int_\eta^y [b(\xi)\pi(\xi)]^{-1}\, d\xi + \int_y^\infty [b(\xi)\pi(\xi)]^{-1}\, d\xi \int_z^y \pi(\eta)\, d\eta\right]$ [a]	
$M_j(z \mid y)$	$2j\left[\int_z^y d\eta\, \pi(\eta)\int_\eta^y \frac{M_{j-1}(z \mid \xi)}{b(\xi)\pi(\xi)}\, d\xi + \int_y^\infty \frac{M_{j-1}(z \mid \xi)}{b(\xi)\pi(\xi)}\, d\xi \int_z^y \pi(\eta)\, d\eta\right]$ [a]	

$\pi(y) = \exp\left\{-\int^y \frac{2a(\eta)}{b(\eta)}\, d\eta\right\}$. For A reflecting, $P(x \mid y, \infty)$ is extremal at x satisfying $a(x) = \frac{1}{2}\, db(x)/dx$.

[a] The expressions above are valid only when $\int_A^\infty \pi(\eta)\, d\eta = \infty$.

TABLE 3.11

Expressions for various quantities of interest for a process restricted *only by one imposed boundary* B (reflecting or absorbing), $-\infty < x \leqslant B$

	B absorbing	B reflecting
$P(x \mid y, \infty)$	$0 \quad x < B$	$\left\{b(x)\,\pi(x) \int_{-\infty}^{B} [b(\eta)\,\pi(\eta)]^{-1}\, d\eta\right\}^{-1}$
	$z \leqslant y \leqslant B$	
$R(z \mid y)$ $M_1(z \mid y)$ $M_j(z \mid y)$	See column 2, Table 3.5	See column 2, Table 3.6
	$y \leqslant z \leqslant B$ (independent of the nature of B)	
$R(z \mid y)$	$\int_{-\infty}^{y} \pi(\eta)\, d\eta \Big/ \int_{-\infty}^{z} \pi(\eta)\, d\eta$	
$M_1(z \mid y)$	$2\left[\int_{y}^{z} d\eta\, \pi(\eta) \int_{y}^{\eta} [b(\xi)\,\pi(\xi)]^{-1}\, d\xi + \int_{-\infty}^{y} [b(\xi)\pi(\xi)]^{-1}\, d\xi \int_{y}^{z} \pi(\eta)\, d\eta\right]$ [a]	
$M_j(z \mid y)$	$2j\left[\int_{y}^{z} d\eta\, \pi(\eta) \int_{y}^{\eta} \frac{M_{j-1}(z\mid\xi)}{b(\xi)\,\pi(\xi)}\, d\xi + \int_{-\infty}^{y} \frac{M_{j-1}(z\mid\xi)}{b(\xi)\,\pi(\xi)}\, d\xi \int_{y}^{z} \pi(\eta)\, d\eta\right]$ [a]	

$\pi(y) = \exp\left\{-\int^{y} \frac{2a(\eta)}{b(\eta)}\, d\eta\right\}$. For B reflecting, $P(x \mid y, \infty)$ is extremal at x satisfying $a(x) = \frac{1}{2}\, db(x)/dx$.

[a] The expressions above are valid only when $\int_{-\infty}^{B} \pi(\eta)\, d\eta = \infty$.

When the "built-in" boundary r, finite or infinite, is such that the integral of Eq. (62) diverges (i.e., the limit is ∞), then the limiting procedure $B \to r$ ($A \to r$) is applied to the expressions for a state space with the same boundary A (B), and a reflecting boundary B (A). This latter procedure applies when r is a regular-reflecting, an entrance, or a natural boundary (see Table 3.3).

In case of a state space unbounded in one direction (r infinite), we are interested only in those processes for which the probability to drift without limit is zero; i.e., when r is an entrance or a natural boundary such that $\pi(y)$ is not integrable over the state space. In Tables 3.10 and 3.11 the results for such processes are summarized by taking the above-mentioned limits. Thus in Table 3.10 we take the limit $B \to \infty$ of the corresponding expressions of Table 3.9 when A is reflecting and of Table 3.6 when A is absorbing, while in Table 3.11 we take the limit $A \to -\infty$ of the corresponding expressions of Table 3.5 when B is absorbing and of Table 3.9 when B is reflecting.

Relation (56) [(57)] at an absorbing boundary $A[B]$ is also valid for processes confined by one absorbing boundary $A[B]$ and one "built-in" boundary [or one infinite boundary with condition (62) satisfied]. At a finite "built-in" boundary r a similar relation holds, i.e.,

$$F(r\,|\,y,t) = \lim_{z\to r} J(z\,|\,y,t) \tag{63}$$

whenever this boundary is accessible.

References

Abramowitz, M., and Stegun, I. A., eds. (1964). "Handbook of Mathematical Functions." Nat. Bur. Stand., Washington, D.C.

Buff, F. P., and Goei, N. S. (1969). Electrostatics of diffuse anisotropic interfaces. II. Effects of long-range diffuseness. *J. Chem. Phys.* **51**, 5363.

Clay, J. R., Goel, N. S., and Buff, F. P. (1972). Electrostatics of diffuse anisotropic interfaces. III. Point charge and dipole image potentials for air-water and metal-water interfaces. *J. Chem. Phys.* **56**, 4245.

Cox, D. R., and Miller, H. D. (1965). "The Theory of Stochastic Processes." Wiley, New York.

Crow, J. F., and Kimura, M. (1970). "An Introduction to Population Genetics Theory." Harper, New York.

Doob, J. L. (1953). "Stochastic Processes." Wiley, New York.

Feller, W. (1952). The parabolic differential equations and the associated semi-groups of transformations. *Ann. of Math.* **55**, 468.

Ito, K. (1944). Stochastic Integral. *Proc. Imp. Acad.* (*Tokyo*) **20**, 519.

Keilson, J. (1965). A review of transient behavior in regular diffusion and birth-death processes. Part II. *J. Appl. Probability* **2**, 405.

Magnus, W., Oberhettinger, F., and Soni, R. P. (1966). "Formulas and Theorems for the Special Functions of Mathematical Physics." Springer-Verlag, Berlin and New York.

Mortensen, R. E. (1969). Mathematical problems of modeling stochastic nonlinear dynamic systems. *J. Statist. Phys.* **1**, 271.

Stratonovich, R. (1963). "Topics in the Theory of Random Noise." Gordon & Breach, New York.

Stratton, J. A., Morse, P. M., Chu, L. J., Little, J. D. C., and Corbató, F. J. (1956). "Spheroidal Wave Functions." Technol. Press of MIT, Cambridge, Massachusetts and Wiley, New York.

Titchmarsh, E. C. (1962). "Eigenfunction Expansions associated with Second-Order Differential Equations." Oxford Univ. Press (Clarendon), London and New York.

Additional References

Bailey, N. T. J. (1964). "The Elements of Stochastic Processes with Application to the Natural Sciences." Wiley, New York.

Bharucha-Reid, A. T. (1960). "Elements of the Theory of Markov Processes and Their Applications." McGraw-Hill, New York.

References

Feller, W. (1954). Diffusion processes in one dimension. *Trans. Amer. Math. Soc.* **77**, 1.

Feller, W. (1959). The birth and death processes as diffusion processes. *J. Math. Pures Appl.* **38**, 301.

Karlin, S. (1966). "A First Course in Stochastic Processes." Academic Press, New York.

Keilson, J. (1964). A review of transient behavior in regular diffusion and birth-death processes. *J. Appl. Probability* **1**, 247.

Keilson, J. (1966). A technique for discussing the passage time distribution for stable systems. *J. Roy. Statist. Soc. Ser. B* **28**, 477.

Wax, N., ed. (1954). "Noise and Stochastic Processes." Dover, New York.

4

Population Growth and Extinction

The study of the growth of human population is several centuries old, and is perhaps the oldest branch of biology which has been studied quantitatively. The first law for population growth was given by Thomas Robert Malthus in 1798 when he observed that population growth follows a geometrical progression in contrast to its means of subsistence which tends to grow in an arithmetical progression. Thus he envisioned a situation in which the world's population would soon outstrip its means of support. Under conditions of unlimited sources, Malthus' geometrical growth can be expressed in the form of the equation

$$dn/dt = rn \tag{1}$$

where $n(t)$ is the instantaneous population of the species under consideration and r is some proportionality constant. Equation (1) is known as Malthus' law, according to which the population will grow exponentially, i.e.,

$$n(t) = n(0) \exp(rt) \tag{2}$$

Malthus did not take into account the fact that in a given environment, the growth may stop due either to the tendency of one organism to destroy others or to the limit on the density of population which the environment can hold (limited food). Verhulst took into account the limitation on the population due to the latter effect, and postulated that the rate of population growth was proportional to the product of the existing population and the difference between the total available resources and the resources used by the present population. If K^* is the maximum population that a given amount of food can support, then according to Verhulst,

$$dn/dt = rn(1 - n/K^*) \tag{3}$$

When $n \ll K^*$ the growth is according to Malthus' law, and as n approaches K^*, the rate of population growth decreases rapidly. Equation (3) can be integrated to give

$$f(t) = f(0)\{f(0) + (1 - f(0))e^{-rt}\}^{-1} \tag{4}$$

where

$$f(t) \equiv n(t)/K^* \tag{5}$$

Equation (4) is the equation of the so-called logistic curve which has a characteristic S-like shape. This equation can be fitted to the populations of many countries by choosing the parameters appropriately. Since there are three parameters [$f(0)$, K^*, and r] in Eq. (4), the population sizes at three different times determine all the parameters.

In addition to the Verhulst equation (3), there are other equations which give rise to an S-shaped curve. Some of these may fit the population data better than Verhulst's law. One such equation is

$$dn/dt = -rn \ln(n/K^*) \tag{6}$$

This equation is known as Gompertz' equation and was first proposed (with $r < 0$) in connection with mortality analysis of elderly people.

Another equation which is a generalization of the Verhulst and Gompertz equation is

$$dn/dt = rn[1 - (n/K^*)^\alpha]/\alpha \tag{7}$$

For $\alpha = 0$ it reduces to Eq. (6) and for $\alpha = 1$ it reduces to Eq. (3). The solution of Eq. (7) is

$$f(t) = f(0)\{[f(0)]^\alpha + e^{-rt}(1 - [f(0)]^\alpha)\}^{-1/\alpha} \tag{8}$$

In any of the deterministic models (3), (6), or (7), with $r > 0$, a population of size n, $0 < n < K^*$, keeps increasing as long as $n < K^*$, and then stays at level K^*. Similarly, a population starting with $n > K^*$ decreases to the value K^*, and then stays there. Extinction of a population is possible within these models only if the net growth rate per individual for a small-sized population r is negative, and then any population starting with any number of individuals less than K^* keeps decreasing until it becomes extinct. Thus the net growth rate determines the behavior of the population, including its success or failure in establishing a long-lasting community. In a realistic situation the basic mechanisms and factors in population growth, e.g., the amount of resources, birth and death rates, immigration and emigration rates, etc., change in a non-deterministic fashion. In this chapter we discuss the effects of random changes in the factors affecting population growth.

For populations with a positive net growth rate, the introduction of these random elements have two important consequences. A population can become extinct in spite of a positive net growth rate or it can grow above and fluctuate around the saturation level K^*. The consequences of the random factors can be studied by looking on the size of the population as a random variable, which is confined to some state space with specified transition probability rates which are the probabilistic counterparts of the growth rate in the deterministic model. The state space can be chosen to be either discrete or continuous. A discrete space is natural to the problem since population size changes discretely, while a continuous space is a good approximation to the growth of a large population where changes are small compared to its size. For investigating the problem of the extinction of a species, where the behavior of the population when its size is small is very crucial, a discrete state space is a better choice. On the other hand, for investigating the fluctuation of population around its saturation level, when this level is not too small, continuous state space is more desirable because of the better tractability of the mathematical analysis.

In the next section we study the evolution of a species colonizing on a limited source of food (i.e., an island), its success in colonizing the island, and if it fails, the length of time it persists before becoming extinct, by taking the growth process as a birth and death process and using the mathematical analysis of Chapter 2. In Section 4.2, we study the fluctuations in a large-sized population, using the mathematical analysis of Chapter 3.

4.1 Extinction of a Colonizing Species

A somewhat simple analysis of the problem of the extinction of a colonizing species due to random elements has been carried out in terms of the birth and death process by MacArthur and Wilson (1967). They take the growth process (assuming no immigration or emigration) as a stochastic birth and death process with λ_n and μ_n as the probability rates of birth and death,† respectively,

† For brevity, in this section we refer to the quantities λ_n and μ_n as birth and death rates.

when the size of the population is n. In other words, λ_n and μ_n are the probability rates of transition from state n to $n+1$ and from state n to $n-1$, respectively. MacArthur and Wilson assume that λ_n and μ_n depend linearly on n and that for $n = K^*$, either the birth rate is zero or the death rate is infinite. Using simple mathematical analysis, they calculate the average time for extinction and then attempt to decide on a strategy for successful colonization by a species.

Recently, the present authors (Richter-Dyn and Goel, 1972) reinvestigated this problem by taking more realistic dependence of λ_n and μ_n on n, and using the mathematical analysis of Chapter 2. We now present the salient features of our analysis, and conclude the section by summarizing the important characteristics of the process of colonization and extinction.

For the problem of colonization and extinction, the process is confined between the state 0 and some upper limit state K (which can be infinite). In the language of birth and death processes, the state 0 is an absorbing state because if the process reaches the state 0, the process is trapped forever. The state K is a reflecting state because if the process reaches this state, it is retracted to a previously occupied state. (In case $K = \infty$ we deal only with processes in which the probability of drifting to ∞ is zero.) Thus $\lambda_0 = \mu_0 = 0$ and $\lambda_K = 0$, $\mu_K > 0$.

We now consider three models, i.e., three dependences of λ_n and μ_n, which may reasonably describe the growth of the population. For convenience let us write

$$\lambda_n = n\lambda(n), \qquad \mu_n = n\mu(n) \tag{1}$$

so that $\lambda(n)$ is the birth rate per individual and $\mu(n)$ the death rate per individual when the population consists of n individuals. The three models are as follows:

1. *MacArthur and Wilson's (MW) model.* In this model

$$\lambda(n) = \lambda, \qquad n < K^*, \qquad \lambda(K^*) = 0; \qquad \mu(n) = \mu \quad \text{for all } n, \qquad \lambda > \mu \tag{2}$$

This dependence of $\lambda(n)$ and $\mu(n)$ corresponds to the assumption that birth but not death is density dependent.[†] As in the deterministic model, K^* represents either the maximal possible population level (i.e., $\lambda_{K^*}/\mu_{K^*} = 0$) or the population size where the net growth rate $(\lambda_n - \mu_n)$ changes sign. However,

† This model is mathematically identical with the second model proposed by MacArthur and Wilson where the controls act via the death rate i.e.,

$$\lambda(n) = \lambda \quad \text{for all } n; \qquad \mu(n) = \mu, \qquad n \leqslant K^*, \qquad \mu(K^*+1) = \infty \tag{3}$$

since in this case K^* is also a reflecting state, as the process can never stay in state K^*+1, but is retracted back to state K^* in no time.

the net growth rate per individual $\lambda(n) - \mu(n)$ is independent of the size of the population below the saturation level and then abruptly changes its sign from positive to negative. This is contrary to the deterministic model where the net growth rate per individual decreases gradually with the size of the population, vanishes only when the population reaches its saturation level, and then becomes negative. Because of the coarse form of density-dependent controls, this model will model populations that are allowed to grow without homeostatic controls until they reach the maximum size (K^*) permitted by the environment, and are then stopped abruptly. Further, as in the deterministic model, the population cannot evolve to levels above the level where the net growth rate becomes negative.

2. *Model A.* In this model, we choose

$$\lambda(n) = \lambda(1-(n/K)^\alpha), \qquad \mu(n) = \mu(1+(n/K)^\alpha), \qquad \lambda > \mu \tag{4}$$

so that the net growth rate is

$$\lambda_n - \mu_n = (\lambda-\mu)n(1-(n/K^*)^\alpha) \tag{5}$$

where

$$K^* = \left(\frac{\lambda-\mu}{\lambda+\mu}\right)^{1/\alpha} K, \qquad K^* < K \tag{6}$$

Though the form of the net growth rate (5) is identical with the corresponding deterministic rate [Eq. (4.0.7)], there is a difference between the two models which is introduced by the stochastic approach. In the deterministic model, population never exceeds the size K^*, whereas in the stochastic model A it may exceed K^* (although $\lambda_n < \mu_n$ for $n > K^*$), but never exceeds the size K where $\lambda_K = 0$. This is reasonable because fluctuations may change the size of the population beyond its deterministic saturation value. It may be noted that as $\alpha \to \infty$, model A approaches the MW model, since for all $n < K$, λ_n and μ_n become linear in n and $K \to K^*$. In this model the controls act via the birth and death rates.

3. *Model B.* In this model we choose

$$\lambda(n) = \lambda, \qquad \mu(n) = \mu\left(1 + \frac{\lambda-\mu}{\mu}\frac{n}{K^*}\right), \qquad n > 0, \qquad \lambda > \mu \tag{7}$$

so that the net growth rate is

$$\lambda_n - \mu_n = (\lambda-\mu)n(1-n/K^*) \tag{8}$$

This net growth rate has the same characteristics as those for model A (with $\alpha = 1$) and the deterministic model (with $\alpha = 1$). However, in model B there is no built-in requirement on the population to stay below some finite size.

Although the population size is not limited ($\lambda_n > 0$ for all n), the probability that the population will increase without limit is zero since

$$\frac{\mu_1 \cdots \mu_n}{\lambda_1 \cdots \lambda_n} \cdot \frac{\mu_{n+1}}{\lambda_{n+1}} > \frac{\mu_1 \cdots \mu_n}{\lambda_1 \cdots \lambda_n} \qquad \text{for all} \quad n \geqslant K^* \tag{9}$$

so that

$$\sum_{i=1}^{\infty} \frac{\mu_1 \cdots \mu_i}{\lambda_1 \cdots \lambda_i} = \infty \tag{10}$$

and condition (2.2.48) is satisfied.

One can generalize model B by replacing n/K^* in Eqs. (7) and (8) by $(n/K^*)^\alpha$, but this generalization does not provide any new insight and the minor differences are similar to those for model A with $\alpha = 1$ and $\alpha \neq 1$. Further, instead of choosing the death rate to be density dependent, one can choose the birth rate to be density dependent but, as shown later in this section, the most important features depend on the ratio μ_n/λ_n, and decreasing λ_n is equivalent to increasing μ_n.

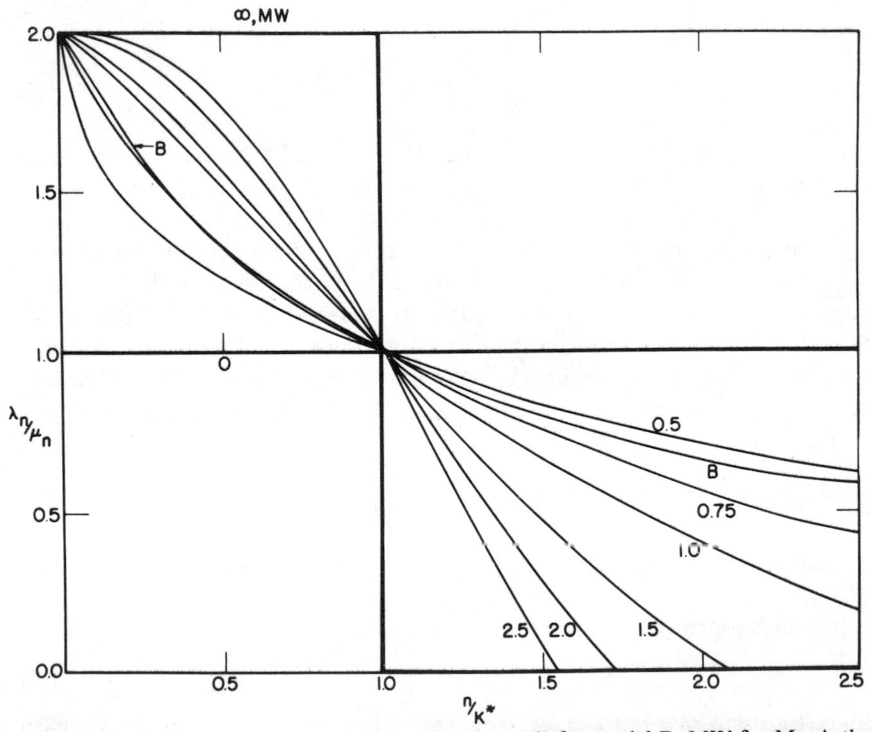

Fig. 4.1 λ_n/μ_n vs. n/K^* for various models. B stands for model B, MW for MacArthur and Wilson's model. The numbers near the curves represent the value of α for model A. For all models $\lambda/\mu = 2$.

In Fig. 4.1 we have plotted λ_n/μ_n vs. n/K^* for the MW model, model A for a variety of values of the parameter α, and model B ($\lambda/\mu = 2$ for all models). In all the models λ_n/μ_n is identical for $n/K^* = 0$ and 1, and the values λ_n/μ_n for the MW model and model A with $\alpha = 0$ are bounds for λ_n/μ_n for all the other models.

For analyzing the problem of the extinction of a species, by any of the above models, the most central quantity is $R_{N,m}(0)$, the probability of population reaching the size N, starting with the size m, without reaching the size 0, i.e., before becoming extinct. From Table 2.6, this probability is given by

$$R_{N,m}(0) = R_{N,1}(0)/R_{m,1}(0) \tag{11a}$$

where

$$R_{N,1}(0) = \left[1 + \sum_{i=1}^{N-1} \frac{\mu_1 \cdots \mu_i}{\lambda_1 \cdots \lambda_i}\right]^{-1} \tag{11b}$$

From Eqs. (1) and (3) for the MW model (with K^* as the saturation value of the population)

$$R_{N,m}(0) = \begin{cases} \dfrac{1-(\mu/\lambda)^m}{1-(\mu/\lambda)^N}, & N \leqslant K^* \\ 0, & N > K^* \end{cases} \tag{12}$$

This result can be derived in a simpler manner by noting that for the MW model, Eq. (2.1.7) with $v = 0$ becomes

$$\frac{\partial G}{\partial t}(t, \mu/\lambda) = 0$$

i.e.,

$$\frac{d}{dt} \sum P_{n,m}(t) \left(\frac{\mu}{\lambda}\right)^n = 0$$

Thus the average value of $(\mu/\lambda)^{n(t)}$ is independent of t (i.e., $\{(\mu/\lambda)^{n(t)}\}$ is a martingale) and is therefore equal to $(\mu/\lambda)^m$. Starting with a population of size m, the probability of reaching size N without first becoming extinct is the same as the probability of being absorbed at state N, since whatever happens after state N is reached does not affect this quantity. For this modified process as $t \to \infty$, either state 0 or state N is occupied and the average of $(\mu/\lambda)^{n(t)}$ is given by

$$(\mu/\lambda)^m = R_{N,m}(0)\,(\mu/\lambda)^N + [1 - R_{N,m}(0)]\,(\mu/\lambda)^0$$

from which we recover Eq. (12).

From Eq. (12) the probability $R_{N,m}(0)$ is independent of K^* for all $N \leqslant K^*$. In Fig. 4.2 we have plotted this probability for $m = 1$ for a variety of values of

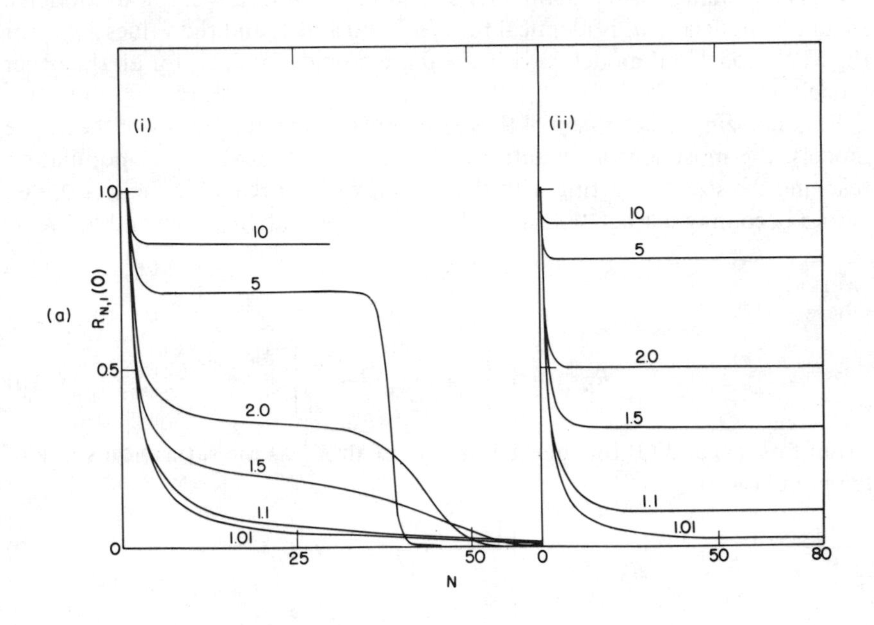
(i)
(ii)
(a)
$R_{N,1}(0)$
1.0
0.5
0
10
5
2.0
1.5
1.1
1.01
25
50
0
50
80
N

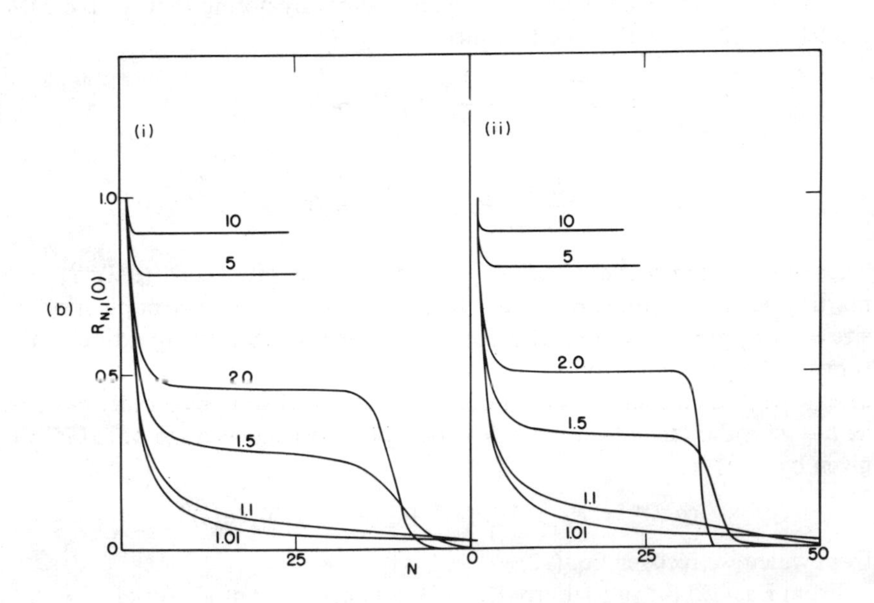
(i)
(ii)
(b)
$R_{N,1}(0)$
1.0
0.5
0
10
5
2.0
1.5
1.1
1.01
25
N
0
25
50

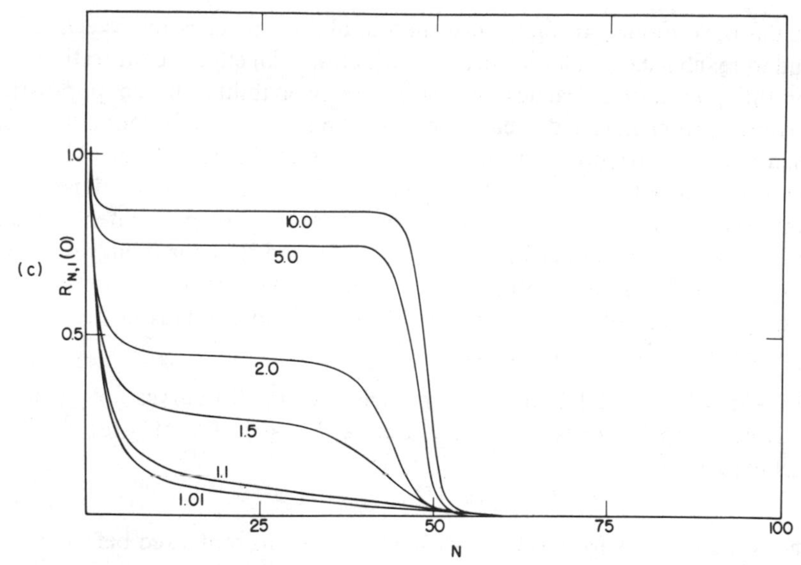

Fig. 4.2 The probability $R_{N,1}(0)$ of reaching the state N starting from state 1 without going extinct versus N for various models. The numbers near the curves represent the values of λ/μ. (a) (i) Model A, $\alpha = 0.5$, $K^* = 20$; (ii) MW model. (b) (i) Model A, $\alpha = 1.0$, $K^* = 20$; (ii) model A, $\alpha = 2.0$, $K^* = 20$. (c) Model B, $K^* = 20$.

λ/μ. Note that for $\lambda/\mu > 1$, the probability drops from the value 1 at $N = 1$ to the value $(1 - \mu/\lambda)$, and then remains practically constant. From Eq. (12), the value n_c such that for $N > n_c$, $R_{N,1}(0)$ does not change significantly, is the value of N for which $(\mu/\lambda)^N$ is smaller than a preassigned small number, say 10^{-3}, i.e.,

$$n_c \approx 3/\log(\lambda/\mu) \tag{13}$$

From Eq. (11a) we note that if the initial size of the population is n_c, the probability of population reaching the size K^* before becoming extinct is practically unity. Thus, according to the MW model, if the population either reaches the critical size n_c or initially starts with the critical size, it is going to colonize, almost surely.

This qualitative behavior in large part is also true for both models A and B, as can be seen in Fig. 4.2 where we have also plotted $R_{N,1}(0)$ for both these models by using Eqs. (11), (4), and (7). However, there are some differences that are worth noting.

(a) The curves for both models A and B are slightly below those for the MW model, for $N < K^*$.

(b) The probability of a species reaching a size greater than K^* is not zero as in the MW model. In fact, once the population reaches the size K^*, it is bound to reach a certain larger size $K_e{}^*$, which we call effective saturation size. After the population reaches this size, the probability of the population increasing further in size decreases sharply. This is so even in model B where we had put no restrictions on the maximum size of the population. The exact value of $K_e{}^*$, which is the end of the plateau of the $R_{N,1}(0)$ curve, depends on the model and the values of the parameters. It can be shown (Richter-Dyn and Goel, 1972) that for model B, $2 < K_e{}^*/K^* < e$, and that for model A, $\alpha \geqslant 1$, $1 < K_e{}^*/K^* < 2$, while for other values of α, e is the upper limit for $K_e{}^*/K^*$. For smaller values of K^*, e.g., $K^* = 5$, there are no plateaus in the $R_{N,1}(0)$ curves, even for λ/μ as large as 100.

For $\lambda/\mu = 1.01$ and 1.1 in all the models the $R_{N,1}(0)$ curves are identical. This is not surprising because, for $\lambda/\mu \simeq 1+\theta$, $\theta \ll 1$, Eq. (11) for the MW model reduces to

$$R_{N,1}(0) \simeq 1/N, \qquad N \leqslant K^*$$

which is the same as for model A with $\alpha = 0$, and, as remarked before, for all the models λ_n/μ_n are bounded between λ_n/μ_n of the MW model and model A, $\alpha = 0$.

The critical value n_c, as calculated by the MW model, provides an exceedingly good estimate for models A and B. This is expected since for $n \ll K^*$, all the models are basically the same. Since this is so, the average time to extinction (when it is assumed a priori that the critical value n_c is never reached) as calculated by using the MW model should also provide a good estimate for the corresponding quantity in models A and B. Since n_c is characterized by $(\lambda/\mu)^{n_c} \ll 1$, $R_{0,m}(N)$ of Table 2.6 is given by

$$R_{0,m}(N) \simeq (\mu/\lambda)^m, \qquad N \geqslant n_c \tag{14}$$

Substituting λ_n and μ_n from Eqs. (1), (3), and (4) into the expression for $M^*_{0,m}(n_c)$ [Table 2.6 and Eq. (2.2.23) with $N = n_c$], the conditional average time for extinction for the MW model, when colonization fails, is given by

$$M^*_{0,1}(n_c) \equiv T_e = \frac{1}{\mu}\sum_{l=1}^{n_c}\frac{(\mu/\lambda)^l}{l} \simeq -\frac{1}{\mu}\ln\left(1-\frac{\mu}{\lambda}\right) \simeq \frac{1}{\lambda}\left(1+\frac{\mu}{2\lambda}+\cdots\right) \tag{15}$$

Thus $\lambda T_e \simeq 1$ for all the models. In the work of Richter-Dyn and Goel (1972), this result was verified by direct computation, where it was also shown that the coefficient of variation (ratio of the square root of variance to the average) is approximately 1 or less.

By the same reasoning as above for calculating T_e, we can use the MW model to calculate $M^*_{n_c,1}(0)$ also, the conditional average time for the population to reach the size n_c without becoming extinct, when a priori it is known that the population is going to reach the size n_c (i.e., colonize). Putting $N = n_c$, $m = 1$, and $L = 0$ in the expression for $M_{N,m}(L)$ given in Table 2.6, the conditional average time is

$$M^*_{n_c,1}(0) \equiv T_{n_c} = \left[R_{n_c,1} \sum_{i=1}^{n_c-1} \sum_{n=1}^{i} \lambda_n^{-1} \Pi_{n+1,i} R_{n_c,n} \right] \Big/ R_{n_c,1}$$

Using the defining relation (2.2.34) for $\Pi_{n+1,i}$, and using Eq. (12) with $(\mu/\lambda)^{n_c} \ll 1$, we obtain

$$\begin{aligned} M^*_{n_c,1}(0) \equiv T_{n_c} &= \sum_{i=1}^{n_c-1} \sum_{n=1}^{i} \frac{1}{\lambda n} \left(\frac{\mu}{\lambda}\right)^{i-n} \left[1 - \left(\frac{\mu}{\lambda}\right)^n\right] \\ &= \frac{1}{\lambda} \sum_{n=1}^{n_c-1} \frac{1}{n} \left[\left(\frac{\mu}{\lambda}\right)^{-n} - 1\right] \sum_{i=n}^{n_c-1} \left(\frac{\mu}{\lambda}\right)^i \\ &= \left\{ \sum_{n=1}^{n_c-1} \frac{1}{n} - \left[\sum_{n=1}^{n_c-1} \frac{1}{n_c - n} \left(\frac{\mu}{\lambda}\right)^n + \sum_{n=1}^{n_c-1} \frac{1}{n} \left(\frac{\mu}{\lambda}\right)^n \right] \right\} (\lambda - \mu)^{-1} \\ &= \left[\sum_{n=1}^{n_c-1} \frac{1}{n} - \frac{\mu}{\lambda} \frac{n_c}{n_c - 1} - \frac{1}{2} \left(\frac{\mu}{\lambda}\right)^2 \frac{n_c}{n_c - 2} - \cdots \right] (\lambda - \mu)^{-1} \end{aligned} \tag{16}$$

For $\lambda/\mu = 5$, $n_c \simeq 5$, $\lambda T_{n_c} \simeq 2.25$ as compared to $\lambda T_e \simeq 1.01$.

Knowing T_{n_c}, we can also derive an approximate expression for the probability density $F_{0,1}(n_c)$ of species becoming extinct in time t before reaching size n_c. As we have shown above, the extinction of an unsuccessful population is determined by its behavior for small values of n ($n < n_c$). In this regime, the linear model with $\lambda_n = n\lambda$, $\mu_n = n\mu$ well represents the growth of the population. For this model, from Table 2.1, the probability density $P_{n,m}(t)$ is given by

$$\begin{aligned} P_{n,m}(t) &= \left(\frac{\lambda}{\mu}\right)^n \sum_{k=0}^{\min\{m,n\}} (-1)^k \binom{m+n-k-1}{n-k} \binom{m}{k} \\ &\quad \times \left(\frac{e^{(\lambda-\mu)t} - 1}{(\lambda/\mu) e^{(\lambda-\mu)t} - 1}\right)^{m+n-k} \left(\frac{(\mu/\lambda) e^{(\lambda-\mu)t} - 1}{e^{(\lambda-\mu)t} - 1}\right)^k \end{aligned} \tag{17}$$

If initially there is only one colonizer,

$$P_{0,1}(t) = \frac{e^{(\lambda-\mu)t} - 1}{(\lambda/\mu) e^{(\lambda-\mu)t} - 1} \tag{18a}$$

$$P_{n,1}(t) = \left(\frac{\lambda}{\mu} - 1\right)^2 \frac{(e^{(\lambda-\mu)t} - 1)^{n-1}}{((\lambda/\mu) e^{(\lambda-\mu)t} - 1)^{n+1}} e^{(\lambda-\mu)t}, \qquad n \geqslant 1 \tag{18b}$$

Equations (18) are valid for $n < n_c$ and $t < T_{n_c}$.

The probability $F_{0,1}(t)\,\Delta t$ of first reaching state $n = 0$ in the time interval $(t, t+\Delta t)$ is related to $P_{n,1}$ by the relation

$$F_{0,1}(t)\,\Delta t = P_{1,1}(t)\,\mu\,\Delta t \tag{19}$$

Therefore from Eq. (18b) we obtain

$$F_{0,1}(t) = \mu\left(\frac{\lambda}{\mu} - 1\right)^2 \frac{e^{(\lambda-\mu)t}}{[(\lambda/\mu)\,e^{(\lambda-\mu)t} - 1]^2} \tag{20}$$

This is the expression for the probability density of a species becoming extinct, for small values of t, and is a monotone decreasing function of t.

Having discussed the details of the process of extinction, we now discuss the best strategy for successful colonization. For a population with a birth rate reasonably larger than the death rate, the existence of a critical value n_c such that for all $m > n_c$ the colonization is practically a certain event, points out the best strategy for successful colonization. The colonizing group which arrives at the island should consist of $m > n_c$ individuals where $n_c \simeq 10$ to 20. When this is not possible the best strategy is to increase the probability of reaching n_c without becoming extinct. For an initial population of size m this probability is $\simeq 1-(\mu/\lambda)^m$. Thus if m cannot be increased, μ/λ should be decreased. This goal can be achieved by either decreasing μ, the death rate per individual for a low density population, or by increasing λ, the birth rate per individual for a low density population. However, the first alternative is favorable since by decreasing μ the average time for extinction of a failing population $T_e \simeq \mu^{-1}\,|\ln(1-\mu/\lambda)|$ increases, and there is more time for an unpredictable favorable event, not incorporated in the model, to change the fate of the population. It may be noted that the strategy mentioned above for population with $m < n_c$ was also suggested by MacArthur and Wilson (1967). Their strategy was aimed at decreasing the probability of quick extinction μ/λ and increasing the average time of extinction $M_{0,1}$, which includes the average time of extinction for both the successful and unsuccessful populations. The relation between these three averages in case the two events of success and failure are practically exclusive (no extinction when the size is moderate but only after it reaches K^* and above) is

$$T_1 = M_{0,1} \simeq T_e(\mu/\lambda) + T_s(1-\mu/\lambda), \qquad T_s \gg T_e \tag{21}$$

where T_s is the average time for extinction of a successful population. This relation follows from the observation that the probability of failure of the population to colonize is $\simeq \mu/\lambda$ and the probability to colonize is $\simeq (1-\mu/\lambda)$.

The average time T_s for extinction of a successful population can be approximately calculated from values of T_1 by using Eq. (21) and noting that $T_s \gg T_e$ if $\mu/\lambda \lesssim 1-\mu/\lambda$. Thus

$$T_s \simeq T_1(1-\mu/\lambda)^{-1} \tag{22}$$

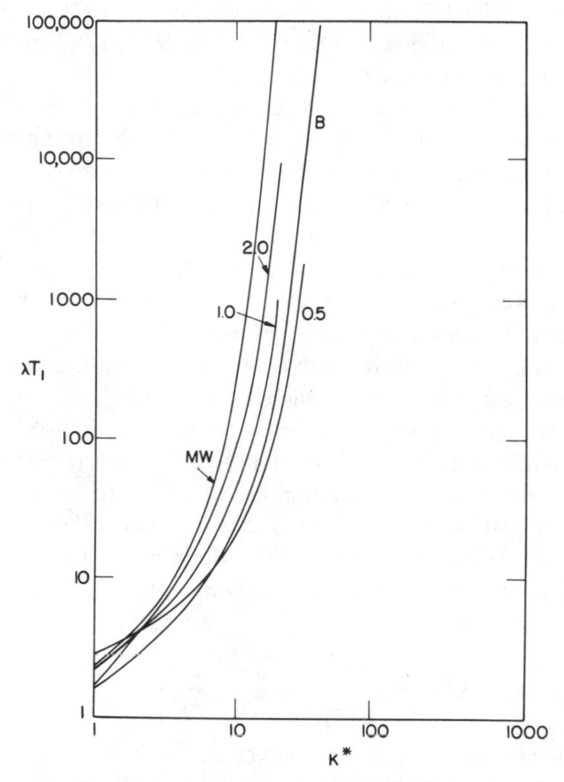

Fig. 4.3 The average time of extinction T_1 (when initially the population has one colonizer) times λ versus K^* for various models with $\lambda/\mu = 2$. B stands for model B, MW for MacArthur and Wilson's model. The numbers near the curves represent the value of α for model A.

In Fig. 4.3 we have plotted λT_1 vs. K^* for various models for $\lambda/\mu = 2$ by using the expression for $M_{0,1}$ given in Table 2.4. The calculated values of T_1 and therefore of T_s demonstrate that a successful population will persist on the average an exceedingly long time before becoming extinct for $\lambda/\mu \gtrsim 1.5$ and $K^* \simeq 40$. This can be seen from Fig. 4.3, where λT_1 increases sharply with K^*. For $K^* = 20$, λT_1 varies from 1.3×10^2 for $\alpha = 0.5$ to 1.1×10^5 in the MW model, and for $K^* = 40$, it varies from 1×10^4 to 5.6×10^{10}. Since the birth rate per individual is typically several offspring per generation, T_s as given by Eq. (22) for K^* as small as 40 is between tens of thousands to millions of generations, an exceedingly long time. Since T_s is the average time for extinction when it is known a priori that the population will colonize, it is practically independent of initial size of the population. On the other hand, small changes in λ/μ yield large changes in T_s, e.g., for $\lambda/\mu = 10$, $T_1 \simeq 6 \times 10^5$

for model B with $K^* = 20$, compared to 300 for $\lambda/\mu = 2$. In this example a fivefold increase in λ causes a 100-fold increase in T_1 (and greater increases result if μ is decreased and λ is kept constant).

It may be noted that T_1 in the MW model is greater than T_1 in any of the other models for $K^* > 3$. This difference becomes more prominent when K^* in the MW model is regarded as the maximal possible size of the population, and taken equal to K, the maximal size of the population in model A.

For the case in which the $R_{N,1}(0)$ curve does not have a plateau (i.e., K^* is small or $\lambda/\mu \simeq 1$), the average time for extinction T_1 is not an exceedingly large time. Therefore, the population can become extinct in a reasonable time, even after it reaches a level comparable to the "capacity" of the island (i.e., $n \simeq K^*$). In this case, both the average time for extinction T_s, when it is a priori known that the population reaches the level K^*, and the average time for extinction $M^*_{0,1}(K^*)$ when it is a priori known that it does not reach this level K^*, are relevant to the choice of a strategy of colonization. In case $\lambda/\mu \simeq 1$, λ_n/μ_n for $n < K^*$ is approximately the same for all the models, and $M^*_{0,1}(K^*)$ can be calculated by using Table 2.6, with $N = K^*$ for the MW model. Substituting Eqs. (1) and (2), the defining equations for the MW model, into this expression, using Eq. (12), and carrying out the summation within the approximation $\mu/\lambda \simeq 1-\theta$, $\theta \ll 1$, we obtain

$$M^*_{0,1}(K^*) = \frac{1}{\lambda}\left[\sum_{l=1}^{K^*}\frac{1}{l} - \frac{3}{2} + \frac{1}{2K^*}\right] \Big/ R_{0,1}(K^*) \tag{23}$$

For $K^* = 5$ to 100, the term in the bracket varies from 1 to 3.7.

To calculate T_s, we note that by arguments similar to those used in the derivation of relation (21),

$$T_1 = M^*_{0,1}(K^*)\,R_{0,1}(K^*) + T_s\,R_{K^*,1}(0)$$

Using the numerical value of λT_1 (which varies between 4.0 and 7.3 for $K^* = 5$ to 100), the relation $R_{K^*,1}(0) \simeq 1/K^*$, noted earlier in this section, and Eq. (23), we get for model A, $\alpha = 1$, and for model B

$$\lambda T_s \simeq cK^*, \qquad 3 < c < 3.6, \qquad 5 < K^* < 100$$

As expected, T_s is much smaller than calculated earlier in this section for $\lambda/\mu = 2$, and is approximately a linear function of K^*.

For a population with $\lambda/\mu \simeq 1$, the best strategy for colonization is to decrease $R_{0,m}(K^*) = 1 - R_{K^*,m}(0)$, and to increase T_s. Thus it is desirable to increase m and decrease μ/λ by decreasing μ rather than by increasing λ, so as to increase T_s.

Summarizing our results, we have shown that for a given growing population, with birth rate per individual reasonably greater than the death rate per individual for a small-sized population ($\lambda/\mu \gtrsim 1.5$), and which becomes comparable to the death rate for a moderate population size ($\geqslant 40$), there is a critical size n_c ($n_c \lesssim 20$) such that if the population reaches this size, it is going to persist for an immensely long time. The maximum size of a population, once it escapes extinction and colonizes, is between K^* and eK^* where K^* is the size of the population for which the birth rate is equal to the death rate. Thus the relevant average time of extinction is the average time of extinction T_e when it is known a priori that n never reached the critical value n_c. For an initial population of one propagule λT_e is shown to be approximately 1, with coefficient of variation approximately equal to 1, independent of the details of the model. A corollary of these results is that if the initial population m of a species is greater than n_c and the birth rate per individual is reasonably larger than the death rate per individual, the species will become extinct in immensely large times or for practical purposes, will colonize. Since for small n, MacArthur and Wilson's linear model represents the growing population well, the critical value n_c, the probability of a population reaching this value without becoming extinct, and the average time for extinction when n_c is never reached can be calculated rather accurately using this model. For a population with $K^* \approx 10$ or $\lambda/\mu \gtrsim 1$, the time for extinction T_s when a priori it is known that the population reaches the level comparable to the one which can be supported by the limited food, is not large and such populations can become extinct in a short period even after colonization.

In the analysis carried out so far, we have been assuming that all the individuals of the population interact with each other. We now generalize our discussion to populations where this is not true, using the mathematical analysis already at hand. We consider an ensemble of local subpopulations of the same species which constitute the whole population. The subpopulations are generated by migration from the already established ones, while existing subpopulations become extinct. We will show that there exists a critical number of subpopulations above which the ensemble is going to persist for a long time, and that by choosing a good strategy of migration the population can increase its prospects to persist when decomposed into a number of subpopulations.

As a first approximation, we ignore differences among subpopulations and assume that in all the subpopulations the birth and death rates are the same. Let N_0 denote the available number of sites for colonization, Λ_n the probability per unit time of the creation of a new subpopulation, and ν_n the probability per unit time of the extinction of one subpopulation when n such subpopulations are present. A reasonable set of forms for Λ_n and ν_n is

$$\Lambda_n = \Lambda n(1-n/N_0), \qquad \nu_n = \nu n \tag{24a}$$

so that

$$\Lambda_{N_0^*} = \nu_{N_0^*} \quad \text{at} \quad N_0^* = (1-\nu/\Lambda)\, N_0 \tag{24b}$$

Here we have made the assumption that the probability of initiating a new subpopulation is proportional to the number of already established subpopulations and the fractions of empty sites for colonization. We also assume that the sites are isolated so that the probability of extinction for one subpopulation is independent of the number of existing subpopulations.

From the results presented earlier in this section, we conclude that if $\Lambda/\nu > 1.5$ and $N_0^* \gtrsim 40$, the probability of a population eventually having n subpopulations before becoming extinct has a plateau starting at some critical number n_c of subpopulations. Once this number is reached, the species will persist for an immensely long time in spite of local extinction of subpopulations. The number of established colonies N_e^* will exceed N_0^* but for moderate values of Λ/ν it will not reach N_0.

These results differ from those for the corresponding deterministic model, where $n(t)$ satisfies the differential equation

$$dn/dt = \Lambda n(1-n/N_0) - \nu n \tag{25}$$

Carrying out the simple integration, the number of subpopulations at time t is given by

$$n(t) = N_0(1-\nu/\Lambda)\left\{1 - \exp(-(\Lambda-\nu)t)\left[1 - \frac{N_0}{n(0)}(1-\nu/\Lambda)\right]\right\}^{-1} \tag{26}$$

For $\Lambda > \nu$, $n(t)$ monotonically approaches the steady-state value $N_0(1-\nu/\Lambda)$, which is the N_0^* defined in Eq. (24b), and for $\nu > \Lambda$ the population becomes extinct.

Our analysis of colonization by a subpopulation provides further insight into the dynamics of an ensemble of subpopulations. For a starting subpopulation, the probability of becoming extinct before getting established is μ/λ by Eq. (14), and from Eq. (15) the average time for such extinction is $T_e \simeq (1/\lambda)(1+\mu/2\lambda)$. Here we assume that in all the subpopulations the birth and death rates are the same, i.e., the same λ and μ such that $\lambda/\mu > 1.5$ and such that K^* in each subpopulation is above 40. Therefore, the probability rate of extinction of a single subpopulation ν (neglecting the rare extinction of successful subpopulations) is given by

$$\nu \cong \left(\frac{\mu}{\lambda}\right) \bigg/ \frac{1}{\lambda}\left(1 + \frac{\mu}{2\lambda}\right) = \mu\left(1 + \frac{\mu}{2\lambda}\right)^{-1} \tag{27}$$

To get an estimate of Λ, we note that new subpopulations are initiated by individuals migrating from successful subpopulations. Since the probability of a subpopulation to colonize successfully is $1-\mu/\lambda$ (assuming $\lambda/\mu > 1.5$)

$$\Lambda = (1-\mu/\lambda)\tilde{\Lambda} \tag{28}$$

where $\tilde{\Lambda}$ is the rate of immigration characterizing successful subpopulations.

Therefore, the condition $\Lambda/\nu > 1.5$ becomes

$$\tilde{\Lambda} > \tfrac{3}{2}\mu\left\{\left(1 + \frac{\mu}{2\lambda}\right)\left(1 - \frac{\mu}{\lambda}\right)\right\}^{-1} \tag{29}$$

Now the probability that the species will establish more subpopulations than the critical number n_c is given by

$$1 - \frac{\nu}{\Lambda} = 1 - \frac{\lambda/\tilde{\Lambda}}{(1+\mu/2\lambda)(1-\mu/\lambda)}\frac{\mu}{\lambda} \tag{30}$$

From Eq. (30), the probability of establishing a long-lasting ensemble of subpopulations by the species is greater than the probability of success within each subpopulation ($\simeq 1-\mu/\lambda$) if

$$\frac{\lambda/\tilde{\Lambda}}{(1+\mu/2\lambda)(1-\mu/\lambda)} < 1 \tag{31}$$

Thus for $\tilde{\Lambda}$ satisfying the inequality [consistent with condition (29)]

$$\tilde{\Lambda} > \lambda[(1+\mu/2\lambda)(1-\mu/\lambda)]^{-1} \tag{32}$$

the prospects of the species to persist as an ensemble of subpopulations are better than its prospects of existence as one subpopulation.

There are many important features of the population, e.g., age distribution and differences in the genetic consitution, which we have not included in our model. The age distribution can, in principle, be incorporated by splitting the population into a few subgroups, each corresponding to a range of ages characterized by approximately the same λ and μ. The stochastic model resulting from this approach will lead to a process with several random variables, each representing the number of individuals in a particular subgroup. To pursue this approach further one first has to derive expressions analogous to those given in Chapter 2 for the multivariate "birth and death" processes, which is beyond the scope of this monograph.

Before concluding this section, we point out that the analysis presented in this section can also be used to discuss strategies for pest control. In the case of pests, the best strategy is aimed at the extinction of the pest population in the shortest possible period before extensive damage is done to the crop. By the analysis of this section, for a small-sized pest population, with λ and μ as the stochastic birth and death rates per individual, the best strategy is to increase μ/λ by increasing μ rather than by decreasing λ. This strategy will simultaneously increase the probability of extinction of pest population and decrease the average time for extinction when this event occurs. When the pest population consists of subpopulations, e.g., due to crops in several neighboring areas, the pest population can be reestablished by immigration, even after extinction. For this case the best strategy should be aimed at increasing ν/Λ, the ratio of extinction and immigration probability rates. This can be achieved by either increasing the extinction rate of subpopulations or by decreasing the immigration rate (e.g., by increasing the distances among areas of crops). The first measure is preferred since it also decreases the average time for extinction of the collection of subpopulations.

4.2 Population Growth in a Random Environment

In this section, we study the effects of randomly fluctuating environment on a successful colonized population (i.e., a population with size much larger than the critical size considered in Section 4.1). Since the population size is large, the relative changes in the size of the population are small, and we can treat the population size x as a continuous variable. In the absence of fluctuating environment, the population is assumed to grow according to Eq. (4.0.7), i.e., according to

$$dx/dt = rx(1-(x/K^*)^\alpha)/\alpha, \qquad r > 0 \tag{1}$$

For simplicity we will take α to be 1 (Verhulst model for population growth) and leave the case of $\alpha \neq 1$ as an exercise for the reader. The fluctuating environment may affect the growth in several different ways, some of which we discuss below.

a. *Changes in net growth rate.* The fluctuating environment may introduce a stochastic element in the growth rate r. A stochastic form of r may be

$$r = \bar{r} + \sigma F(t) \tag{2}$$

where $F(t)$ is some noise. Since environmental changes are due to many factors, and are fast compared to the time scale of population growth, we can approximate the noise by a white noise with the characteristics defined by Eqs. (3.0.17a) and (3.0.17b). Here σ is a constant and $\langle r \rangle = \bar{r}$. The stochastic differential equation describing the growth of the population is then

$$dx/dt = \bar{r}x(1-x/K^*) + \sigma x(1-x/K^*)\, F(t) \tag{3}$$

and the coefficients of the corresponding Fokker–Planck equation, according to Eq. (3.0.30b), are

$$a(x) = x(1-x/K^*)[\bar{r}+(\sigma^2/2)(1-2x/K^*)] \tag{4a}$$

$$b(x) = \sigma^2 x^2 (1-x/K^*)^2 \tag{4b}$$

Since $b(0) = b(K^*) = 0$, the boundaries $x = 0$, $x = K^*$ are singular. Near $x = 0$, $a(x) \simeq (r+\sigma^2/2)x$ and $b(x) \simeq \sigma^2 x^2$, and by the classification of singular boundaries given in Section 3.1, the boundary $x = 0$ is an inaccessible natural boundary and can never be reached. Similarly, the boundary $x = K^*$ is a natural boundary and cannot be reached in a finite time. This process therefore can describe a population which is far from extinction and which fluctuates about some average value ($< K^*$) due to fluctuations in the net growth rate.

The steady-state probability density function, from Eqs. (3.0.56a) and (4), is given by

$$P(x, \infty) = Cx^{(2\bar{r}/\sigma^2 - 1)}(1 - x/K^*)^{-(2\bar{r}/\sigma^2 + 1)} \tag{5}$$

where C is a normalization constant. This density function is slightly different from that derived by Levins (1969) using the same stochastic differential equation (3). In his probability density function $2\bar{r}/\sigma^2 - 1$ is replaced by $2\bar{r}/\sigma^2 - 2$ and $2\bar{r}/\sigma^2 + 1$ is replaced by $2\bar{r}/\sigma^2 + 2$. This difference is due to the Ito calculus used by Levins as compared to the Stratonovich calculus used by us (see Section 3.0). However, qualitatively the two distributions are similar. If $2\bar{r}/\sigma^2 < 1$, the density function (5) is U shaped, indicating the tendency of the population to be either near 0 (of very small size) or near K^*. For $2\bar{r}/\sigma^2 > 1$, the density (5) is monotonically increasing and is J shaped, indicating accumulation of the population near K^*. The minimum of the U-shaped distribution is given by (see Table 3.9)

$$a(x) = \frac{1}{2}\frac{db}{dx}$$

i.e., by

$$x = (1 - 2\bar{r}/\sigma^2)\, K^*/2$$

To derive the time-dependent probability density, we introduce the variable

$$z = \frac{1}{\sigma} \ln[x/(1 - x/K^*)], \qquad dz = [\sigma x(1 - x/K^*)]^{-1}\, dx \tag{6}$$

Equation (3) then becomes

$$dz/dt = (\bar{r}/\sigma) + F(t) \tag{7}$$

which is the stochastic differential equation for an unrestricted Wiener process, and hence can be analyzed by the methods of Chapter 3. The Fokker–Planck equation, satisfied by the probability density function of z, $g(z \mid z_0, t)$, is

$$\frac{\partial g}{\partial t} = -\frac{\partial}{\partial z}((\bar{r}/\sigma)g) + \frac{1}{2}\frac{\partial^2 g}{\partial z^2} \tag{8a}$$

with the boundary conditions

$$\lim_{z \to \pm\infty} g(z \mid z_0, t) = 0 \tag{8b}$$

corresponding to the inaccessible boundaries $x = 0$ ($z = -\infty$) and $x = K^*$ ($z = +\infty$). From Table 3.4, row 1, $g(z\,|\,z_0, t)$ is given by

$$g(z\,|\,z_0, t) = \frac{1}{(2\pi t)^{1/2}} \exp\left\{-\frac{1}{2t}(z - z_0 - \bar{r}t/\sigma)^2\right\} \tag{9}$$

which is a Gaussian distribution peaked around $z = z_0 + \bar{r}t/\sigma$. Substituting Eq. (6) into (9) and using relation (3.0.32c), we get

$$P(x\,|\,y, t) = \frac{1}{\sigma(2\pi t)^{1/2} x(1 - x/K^*)} \times \exp\left[-\frac{1}{2t}\left(\frac{1}{\sigma}\ln x/y - \frac{1}{\sigma}\ln\frac{(1-x/K^*)}{(1-y/K^*)} - \bar{r}t/\sigma\right)^2\right] \tag{10}$$

which is the density function determining the behavior of the population. Using this probability density function various moments of x can be calculated by numerical integration. When $K^* \to \infty$, i.e., either the population is far from saturation or the supply of food is unlimited, the integration can be calculated analytically. For this case, from Eq. (6),

$$x = e^{\sigma z} \tag{11}$$

and

$$\langle x \rangle = \int_0^{K^*} xP(x\,|\,y, t)\, dx = \int_{-\infty}^{\infty} e^{\sigma z} g(z\,|\,z_0, t)\, dz$$

$$= \exp(\sigma^2 t/2) \exp\{\sigma(z_0 + \bar{r}t/\sigma)\} = ye^{\bar{r}t} \exp(\sigma^2 t/2) \tag{12}$$

as compared to

$$x = ye^{\bar{r}t} \tag{13}$$

for the deterministic case in the absence of random fluctuations. Similarly,

$$\operatorname{var}(x) = \langle x^2 \rangle - \langle x \rangle^2 = y^2 e^{2\bar{r}t} \exp(\sigma^2 t)(\exp(\sigma^2 t) - 1) = \langle x \rangle^2 [\exp(\sigma^2 t) - 1] \tag{14a}$$

as compared to zero variance for the deterministic case. The coefficient of variation is given by

$$\frac{[\operatorname{var}(x)]^{1/2}}{\langle x \rangle} = (\exp(\sigma^2 t) - 1)^{1/2} \tag{14b}$$

As t increases, the coefficient of variation is increasing, and already for moderate times the average cannot describe the growth of the population.

b. *Changes in the death rate.* If the number of zygotes (a cell formed by the union of two germ cells) produced is large compared to the adult population, and is therefore relatively less subject to random fluctuations, then random variation will occur mostly in the death of adults. We now derive a continuous model approximating this situation.

Let the deterministic birth rate be λ_x (when the population size is x) and let $\mu\,\Delta t$ be the probability for a given individual to die in time Δt (independent of the age of the individual). If $x(t)$ is the number of individuals at time t, the number of individuals at time $(t+\Delta t)$ is a random variable which takes the value

$$x(t+\Delta t) = x(t) + \lambda_x\,\Delta t - i \tag{15}$$

with probability

$$\binom{x(t)}{i}\mu^i(\Delta t)^i(1-\mu\,\Delta t)^{x-i} \tag{16}$$

Up to first order in Δt the number i is Poisson distributed since

$$\langle i\rangle = \mu x\,\Delta t \tag{17a}$$

$$\langle i^2\rangle = \mu x\,\Delta t(1-\mu\,\Delta t) + (x\mu\,\Delta t)^2 \tag{17b}$$

Thus from Eqs. (15) and (17)

$$\langle x(t+\Delta t) - x(t)\rangle \equiv \langle \Delta x(t)\rangle = (\lambda_x - \mu x)\,\Delta t \tag{18a}$$

$$\langle[\Delta x(t)]^2\rangle = \lambda_x(\Delta t)^2 + 2\lambda_x x\mu(\Delta t)^2 + x\mu\,\Delta t(1-\mu\,\Delta t) + (x\mu\,\Delta t)^2 \tag{18b}$$

Therefore,

$$\lim_{\Delta t\to 0}\frac{1}{\Delta t}\langle\Delta x(t)\rangle = \lambda_x - \mu x \tag{19a}$$

$$\lim_{\Delta t\to 0}\frac{1}{\Delta t}\langle[\Delta x(t)]^2\rangle = \mu x \tag{19b}$$

If λ_x is chosen to be of the same form as λ_n of model A with $\alpha = 1$ of Section 4.1, i.e.,

$$\lambda_x = \lambda x(1-x/K) \tag{20}$$

then from Eqs. (3.0.3), (3.0.6a), (3.0.6b), (19a), and (19b), the Fokker–Planck equation for the continuous model is

$$\frac{\partial P}{\partial t} = -\frac{\partial}{\partial x}(a(x)P) + \frac{1}{2}\frac{\partial^2}{\partial x^2}(b(x)P) \tag{21}$$

with

$$a(x) = rx(1-x/K^*), \qquad r \equiv \lambda - \mu, \qquad K^* = K(1-\mu/\lambda) \tag{22a}$$

$$b(x) = \mu x \tag{22b}$$

We may remark that if we had assumed the probability for a given individual to die in time Δt to be not constant ($\mu\,\Delta t$) but of the form $(1+x/K)\,\Delta t$, K^* in Eq. (22a) would have been related to K in the same fashion as in model A of Section 4.1, Eq. (4.1.6), but $b(x)$ would have become $\mu x(1+x/K)$. Thus in deriving Eqs. (22), we have incorporated the reduction in fertility due to the limitation of food, but neglected the increase in the probability of death. However, in the regime $x, y \ll K$ both models can be approximated by taking $K \to \infty$, as is done later in this section.

To determine the steady-state distribution, we note that the variable x is confined to $x \geqslant 0$, with the boundary $x = 0$ being singular $[b(0) = 0]$. Near $x = 0$, $b(x) = \mu x$ and $a(x) \simeq rx$, and the boundary is an exit boundary, i.e., whatever reaches the boundary $x = 0$ is trapped there forever, corresponding to the extinction of the population. For large values of x ($x \gg K^*$), $a(x) \simeq -(r/K^*)x^2$, and $x = \infty$ is an entrance boundary. Thus eventually extinction is bound to occur and the steady-state probability density function is zero for all $x > 0$. We may point out that Levins (1969), without analyzing the nature of boundaries, used Eq. (3.0.56a) and the expressions (22a) and (22b) for $a(x)$ and $b(x)$ to derive a nonzero steady-state probability density

$$P(x, \infty) = (C/x)\exp(r(2x - x^2/K^*)/\mu) \qquad \left[\int_0^\infty P(x,\infty)\,dx = \infty\right]$$

which is clearly wrong.

To determine $P(x \mid y, t)$ we transform the Fokker–Planck equation (21), which is of form (3.0.29), into one of the type (3.0.27) by making the transformation (3.0.32), i.e.,

$$dz = dx/(\mu x)^{1/2}, \qquad z = 2(x/\mu)^{1/2} \tag{23a}$$

$$P(x \mid y, t) = g(z \mid z_0, t)/(\mu x)^{1/2} \tag{23b}$$

$$a(z) = [rx(1 - x/K^*) - \mu/4]/(\mu x)^{1/2} \tag{23c}$$

The resulting equation is

$$\frac{\partial g}{\partial t} = -\frac{1}{2}\frac{\partial}{\partial z}\left(rz - \frac{rz^3\mu}{4K^*} - \frac{1}{z}\right)g + \frac{1}{2}\frac{\partial^2 g}{\partial z^2} \tag{24}$$

An analytical solution to this equation is not known. However, by simple transformations

$$\xi = z(r\mu/4K^*)^{1/4}, \qquad \alpha \equiv 2(rK^*/\mu)^{1/2}, \qquad \tau = \tfrac{1}{4}t(r\mu/K^*)^{1/2} \tag{25}$$

this equation reduces to

$$\frac{\partial g}{\partial \tau} = \frac{\partial}{\partial \xi}[(\xi^{-1} - \alpha\xi + \xi^3)g] + \frac{\partial^2 g}{\partial \xi^2} \tag{26}$$

An equation similar to the above equation, with ξ^{-1} replaced by $-\xi^{-1}$, has been studied in the theory of noise in the laser (Risken and Vollmer, 1967). The slightly different equation has been solved numerically and also by using quantum mechanical perturbation theory. It may be noted that the formal solution of Eq. (26) is

$$g(\xi \mid \xi_0, t) = \sum_{m=0}^{\infty} A_m \xi^{-1/2} \exp\left(-\frac{\xi^4}{8} + \frac{\alpha\xi^2}{4}\right) \psi_m \exp(-\lambda_m \tau) \tag{27}$$

where λ_m and ψ_m are the eigenvalue and eigenfunction of the equation

$$d^2\psi_m/d\xi^2 + (\lambda_m - U(\xi))\psi_m = 0 \tag{28a}$$

with

$$U(\xi) \equiv \frac{\xi^6}{2} - \frac{\alpha\xi^4}{2} + \left(\frac{\alpha^2}{4} - 1\right)\xi^2 + \frac{3}{4\xi^2} \tag{28b}$$

Approximate solutions to Eq. (21) can be derived for two regimes: the Malthusian regime when the initial population size y is small ($y \ll K^*$, or equivalently $K^* \to \infty$) and the regime in which $y \simeq K^*$. For the Malthusian regime $a(x) \simeq rx$, so that from Table 3.4, the solution of Eq. (21) is

$$P(x \mid y, t) = \frac{2r}{\mu} \frac{(y/x)^{1/2}}{e^{rt/2} - e^{-rt/2}} \exp\left[-\frac{2r}{\mu} \frac{x + ye^{rt}}{e^{rt} - 1}\right] I_1\left[\frac{4r}{\mu} \frac{(xy)^{1/2}}{e^{rt/2} - e^{-rt/2}}\right] \tag{29}$$

[This equation was derived by Feller (1951) for the continuous analog of the branching process.] According to the density (29), the lth moment of $x(t)$ is given by

$$\begin{aligned} \langle x^l(t) \rangle &= \int_0^{\infty} x^l P(x \mid y, t)\, dx \\ &= ye^{rt}\left[\frac{\mu(e^{rt} - 1)}{2r}\right]^{l-1} \exp\left[-\frac{2r}{\mu}\frac{y}{(1 - e^{-rt})}\right] \\ &\quad \cdot \Gamma(l+1) F\left[l+1; 2; \frac{2ry}{\mu(1 - e^{-rt})}\right] \end{aligned} \tag{30}$$

where $F(a; c; z)$ is the confluent hypergeometric function and where we have used the standard formula (Magnus *et al.*, 1966, p. 275)

$$\int_0^{\infty} t^{a-1}(zt)^{(1-c)/2} e^{-t} I_{c-1}[2(zt)^{1/2}]\, dt = \frac{\Gamma(a)}{\Gamma(c)} F(a; c; z), \qquad a > 0 \tag{31a}$$

In particular, by using the standard formula (Magnus *et al.*, pp. 285, 267)

$$F(a; a; z) = e^z \tag{31b}$$

$$F(a+1; a; z) = \frac{a+z}{a} F(a; a; z) \tag{31c}$$

we obtain

$$\langle x(t)\rangle = ye^{rt} \tag{32a}$$

$$\langle x^2(t)\rangle - \langle x(t)\rangle^2 = \frac{\mu}{r}\langle x(t)\rangle(e^{rt}-1) \tag{32b}$$

The average value given by Eq. (32a) is identical to the Malthusian deterministic behavior $[b(x)=0]$. This is expected since $a(x)$ is linear in x (see footnote on p. 46), and the term in the square brackets in Eq. (3.0.44) vanishes at the boundaries. From Table 3.10, the probability of population growing without limit is

$$R(\infty \mid y) = \int_0^y \pi(\eta)\,d\eta \bigg/ \int_0^\infty \pi(\eta)\,d\eta \tag{33}$$

where

$$\pi(\eta) = \exp\left(-2\int^\eta \frac{a(y)}{b(y)}\,dy\right) \tag{34}$$

Substituting for $a(y)$ and $b(y)$ from Eq. (22), with $K^* = \infty$, we get

$$R(\infty \mid y) = 1 - e^{-2ry/\mu} \tag{35}$$

Thus for $r>0$, the population will keep growing with probability $1-e^{-2ry/\mu}$ and will become extinct with probability $e^{-2ry/\mu}$, while for $r<0$, the fate of the population is extinction. Hence the probability of population having any finite size greater than zero, at infinite time, is zero, as can also be seen by taking the limit $t\to\infty$ in Eq. (29). Of course, when Eq. (29) is regarded as an approximate solution of Eq. (21) in the Malthusian regime, the validity of the approximation is limited to short times, and the limit $t\to\infty$ is of no meaning in this context.

In the regime $y \simeq K^*$, $(x-y)/K^* \ll 1$ and we can take

$$a(x) \simeq r(K^*-x), \qquad r>0 \tag{36}$$

Let us define a parameter η by

$$\eta \equiv 2rK^*/\mu \tag{37}$$

For $\eta \geqslant 1$, the boundary $x=0$ is an entrance boundary, while for $0<\eta<1$ it is a "regular" boundary which is reflecting. For all $\eta>0$, $x=\infty$ is a natural inaccessible boundary. The steady-state probability density exists, and from Eqs. (3.0.56a), (36), and (22b), is given by

$$P(x \mid y, \infty) = cx^{\eta-1}\exp\{-\eta(x/K^*)\} \tag{38a}$$

where c is a normalization constant and is given by

$$c = (\eta/K^*)^\eta/\Gamma(\eta) \tag{38b}$$

The most probable steady-state size of the population is (see Table 3.10)

$$x_0 = K^* - \mu/2r = K^*(1-\eta^{-1}) \tag{39a}$$

which is positive and below K^* for $\eta > 1$. As η increases, this size tends to K^*. The lth moment of the population size at steady state is

$$\langle x^l \rangle = (K^*/\eta)^l \Gamma(l+\eta)/\Gamma(\eta), \qquad \langle x \rangle = K^* > x_0 \tag{39b}$$

The time-dependent probability density is (see Table 3.4)

$$P(x|y,t) = \frac{2r}{\mu}\left(\frac{x}{y}\right)^{(\eta-1)/2} \frac{e^{\eta rt/2}}{e^{rt/2}-e^{-rt/2}} \exp\left[-\frac{2r}{\mu}\frac{x+ye^{-rt}}{1-e^{-rt}}\right] \cdot I_{\eta-1}\left[\frac{4r}{\mu}\frac{(xy)^{1/2}}{e^{rt/2}-e^{-rt/2}}\right] \tag{40}$$

From Eqs. (38a) and (40), we conclude that for all $t > 0$ the probability of having a large population ($x \gg K^*$) is very small for all $\eta > 0$ [$I_\nu(z) \sim e^z/(2\pi z)^{1/2}$ for $z \to \infty$] while the probability of having population of small size is low when $\eta > 1$ and very high for $\eta < 1$. Therefore, the case $\eta > 1$ describes a population which fluctuates around an average far from zero, and only in this case is the approximation leading to Eq. (36) a good approximation to the original process (22). Using Eq. (31a) the lth moment is given by

$$\langle x^l \rangle = \left(\frac{\mu}{2r}\right)^l (1-e^{-rt})^l \exp\left[-\frac{2ry}{\mu(e^{rt}-1)}\right] \frac{\Gamma(l+\eta)}{\Gamma(\eta)} F\left[l+\eta;\eta;\frac{2ry}{\mu(e^{rt}-1)}\right] \tag{41a}$$

For $l = 1, 2$ using the identities (31b) and (31c) we get from the above equation

$$\langle x \rangle = K^*(1-e^{-rt}) + ye^{-rt} = K^* - (K^*-y)e^{-rt} \tag{41b}$$

As expected the average given by this equation is identical to the deterministic behavior obtained by solving the deterministic equation

$$dx/dt = r(K^*-x)$$

For the variance we get

$$\langle x^2 \rangle - \langle x \rangle^2 = \frac{\mu}{2r}[\langle x \rangle + ye^{-rt}](1-e^{-rt}) \tag{41c}$$

as compared to the zero value in the deterministic approach.

c. *Changes in the upper limit to growth.* Another possible effect of fluctuating environment on the growth of a population is the random variations in the deterministic saturation level (Levins, 1969). In order to incorporate randomness in K^*, we define

$$M(t) = 1/K^*(t) \tag{42a}$$

and assume $M(t)$ to be approximated by

$$M(t) = \overline{M} + \sigma F(t) \tag{42b}$$

where $\langle M(t)\rangle = \overline{M}$, σ is a constant, and $F(t)$ is the same white noise as in Eq. (3.0.17). The dynamical equation for x [Eq. (1) with $\alpha = 1$] then becomes

$$dx/dt = rx - rx^2[\overline{M} + \sigma F(t)] \tag{43}$$

The coefficients of the corresponding Fokker–Planck equation, according to Eq. (3.0.30b), are given by

$$a(x) = rx - r\overline{M}x^2 + \sigma^2 r^2 x^3 \tag{44a}$$

$$b(x) = \sigma^2 r^2 x^4 \tag{44b}$$

The boundary $x = 0$ is a singular boundary, and since near $x = 0$, $a \simeq rx$, and $b = r^2\sigma^2 x^4$, this boundary is a natural† boundary for $r > 0$. The case $r < 0$ is uninteresting because the saturation level has a biological meaning only for a population that is growing and not decaying. At $x = \infty$, $a(x) \sim \sigma^2 r^2 x^3$, $b(x) = \sigma^2 r^2 x^4$, $\pi(x) = x^{-2}$, and the boundary is regular so that the condition $P(\infty \mid y, t) = 0$ can be imposed. To solve the Fokker–Planck equation with the coefficients (44), we make the transformation

$$z = (1/x - \overline{M})/\sigma r \tag{45}$$

so that Eq. (43) becomes

$$dz/dt = -rz + F(t) \tag{46}$$

which is the stochastic differential equation for an Ornstein–Uhlenbeck process. The Fokker–Planck equation satisfied by the probability density function of z is

$$\frac{\partial g(z,t)}{\partial t} = \frac{\partial}{\partial z}(rzg) + \frac{1}{2}\frac{\partial^2 g}{\partial z^2} \tag{47}$$

and the steady-state probability density function, from Eq. (3.0.56), is

$$g(z, \infty) = c\exp(-rz^2) \tag{48}$$

where c is a normalization constant. Substituting Eq. (45) into (48) we get for $\xi = 1/x$,

$$P(\xi, \infty) = c'\exp[-(\xi - \overline{M})^2/\sigma^2 r] \tag{49}$$

Thus $\xi = 1/x$ is distributed normally with mean $\overline{M}$ and variance $1/(2\sigma^2 r)$ and therefore the harmonic mean of K^* $(=\overline{M})$ is the median‡ of x (which equals

† This can be verified easily by considering the corresponding boundary after the transformation (45).

‡ The median x_m is defined by the relation $\text{prob}[x \geqslant x_m] = \text{prob}[x \leqslant x_m]$.

the median of ξ). The most probable size ($\simeq 1/\overline{M} - \sigma^2 r/\overline{M}^3$) is reduced by the variations in K^*, and this effect increases with r and σ.

To determine $P(x|y,t)$ we note that by Eq. (45), the boundaries $x = 0$ and $x = \infty$ correspond to the boundaries $z = \infty$ and $z = -\overline{M}/\sigma r$ which are natural and regular boundaries, respectively. For these boundaries, the solution to Eq. (47) is analytically complicated, and is not discussed here. For $\overline{M} \ll 1$ (very large saturation level) we can approximate the boundary $z = -\overline{M}/\sigma r$ by $z = 0$ and take it to be a reflecting boundary. Then $P(x|y,t)$ can be calculated from Table 3.4, but since for a very large saturation level, the fluctuations are important, this case is biologically meaningless.

References

Feller, W. (1951). Diffusion processes in genetics. *Proc. Symp. Math. Statist. Probability, 2nd, Berkeley, California*, 1951, p. 227.

Levins, R. (1969). The effect of random variations of different types on population growth. *Proc. Nat. Acad. Sci. U.S.A.* **62**, 1061.

MacArthur, R. H., and Wilson, E. O. (1967). "The Theory of Island Biogeography." Princeton Univ. Press, Princeton, New Jersey.

Magnus, W., Oberhettinger, F., and Soni, R. P. (1966). "Formulas and Theorems for the Special Functions of Mathematical Physics." Springer-Verlag, Berlin and New York.

Richter-Dyn, N., and Goel, N. S. (1972). On the extinction of a colonizing species. *J. Theoret. Pop. Biol.* **3**, 406.

Risken, H., and Vollmer, H. D. (1967). The influence of higher order contributions to the correlation function of the intensity fluctuation in a laser near threshold. *Z. Physik.* **201**, 323.

Additional References

Baker, H. G., and Stebbins, G. L. (1965). "The Genetics of Colonizing Species." Academic Press, New York.

Goel, N. S., Maitra, S. C., and Montroll, E. W. (1971). On the Volterra and other nonlinear models of interacting populations. *Rev. Modern Phys.* **43**, 231; "On the Volterra and Other Nonlinear Models of Interacting Populations." Academic Press, New York.

Levins, R. (1968). Evolution in changing Environments. Princeton Univ. Press, Princeton, New Jersey.

Levins, R. (1970). Extinction, *in* "Some Mathematical Questions in Biology" (M. Gerstenhaber, ed.), Vol. 2, p. 77. Amer. Math. Soc., Providence, Rhode Island.

Pielou, E. C. (1969). "An Introduction to Mathematical Ecology." Wiley, New York.

Skellam, J. G. (1951). Random dispersal in theoretical populations. *Biometrika* **38**, 196.

5

Population Growth of Two-Species Systems

In the preceding chapter we discussed the population growth of a single species, ignoring its interactions with other species. In natural systems, however, there are interactions of various types among the species, the most common of which are the prey–predator, the host–parasite, and the competitors (for the same food) interactions. In this chapter we consider systems that consist of only two species, one of which is the host for the other, and in the next chapter we consider more complex systems consisting of many interacting biological species.

For a prey–predator system, on the basis of experimental observations of D'Ancona on the dynamics of two species of fish, Volterra in 1928 [Volterra (1928, 1931, 1937), also, D'Ancona (1954)] proposed the following equations describing the dynamics of a prey–predator system:

$$dn/dt = rn - snn' \tag{1a}$$

$$dn'/dt = -r'n' + s'nn' \tag{1b}$$

where n and n' are the sizes of prey and predator species, respectively. The term $-snn'$ represents the loss rate of prey due to "collisions" with predator, and $s'nn'$ the growth rate of the population of the predator species through the same collisions.† The first term in Eq. (1a) describes the growth of prey in the absence of predator (assumed for simplicity to be Malthusian), while the first term in Eq. (1b) implies that predator would die out exponentially without the availability of prey. These deterministic equations have been investigated extensively, and for review we refer to Goel *et al.* (1971). In a more general form of Eqs. (1), the coefficients r and r' can be density dependent, i.e., $r = r(n)$, $r' = r'(n')$; in particular, the form can be the same as the Verhulst form of the growth rate [Eq. (4.0.3)].

Equations (1) also describe the host–parasite interactions where n is the number of hosts and n' is the number of parasites. For the competitive interactions, the equation for the second species is of the same form as for the first species, i.e.,

$$dn'/dt = r'n' - s'nn' \tag{1c}$$

since each species in the absence of the other is growing according to a certain law (given by the form of r or r') while, in the presence of the other species, this growth is reduced due to decrease in the amount of food.

As in the case of the single-species growth, the coefficients in Eqs. (1) are not constant, but may fluctuate due to random elements, e.g., environment, immigration and emigration of species, amount of resources, etc. Thus, more realistically, the number of individuals of each species should be taken as a random variable. One then has two approaches at hand—the discrete approach, where these random variables are assumed to change discretely, and the continuous approach, where changes are continuous. We describe the former approach, and use it in the succeeding sections.

The discrete approach is due originally to Chiang (1954) who introduced $P(n, n'; m, m'; t)$, which is the joint probability function for the size of species, to be n and n' at time t, given that initially the sizes were m and m', respectively. This joint probability function, which for simplicity in notation we denote by $P(n, n'; t)$, is assumed to satisfy a master equation similar to Eq. (2.0.3), i.e.,

$$\begin{aligned}\frac{dP(n,n';t)}{dt} &= -(\lambda_{n,n'} + \mu_{n,n'} + \lambda'_{n',n} + \mu'_{n',n})P(n,n';t) \\ &\quad + \lambda_{n-1,n'}P(n-1,n';t) + \lambda'_{n'-1,n}P(n,n'-1;t) \\ &\quad + \mu_{n+1,n'}P(n+1,n';t) + \mu'_{n'+1,n}P(n,n'+1;t)\end{aligned} \tag{2}$$

Here $\lambda_{n,n'}$ and $\mu_{n,n'}$ are the probability rates of a birth and a death in the first species when its size is n and the size of the other species is n', and $\lambda'_{n',n}$ and

† A more detailed meaning of the coefficients s and s' in case of prey–predator interactions is given in Section 6.0 in connection with many-species systems.

$\mu'_{n',n}$ are similarly defined. The exact forms of these probability rates depend on the nature of the interactions among the two species. For the prey–predator and host–parasite interactions of the type presented in Eqs. (1), these rates can be taken as the stochastic analogs of the deterministic rates, i.e.,

$$\lambda_{n,n'} = \lambda n, \qquad \mu_{n,n'} = snn' + \mu n, \qquad \lambda - \mu \leftrightarrow r \tag{3a}$$

$$\lambda'_{n',n} = vnn' + \lambda' n', \qquad \mu'_{n',n} = \mu n', \qquad \lambda' - \mu' \leftrightarrow r' \tag{3b}$$

Substituting Eqs. (3) into (2) we get a difference-differential equation involving two variables—the equation for a bivariate process which has not been solved analytically. Approximate solutions for various moments of n and n' have been obtained (Goel *et al.*, 1971) for the case $\lambda' = \mu = 0$.

For some systems the conditions are such that the bivariate process can be reduced to a univariate process and can then be analyzed by the techniques of Chapters 2 and 3. Such a system is discussed in the next section. The system (termed epidemics) consists of a population of susceptibles (the hosts) attacked by an infectious disease. In this system, although the parasites are the bacteria (or virus) that cause the disease, the population which interacts directly with the susceptibles is that of the infectives, who by contacts with susceptibles (analogous to the "collision" in the prey–predator interactions) distribute the epidemic in the population. Thus the number n' is not the number of bacteria, but the number of infectives carrying the bacteria in their bodies.

The other system we discuss consists of the two-species bacteriophage (virus) and the host bacteria. In this case, due to the complicated life cycle of the phages, Eq. (2) cannot describe the interactions between these two species. However, several stages in these interactions can be analyzed via univariate birth and death processes, and these stages are presented in Section 5.2.

5.1 Epidemics

The spread of an epidemic in a population of susceptibles involves interactions at two levels. At the microscopic level, there are interactions between the pathogenic agent (bacteria or virus) and the host. At the macroscopic level, there are interactions between infected and susceptible members of the population, which cause the transmission of infection. Both types of interactions occur simultaneously. When an individual is attacked by a pathogenic agent, e.g., bacteria, there is a certain period, known as the latent period, during which the bacteria multiply in his body, but the infected individual is not yet infectious. At the end of this period, the individual is capable of discharging the contagious bacteria, thus infecting other members of the population. The period immediately following the latent period is thus known as the infectious period, while the interval between the moment of infection and the appearance of symptoms is called the incubation period. The infectious period can terminate either by death or by recovery of the infected individual. After recovery, the individual may be immune or susceptible to infection, depending on the type of pathogenic agent. When the latent period and the incubation period are negligible compared to the time scale of the spread of the disease, the effects at microscopic and macroscopic levels can be modeled separately. We will now model the interactions at these two levels by using the theory of birth and death processes. We start with the macroscopic level.

A. *Spread of an Epidemic in a Finite Population*

The general model for understanding the effect of interactions at macroscopic level, namely, the spread of an epidemic in a finite population, involves two stochastic variables, one representing the number of susceptibles r, and the other representing the number of infectives n. The number of infectives is increased by a contagious contact between an infective and a susceptible, the probability of which, under the assumption of homogeneous mixing within the population, is proportional to nr. The number of infectives is decreased by a removal, due to either death, isolation, or recovery, the probability of which is proportional to n, under the assumption that each infective has the same probability to be removed. On the other hand, the number of susceptibles can increase either by natural birth or by recovery of an infective individual (when the recovered individuals are not immune to infection), and can decrease either by natural death or infection.

In the so-called general epidemic models (Dietz, 1967), natural births and deaths are neglected and recovered individuals are taken to be immune to further infection. The joint probability $P_{n,r}(t)$ of n and r satisfies the equation

$$dP_{n,r}/dt = \nu(n-1)(r+1)P_{n-1,r+1} - n(\nu r+\mu)P_{n,r} + \mu(n+1)P_{n+1,r} \tag{1}$$

where μ is the probability rate per infective of being removed, and ν is the probability rate of a contact between two members of the population. $P_{n,r}(t)$ is the probability of having n infectives and r susceptibles in time t when initially $n(0) = m$ and $r(0) = N-m$. N is the size of the population at $t = 0$. The ranges of values of n and r are $0 \leqslant n < N$, $r \leqslant N-m$, and $r+n \leqslant N$, and therefore $P_{m+1,N-m}(t) \equiv 0$, $P_{n,N-m+1}(t) \equiv 0$. The process described by Eq. (1) is a bivariate process whose analysis is beyond the scope of this book. Interested readers are referred to the review article by Dietz (1967). However, an interesting aspect of the problem, namely, the probability of successful spreading of the disease as a function of the initial number of infectives n, the initial size of the population N, and the other parameters of the process, can be described by a univariate process. At the beginning of an epidemic, defined by the condition $n \ll N$, r can be well approximated by N for populations of moderate size ($N > 50$), and Eq. (1) becomes

$$dP_n/dt = \nu N(n-1)P_{n-1} - (\nu N+\mu)nP_n + \mu(n+1)P_{n+1} \tag{2}$$

where $P_n(t)$ is the probability of having n infectives in the time t when initially their number is small ($1 < m < 5$). Equation (2) describes a simple birth and death process with transition probabilities

$$\lambda_n = (\nu N)n, \qquad \mu_n = \mu n \tag{3}$$

This process was analyzed in Section 4.1 in connection with the extinction of colonizing species. Using this analysis, the probability of having n infectives before their number reduces to zero is given by (see p. 79)

$$R_{n,m}(0) = \frac{1-(\mu/\nu N)^m}{1-(\mu/\nu N)^n} \tag{4}$$

For $\mu/\nu N > 1+\varepsilon$ and n large enough such that $(\mu/\nu N)^n \gg 1$ but $n \ll N$, the probability $R_{n,m}(0)$ decreases to zero as n increases according to

$$R_{n,m}(0) \simeq \left[\left(\frac{\mu}{\nu N}\right)^m - 1\right]\left(\frac{\mu}{\nu N}\right)^{-n} \tag{5}$$

Therefore, in this case the probability of infection spreading over the population is very small. On the other hand, when $\nu N/\mu > 1+\varepsilon$, there exists a critical size n_c of infectives defined by $(\mu/N\nu)^{n_c} \ll 1$, $n_c \ll N$, such that once the number of infectives reaches this value the epidemic is bound to spread, as the probability of having $n > n_c$ infectives is approximately $1-(\mu/\nu N)^m$, independent of n. Therefore, when $\nu N/\mu > 1+\varepsilon$, either there is a small total number of infectives with probability $(\mu/\nu N)^m$, or the number of infectives is bound to increase with probability $1-(\mu/\nu N)^m$. Note that for larger N, ε can be smaller, and as N increases, the transition at $\mu = \nu N$ becomes sharper from one case to the other.

This dependence on the value of $\mu/\nu N$ was found by Bailey (1953) by considering the Laplace transform of $P_{n,r}$ of the general epidemic process for $t \to \infty$ and $n = 0$, i.e., when epidemic is over. He found that the probability of having eventually $w = N - r(\infty)$ removed individuals in the course of the epidemic is J shaped, with the highest point at $w = 0$ when $\mu > \nu N$, and is U shaped when $\mu < \nu N$. The J-shaped curve indicates that the most probable fate of the epidemic is extinction with a very low total number of removals, while the U-shaped curve corresponds to the situation in which two exclusive highly probable outcomes exist: either quick extinction or the spread of the epidemic in a great part of the population. The sharp transition at $\mu = \nu N$ was found for $N > 40$, while for other population sizes the transition from one type of distribution to the other was gradual. Analogous deterministic results were obtained almost half a century ago by Kermack and McKendrick (1927) from a deterministic model and simple arguments. According to them, the rate equation describing the growth of the number of infectives is

$$dn/dt = vrn - \mu n \tag{6}$$

where all quantities occurring in this equation are the deterministic analogs of the corresponding stochastic quantities. For the epidemic to spread, when initially n is very small, dn/dt must be positive, i.e.,

$$vr(0) > \mu \tag{7}$$

Assuming $r(0) \simeq N$, they arrived at the conclusion that for $vN > \mu$, the epidemic is spreading, while for $vN < \mu$, it is dying out. Thus $N = \mu/v$ is the threshold size of population for epidemic spread.

In the case of general epidemic with $vN > \mu$, $m = 1$, the average time for the epidemic to die out, when it is known a priori to regress before reaching n_c infectives, is given by Eq. (4.1.15), and is approximated by

$$M^*_{0,1}(n_c) = \frac{1}{vN}\left(1 + \frac{\mu}{2vN} + \frac{1}{3}\left(\frac{\mu}{vN}\right)^2 + \cdots\right) \simeq \frac{1}{\mu}\ln\left(1 - \frac{\mu}{vN}\right) \tag{8}$$

From this expression and from the probability μ/vN of epidemic dying out before spreading, we conclude that the measures taken against the epidemic should increase the probability μ/vN while decreasing the average time $M^*_{0,1}(n_c)$. For a given μ/vN, $M^*_{0,1}(n_c)$ is decreased by increasing μ. Therefore μ/vN is to be increased by increasing μ rather than by decreasing v (N is constant for a given population), i.e., better isolation of infectives is preferable to a reduction of interactions among the members of the population.

We may remark that Kendall (1956) also observed that for $vN > \mu$, there are two possible outcomes, either quick extinction with probability $(\mu/vN)^m$ or large buildup with probability $1-(\mu/vN)^m$. However, he does not justify his observation that the event of extinction corresponds to quick extinctions only, and not to extinctions which occur after the epidemic has reached a moderate size.

When the rate of removal of infectives is negligible, as compared to the rate of spread of the disease, once again one can approximate the epidemic by a univariate process, since in this case for all t, $n+r = N$. In this case

$$\lambda_n = vn(N-n), \qquad \mu_n = 0 \tag{9}$$

The resulting process is known as *simple epidemic* (Dietz, 1967). Within this model, the epidemic is bound to spread and the more interesting quantities are the duration time, i.e., the time for all susceptibles to become infectives and the number of infectives as a function of time.

The average $M_{N,m}$ and variance $V_{N,m}$ of the duration time (first passage time for the number of infectives to become N) are given in Table 2.3 (with $l = 1$ and $\Pi_{i,j} = 0$ for all $i \leqslant j$). Thus

$$M_{N,m} = \sum_{i=m}^{N-1} M_{i+1,i} = \sum_{i=m}^{N-1} \lambda_i^{-1} = v^{-1} \sum_{i=m}^{N-1} [i(N-i)]^{-1} \tag{10a}$$

$$V_{N,m} = \sum_{i=m}^{N-1} V_{i+1,i} = \sum_{i=m}^{N-1} \lambda_i^{-2} = v^{-2} \sum_{i=m}^{N-1} [i(N-i)]^{-2} \tag{10b}$$

For $m = 1$ and N large, the expressions above are asymptotically given by (Gradshteyn and Ryzhik, 1965, formulas 0.131, 0.233)

$$\begin{aligned} M_{N,1} &= (Nv)^{-1}\left[\sum_{i=1}^{N-1} i^{-1} + \sum_{i=1}^{N-1} (N-i)^{-1}\right] = 2(Nv)^{-1} \sum_{i=1}^{N-1} i^{-1} \\ &= 2(\gamma + \ln N)(Nv)^{-1} + O(N^{-1}) \end{aligned} \tag{11}$$

$$\begin{aligned} V_{N,1} &= (Nv)^{-2}\left\{\sum_{i=1}^{N-1} i^{-2} + \sum_{i=1}^{N-1} (N-i)^{-2} + 4N^{-1} \sum_{i=1}^{N-1} i^{-1}\right\} \\ &= (Nv)^{-2}(\pi^2/3) + O(N^{-2}) \end{aligned} \tag{12}$$

where $\gamma \equiv 0.577$ is the Euler constant. The coefficient of variation for large N is of the form

$$\frac{(V_{N,1})^{1/2}}{M_{N,1}} = \frac{\pi}{2(3)^{1/2}} (\ln N)^{-1} + O[(\ln N)^{-1}] \tag{13}$$

which indicates that even for moderately large N, the coefficient of variation is not negligible. The number of infectives at time t is described by the probability function $P_{n,m}(t)$ which satisfies the forward and backward equations (2.0.3) and (2.0.4). Since $\mu_n = 0$ for all n, the Laplace transform of $P_{n,m}(t)$ is easily obtained, but the inversion of this expression back to the variable t is quite involved and will not be presented here. [For details, see Bailey (1957), Section 5.2.11.] For the moments of n one can use the set of differential equations (2.0.8) and (2.0.11), but since λ_n is quadratic in n, this set cannot be solved exactly without a priori knowing $\langle n \rangle$. Here we use the approximate method presented in Appendix A in a very primitive way by replacing $\langle n^2 \rangle$ in the equation for the first moment by $\langle n \rangle^2$. This substitution is equivalent to the assumption that $\mathrm{var}(n) = 0$; i.e., the process is deterministic. The equation for $d\langle n \rangle/dt$ is

$$d\langle n \rangle/dt = \langle \lambda_n \rangle - \langle \mu_n \rangle = v\langle (N-n)n \rangle = vN\langle n \rangle - v\langle n^2 \rangle \tag{14}$$

which, in light of the above-mentioned approximation, becomes

$$d\langle n \rangle/dt = vN\langle n \rangle - v\langle n \rangle^2, \qquad \langle n \rangle(0) = m \tag{15}$$

The solution to Eq. (15) is

$$\langle n \rangle = \frac{Nm}{m+(N-m)\,e^{-N\nu t}} \tag{16}$$

Thus $\langle n \rangle$ tends to N as $t \to \infty$, as is expected.

A more interesting quantity is the so-called epidemic curve which is the histogram of the number of new cases notified as a function of time, i.e., dn/dt as a function of time. According to Eq. (16), it is given by

$$\frac{d\langle n \rangle}{dt} = \frac{\nu N^2 m(N-m)\,e^{-N\nu t}}{[m+(N-m)\,e^{-N\nu t}]^2} = \nu \langle n \rangle (N - \langle n \rangle) \tag{17}$$

This function increases from the value $\nu m(N-m)$ at $t = 0$ to a maximal value when $\langle n \rangle = N/2$,

$$\left(\frac{d\langle n \rangle}{dt}\right)_{\max} = \frac{\nu N^2}{4} \quad \text{at} \quad t^* = (N\nu)^{-1} \ln[(N-m)/m] \tag{18}$$

and drops to zero as $t \to \infty$.

Another quantity related to the epidemic curve is the time between the occurrence of two consecutive new cases as a function of the number of already notified cases, i.e., dt/dn regarded as a function of n. From the deterministic result [which is identical with Eq. (16)]

$$t(n) = (\nu N)^{-1} \left\{ \ln \frac{N-m}{m} - \ln \frac{N-n}{n} \right\} \tag{19}$$

and

$$\Delta t = t(n+1) - t(n) = (\nu N)^{-1} \ln \frac{n(N-n-1)}{(n+1)(N-n)}$$

$$= (\nu N)^{-1} \ln \left[1 - \frac{N}{(n+1)(N-n)} \right] \tag{20}$$

For $N \gg 1$ and n in the vicinity of $N/2$, $N(n+1)^{-1}(N-n)^{-1} \ll 1$ and

$$\Delta t \simeq \frac{1}{\nu N} \frac{N}{(n+1)(N-n)} \simeq [\nu n(N-n)]^{-1} \tag{21}$$

From the stochastic model, the average time between the occurrence of case n and case $n+1$, $M_{n+1,n}$, is given [see Eq. (10a)] by λ_n^{-1}, i.e., by $[\nu n(N-n)]^{-1}$, which has the same dependence on n as the expression in Eq. (21). Thus, by this interpretation of the epidemic curve, the average behavior of the stochastic model coincides with the deterministic behavior only for moderately large populations and only after the epidemic has already spread to a considerable part of the population. The variance of the time between occurrence of case n and case $n+1$ is also readily available and is given by $\lambda_n^{-2} = [\nu n(N-n)]^{-2}$. The coefficient of variation here is 1 and independent of the parameters, i.e., the fluctuations are of the same order of magnitude as the value of the average, an indication of the poor description of the process by the average or by the deterministic model.

We conclude this section by discussing another type of epidemic that can be described by a univariate birth and death process. This type of epidemic is the one for which the removal of infectives is mainly due to recoveries (no fatality) and the recovered individuals are again susceptible to the disease (so that for all t, $r+n = N$). For this case, the number of infectives is described by a birth and death process with transition probability rates of the form

$$\lambda_n = \nu n(N-n), \qquad \mu_n = \mu n \tag{22}$$

For N moderately large and $n \ll N$ (beginning of the epidemic), λ_n and μ_n of Eq. (22) are of the same form as in the general epidemic model (3). Thus the initiation of an epidemic for the present case will be similar to that for the general epidemic model. The transition probabilities of Eq. (22) are similar in form to those discussed in Section 4.1. Therefore, one can use the analysis presented there for the derivation of the maximum size of a successfully colonizing species and the duration for colonization, to determine the maximum size of the epidemic and the duration of the epidemic.

For $\nu N > \mu$ and for those cases in which the epidemic spreads above the critical number (discussed also in connection with the general epidemic model), the maximum number n^* of infectives is between N^* and N where N^* is defined by

$$\lambda_{N^*} = \mu_{N^*}, \qquad \text{i.e.,} \qquad N^* = N - \mu/\nu = N[1-(\mu/\nu N)] \tag{23}$$

Employing the method used in connection with the effective saturation level of a colonizing species [see Section 4.1 and also Richter-Dyn and Goel (1972, Appendix B)], we arrive at the following estimate

$$n^* \simeq N[1-(\mu/\nu N)^2] = N^*[1+\mu/\nu N] \tag{24}$$

Thus the number of susceptibles stays mostly above $N-n^* = N(\mu/\nu N)^2$, and if $N = \mu/\nu+\eta$ ($\eta \gtrsim N/2$), then the maximal number of infectives is estimated by $n^* = 2\eta(1-\eta/2N) \lesssim 3\eta/2$.

Further, we conclude that once the epidemic spreads over the population, it will persist for a very long time before the ultimate recovery takes place. The number of infectives fluctuates around N^* but does not exceed n^*. (Of course, for this long time, N is assumed to be constant by natural deaths balancing the births.) On the other hand, for $\nu N \simeq \mu$, the average time to extinction of epidemic, even after a great part of the population has been infected, is not exceedingly large and the epidemic can persist or reoccur only by the introduction of infectives from outside.

Having discussed the spreading of an epidemic, we now discuss the process of infection. For convenience in expression, we take a bacterium as the pathogenic agent.

B. Bacterial Growth in Infectives

Let us start with modeling of the incubation period. A simple model is based on the following reasonable assumptions:

1. The number of bacteria introduced into the susceptible body by contact with an infectious individual is Poisson distributed with mean d.
2. Each bacterium in the susceptible's body has a probability $\lambda \Delta t$ of dividing into two bacteria and $\mu \Delta t$ of dying in the time Δt; i.e., the process of bacteria growth is a simple birth and death process with transition probability rates

$$\lambda_n = \lambda n; \qquad \mu_n = \mu n, \qquad \lambda > \mu \tag{25}$$

3. Symptoms appear when the number of bacteria in the host reaches a certain fixed level N which is large compared with the mean initial number d.

For a given initial number of bacteria, the mathematical model is a simple birth and death process confined between two absorbing states $n = 0$ and, $n = N$. The two relevant questions are

(a) What is the probability of positive response; i.e., what is the probability of bacteria reaching size N before becoming extinct?
(b) What is the average incubation time?

The answer to the first question is rather simple. When initially the number of bacteria is m, the probability of this number reaching N before 0 is (see the formula in Table 2.6)

$$R_{N,m}(0) = \frac{1-(\mu/\lambda)^m}{1-(\mu/\lambda)^N} \tag{26}$$

Since the initial number m is assumed to be Poisson distributed, i.e.,

$$\text{prob}[m = i] = e^{-d} d^i / i!, \qquad d \ll N \tag{27}$$

the probability of positive response $R(d)$ is

$$R(d) = \frac{e^{-d}}{1-(\mu/\lambda)^N} \sum_{m=1}^{N-1} \frac{d^m}{m!} [1-(\mu/\lambda)^m] \simeq 1 - \exp\{-d(1-\mu/\lambda)\} \tag{28}$$

where, in obtaining the last expression we have neglected $(\mu/\lambda)^N$ as compared to 1 (by assumption 3). Defining d_0 as the median infectious dose, i.e., the average initial number of bacteria for which $R(d) = \frac{1}{2}$, we obtain

$$d_0 = (\ln 2)/(1-\mu/\lambda) \tag{29}$$

and

$$R(d) = 1 - 2^{-d/d_0} \tag{30}$$

Equation (28) is also valid under a modification of assumption 3, introduced by Grat (1965), which applies to diseases where symptoms appear as local lesions that are supposedly caused by a clone of progeny of an initially single bacterium. In such cases the incubation period terminates when at least one clone of progeny has reached the size N. The probability of a clone becoming extinct before reaching the size N is μ/λ when $(\mu/\lambda)^N \ll 1$, and the probability of all clones becoming extinct before reaching size N is therefore $(\mu/\lambda)^m$, when initially there are m bacteria in the host. Thus the probability of a positive response is $1-(\mu/\lambda)^m$, and by averaging over all possible initial numbers, we arrive at the result given by Eq. (28).

The answer to the second question is analytically more complicated, especially when the probability density of the first passage time to the appearance of symptoms is considered. Saaty (1961) derived an approximate form of this probability density for a fixed initial number m by solving for the generating function and approximating $F_{N,N-1}(t)$ (first passage time to N starting at state $N-1$) for $N \gg 1$ by a Dirac delta function. His result

$$F_{N,m}(t) = (\lambda-\mu)\exp\{-N(1-\mu/\lambda)\exp(-(\lambda-\mu)t)\}\sum_{k=0}^{m-1}\frac{1}{k!}\binom{n}{k+1}\beta^{n-k-1} \times [(1-\beta)^2 N \exp(-(\lambda-\mu)t)]^{k+1} \tag{31}$$

was used by Shortley (1965) to derive the probability density $F_N(t,d)$ of first passage time to N when the initial number m is distributed according to Eq. (27). Shortley arrived at the final simple expression

$$F_N(t,d) = (\lambda-\mu)(ay)^{1/2}e^{-y-a}I_1[2(ay)^{1/2}] \tag{32a}$$

where

$$a = (d/d_0)\ln 2, \qquad y = (1-\mu/\lambda)N\exp(-(\lambda-\mu)t) \tag{32b}$$

The distribution is highly peaked for $d/d_0 \gtrsim 8$, and becomes flat for small values of d/d_0. The peak is higher and occurs at smaller times the higher the ratio d/d_0. Thus for $d/d_0 > 8$, the average first passage time serves as a good approximation for the actual first passage time.

The average first passage time is given according to Table 2.6 and Eqs.

(25)–(27) by

$$M_N(d) = \sum_{m=1}^{N-1} e^{-d} \frac{d^m}{m!} M_{N,m}(0) \simeq \sum_{m=1}^{N-1} e^{-d} \frac{d^m}{m!} \{[1-(\mu/\lambda)^m]\, S(N) - S(m)\} \tag{33}$$

where

$$s(n) = \sum_{i=1}^{n-1} \sum_{j=1}^{i} \frac{1}{\lambda j} \left(\frac{\mu}{\lambda}\right)^{i-j} [1-(\mu/\lambda)^j] \tag{34}$$

In case $d \gg 1$ and $(\mu/\lambda)^d \ll 1$ the distribution of the initial number m is peaked at $m = d$, and the average time is approximated by

$$M_N(d) \simeq M_{N,d} \simeq S(N) - S(d) = \sum_{i=d}^{N-1} \sum_{k=0}^{i-1} [\lambda(i-k)]^{-1} \left(\frac{\mu}{\lambda}\right)^k \tag{35}$$

where we have neglected all terms with powers of (μ/λ) greater than $d-1$. In Shortley (1965), the average and the higher moments of the first passage time are derived directly from the probability density given by Eq. (32a). For large values of d the average time is of the form

$$\langle T_N \rangle \equiv M_N(d) \simeq (\lambda-\mu)^{-1} \ln N/d \tag{36}$$

and this dependence was also verified experimentally.

A similar result can be found more directly by considering the average number of bacteria $\langle n \rangle$ as a function of time. According to Eq. (2.0.8), when initially there are $d \gg 1$ bacteria,

$$\frac{d}{dt} \sum_{n=0}^{N} nP_{n,d}(t) = \frac{d\langle n \rangle}{dt} = \langle \lambda_n \rangle - \langle \mu_n \rangle = (\lambda-\mu) \sum_{n=1}^{N-1} nP_{n,d}(t) \tag{37}$$

or

$$\frac{d\langle n \rangle}{dt} = (\lambda-\mu)\langle n \rangle - (\lambda-\mu)\, NP_{N,d}(t) \tag{38}$$

Since $d \gg 1$, $P_{N,d}(t)$ is highly peaked at $\langle T_N \rangle$. Therefore for t less than $\langle T_N \rangle$, $P_{N,d}(t) \simeq 0$, and Eq. (38) can be approximated by the equation

$$d\langle n \rangle/dt = (\lambda-\mu)\langle n \rangle, \qquad \langle n \rangle(0) = d \tag{39}$$

which has the solution

$$\langle n \rangle = de^{(\lambda-\mu)t} \tag{40}$$

The time to reach N is then estimated by putting $n = N$ in Eq. (40), and is of the form given by Eq. (36).

For small values of d the sum in Eq. (33) can be approximated by the first term, and the average conditional first passage time, when a priori it is known that response occurs, is

$$M_N^*(d) = \frac{e^{-d}d}{1-2^{-d/d_0}}[1-(\mu/\lambda)]S(N) = \frac{e^{-d}(d/d_0)\ln 2}{1-\exp(-(d/d_0)\ln 2)} \cdot S(N) \simeq S(N) \tag{41}$$

where by Eqs. (34) and Table 2.6, $S(N) = N_{N,1}^*(0)$. In deriving Eq. (41) we used the relation (29) and made the approximations $e^{-d} \simeq 1$ and $1-\exp((d/d_0)\ln 2) \simeq (d/d_0)\ln 2$. This result is equivalent to the assumption that for $d \ll 1$ and $d/d_0 \ll 1$ the positive responses are initiated by a single bacterium, a result found also by Shortley (1965). A better approximation, which is not independent of d, results when we keep the factor e^{-d}, i.e.,

$$M_N^*(d) = e^{-d}M_{N,1}^*(0) \simeq M_{N,1}^*(0) - dM_{N,1}^*(0) \tag{42}$$

Thus as d increases the average time decreases linearly in d for small values of d.

5.2 Bacteriophage Growth

A bacteriophage (called phage for short) is composed of one of the two genetic materials, RNA (ribonucleic acid) or DNA (deoxyribonucleic acid), arranged in a single- or double-stranded molecule which is always present in the center of the phage particle, surrounded by a protective coat (shell). There is a variation in the structure of the shell; for simpler phages it consists of a large number of identical protein molecules which are arranged into a regular geometric shape, much smaller in size than bacteria. The parasitic life cycle of phages also varies and depends on the types of phages, but basically the life cycle can be described by five different stages. We describe these stages briefly, and for details refer the reader to an excellent text by Watson (1970). These five stages are as follows:

1. Phages are introduced into a suspension of growing bacteria. Being nonmotile they are subject to Brownian motion in the suspension.
2. One or more phages come into contact with a bacterium and insert the DNA (or RNA) strands into the bacterium cell through its wall. The attachment of phages to the bacterium wall changes the properties of the wall, and eventually inhibits new attachments.
3. The phages' genetic material replicates inside the bacterial cell, using the basic material and enzymes provided by the cell. If the genetic material is made up of DNA, then the DNA serves as a template for its own replication (for details see Section 9.2) and also for phage-specific RNA necessary for the synthesis of specific proteins. On the other hand, if the genetic material is made up of RNA, then RNA has two template roles: first to make more RNA molecules, and second to make the viral-specific proteins. Some of these

proteins act as enzymes necessary for replication of virus, and some as specific enzymes frequently needed to ensure the release of progeny virus particles from their host cell. Within a certain time, known as the eclipse period (less than 30 min, typically 5–10 min) a pool of 40–80 DNA (RNA) strands are formed together with proteins needed for protective coating. Generally not all the proteins are formed simultaneously; instead there is a regular time schedule by which they function.

4. The DNA (RNA) strands in the pool are gradually covered by the protein coats, and can no longer replicate. Meanwhile the uncovered strands continue to replicate, and more protein coats are also produced. In this stage, which lasts typically for 10–17 min, 200–300 mature phages are formed.
5. The bacterium is burst by dissolution (lysis) of the cell wall some 15–60 min after the start of phage infection, with the release of the mature phages and a certain number (40–80) of uncoated strands into the bacterial suspension. The mature phages released are ready to infect the not yet infected bacteria, while the uncoated strands constitute waste matter.

It should be noted that at any time after the infection of a bacterium and until lysis occurs (stages 3–4), the amount of DNA (RNA) in the cell is found experimentally to depend linearly on time.

After briefly outlining the life cycle of the phage, we now apply the mathematical analysis of birth and death processes discussed in Chapter 2 to model stages 2, 3, and 4, which can be described to some extent by one random variable. The later stages, the extinction of the colony of bacteria, and other aspects of these phage–bacterium interactions (such as mutations of the phages or the bacteria, age distribution of phages, etc.) cannot be represented by a univariate process. For a review of these topics we refer the readers to a review paper by Gani (1965) in which multivariate processes with time-dependent transition probability rates are used.

In stage 2, we are interested in the number of phages that are attached to a bacterium. Let $n(t)$ be the random variable representing the number of phages attached to a given bacterium at time t. Initially, $n(0) = 0$. Let $x(t)$ stand for the number of phages in the suspension which are not yet attached to any bacterium. As mentioned above, the attachment of phages inhibits further attachment to the same bacterium. Assuming s to be the maximal number of phages that can be attached before total inhibition occurs, reasonable forms of the transition probability rates for the changes in $n(t)$ are

$$\lambda_n = \lambda x(s-n), \qquad \mu_n = 0 \tag{1}$$

and the forward master equation (2.0.3) becomes

$$dP_{n,0}(t)/dt = \lambda x(s+1-n)\,P_{n-1,0}(t) - \lambda x(s-n)\,P_{n,0}(t), \qquad P_{-1,0}(t) \equiv 0 \tag{2}$$

The variable $x(t)$ is also changing randomly in time and thus the λ_n are random and time dependent. For the present problem of fluctuations in the number of attached phages, the fluctuations of $x(t)$ about its mean are not significant and can be ignored. Thus taking $x(t)$ as a deterministic variable that can be measured experimentally, we can consider stage 2 as a univariate process with time-dependent transition probability rates. In order to eliminate the time dependence of $\lambda_n(t) = \lambda(s-n)x(t)$, we define a new variable v, which is monotonically increasing with time, by $dv = x(t)\,dt$ or by

$$v(t) = \int_0^t x(\tau)\,d\tau \tag{3}$$

so that Eq. (2) becomes

$$dP_{n,0}/x\,dt = dP_{n,0}/dv = \lambda(s+1-n)P_{n-1,0} - \lambda(s-n)P_{n,0} \tag{4}$$

Using the method of generating function (or the results in Table 2.1, row 5, with n replaced by $s-n$, μ replaced by λ, and $m = s$), we obtain

$$P_{n,0}(t) = \binom{s}{n}(1-\rho)^n\rho^{s-n}, \qquad \rho = e^{-\lambda v} = \exp\left\{-\lambda\int_0^t x(\tau)\,d\tau\right\} \tag{5}$$

which is a binomial distribution in the parameter $1-\rho$. The probability of already having the maximum number s of attachments by the time t is given by

$$P_{s,0}(t) = \left[1-\exp\left\{-\lambda\int_0^t x(\tau)\,d\tau\right\}\right]^s \tag{6}$$

and is maximal when $x(t^*) = 0$. If $t^* < \infty$, then $P_{s,0}(t) < 1$ for all t, and among all the bacteria in a suspension of N bacteria, there are approximately $NP_{s,0}(t^*)$ bacteria which are "saturated" with attached phages. We assume $x(0) < N$ so that t^* is small enough to regard the number of bacteria constant during this stage. The probability of a bacterium to remain intact, when all phages are already attached, is

$$P_{0,0}(t^*) = \exp\left\{-\lambda s\int_0^{t^*} x(\tau)\,d\tau\right\} \tag{7}$$

and the number of uninfected bacteria can be approximated by $NP_{0,0}(t^*)$. The average number of phages attached to a bacterium at time t is the average of the binomial distribution of Eq. (5), i.e.,

$$\langle n\rangle(t) = [1-\rho(t)]\,s = \left[1-\exp\left\{-\lambda\int_0^t x(\tau)\,d\tau\right\}\right]s \tag{8}$$

Within the approximation, which regards $NP_{n,0}(t)$ as the number of bacteria in the suspension having n attached phages, the present approach becomes equivalent to the deterministic model of Yassky (1962), which deals with the number of bacteria having $0, 1, \ldots, s$ attached phages. Within this approximation, $x(t)$ satisfies the following equation:

$$x(t) = x(0) - \sum_{n=0}^{s} nNP_{n,0}(t) = x(0) - N\langle n\rangle(t) \tag{9}$$

where $\langle n\rangle(t)$ is given by Eq. (8). To solve Eq. (9), we introduce the variable $v(t)$, defined by Eq. (3), to obtain

$$dv/dt = x(0) - Ns + Nse^{-\lambda v}, \qquad v(0) = 0 \tag{10}$$

The solution to this differential equation is easily found to be

$$v(t) = \lambda^{-1} \ln\left[\frac{Ns - x(0)\exp\{-\lambda[Ns - x(0)]\,t\}}{Ns - x(0)}\right] \tag{11}$$

Differentiating this equation with respect to t and using Eq. (3), we get

$$x(t) = \frac{[Ns - x(0)]\,x(0)}{Ns\exp[\lambda(Ns - x(0))\,t] - x(0)} \tag{12}$$

which is the expression that was given by Yassky. Thus by regarding $NP_{n,0}(t)$ as the actual number of bacteria having n attachments at time t, the time dependence of $x(t)$ is determined, while in the proposed stochastic model for the number of attachments on a bacterium, any function $x(t)$ (found experimentally) is acceptable.

In stage 3 of the life cycle of phages, the number of nucleic acid strands in a bacterium is increasing due to replications. This number, n, can be described by a pure birth process (no deaths) with $\lambda_n = n\lambda(n)$, where $\lambda(n)$ is a density-dependent probability rate of "birth" per individual strand when n strands are present. (Note that by one replication, the number of strands increases by one.) We neglect the possible addition of strands due to further attachments of phages to the same bacterium.

The probability of having n strands at time t, after an infection by a single phage, satisfies the master equation [Eq. (2.0.3)],

$$dP_n/dt = \lambda_{n-1} P_{n-1}(t) - \lambda_n P_n(t), \qquad P_n(0) = \delta_{n,1} \tag{13}$$

and the average number of strands satisfies the differential equation [Eq. (2.0.8)]

$$d\langle n\rangle/dt = \langle\lambda_n\rangle = \langle n\lambda(n)\rangle \tag{14}$$

As was mentioned before, the amount of nucleic acid in the bacterium changes linearly in time during stages 3 and 4. Therefore, $\lambda(n)$ should be chosen to yield a constant growth rate of $\langle n\rangle$. From Eq. (14), the growth rate of $\langle n\rangle$ is constant when $\lambda(n)$ is proportional to n^{-1}. This form is quite reasonable since the greater the number of strands present, the less is the raw material (nucleotides) available for further replications, and the probability of each strand to replicate decreases. Taking $\lambda(n) = \lambda n^{-1}$, we arrive at a process with constant transition probabilities $\lambda_n = \lambda$, which is the Poisson process. For this process (see Table 2.1)

$$\langle n\rangle(t) = 1 + \lambda t \tag{15}$$

and

$$P_n(t) = \frac{e^{-\lambda t}(\lambda t)^{n-1}}{(n-1)!}, \qquad n = 1, 2, \ldots \tag{16}$$

In a model proposed by Ohlsen (1963), λ_n is taken to be of the form $\lambda(t)n$, and the dependence of $\lambda(t)$ on time is determined by imposing a linear dependence of $\langle n\rangle$ on time. It seems to us very difficult to determine experimentally whether the change in the replication rate per individual depends on the time since infection or on the number of strands, as these two variables are linearly related.

Denoting by τ the eclipse period (stage 3), we get, by substituting $t = \tau$ in Eq. (16), the distribution of the number of strands in the pool before the coating process is ready to start. Since at $t = \tau$ there are 40–80 strands in the pool, $\lambda\tau \gg 1$, and the distribution $P_n(\tau)$ is unimodal with maximum value at $n = [\lambda\tau]+1$ ($[\alpha]$ stands for the largest integer smaller than α). Thus the average given by Eq. (15) is almost identical with the most probable n, and can represent the process in most cases.

In stage 4 the replication of the strands continues, but at the same time strands are being coated and so are unable to replicate. The number of uncoated strands in this stage can be represented by a birth and death process, with coated strands regarded as dead strands. However, the more interesting number at lysis is the number of coated strands, i.e., the number of "dead" strands, rather than the number of viable strands. Since the number of coated strands depends on the number of the viable strands, the process which represents this stage is a bivariate process defined by $P_u(n, r, t)$, the probability that at time t after the end of the eclipse period (with u strands in the pool) there are n viable strands and r coated strands in the bacterium. Denoting by μ_n the probability of formation of a new coated strand when there are n uncoated strands in the pool (independent of the number of coated strands), and by $\lambda_n = \lambda$ the replication rate (as in the preceding stage), this probability satisfies the differential equation

$$\frac{d}{dt} P_u(n,r,t) = \lambda P_u(n-1,r,t) - (\lambda+\mu_n) P_u(n,r,t) + \mu_{n+1} P_u(n+1,r-1,t) \tag{17}$$

Since the changes in n are independent of the number of coated strands r, we can sum Eq. (17) for all possible values of r to get for $\sum_{r=0}^{\infty} P_u(n,r,t) \equiv P_{n,u}(t)$

the equation

$$dP_{n,u}(t)/dt = \lambda P_{n-1,u}(t) - (\lambda_n + \mu_n) P_{n,u}(t) + \mu_{n+1} P_{n+1,u}(t), \qquad n \geqslant 0 \tag{18}$$

which is a master equation for a birth and death equation for a univariate process. For this process $P_{0,u}(t) \ll 1$, and since $\mu_0 = 0$, we get from Eq. (2.0.8),

$$d\langle n\rangle/dt = \langle \lambda_n \rangle - \langle \mu_n \rangle = \lambda - \mu + \mu P_{0,u}(t) \simeq \lambda - \mu \tag{19}$$

where we have taken $\mu_n = \mu$ in view of the experimental observation that $\langle n\rangle$ changes linearly in time in stage 4 also. This choice of μ_n implies that the probability of an individual strand to be coated is inversely proportional to the number of uncoated strands [$\mu(n) = \mu n^{-1}$]. Integrating Eq. (19) we get

$$\langle n\rangle = u + (\lambda - \mu) t \tag{20}$$

Since in lysis the total number of uncoated strands is of the same order of magnitude (40–80) as their number at the end of the eclipse period, and, in addition, there are 200–300 mature phages after lysis, the rate λ must be of the same order of μ, so that $(\lambda - \mu)/\mu \ll 1$.

For $\mu_n = \mu$, the number of "dead" strands (coated strands) becomes independent of the number of viable strands as long as the latter number is nonzero, and the number of deaths is described by a Poisson process with parameter μ. Since at $t = 0$ there are no coated strands, the probability of having r coated strands at time t is

$$Q_r(t) = e^{-\mu t} \frac{(\mu t)^r}{r} \tag{21}$$

If lysis occurs when there are R coated mature phages in the bacterium, then the burst time is distributed according to the first passage time distribution to state R, i.e., according to the probability density function

$$F_R(t) = \mu Q_{R-1}(t) = \mu e^{-\mu t} \frac{(\mu t)^{R-1}}{(R-1)!} \tag{22}$$

Let t_p, $\langle t\rangle$, and V_t be the most probable burst time, the average burst time, and the coefficient of variation of burst time, respectively. Then from Eq. (22),

$$t_p = (R-1)/\mu, \qquad \langle t\rangle = R/\mu, \qquad V_t = R^{-1/2} \tag{23}$$

The expressions above indicate that for moderately large R ($R > 25$), the burst-time distribution is sharply peaked ($V_t < 0.2$), and the most probable burst time is close to the average burst time.

Qualitatively such behavior is found experimentally when most of the bacteria are burst in a small time interval. In Fig. 5.1, taken from Luria and Darnell (1967), the burst time occurs between 14 and 20 min after infection.

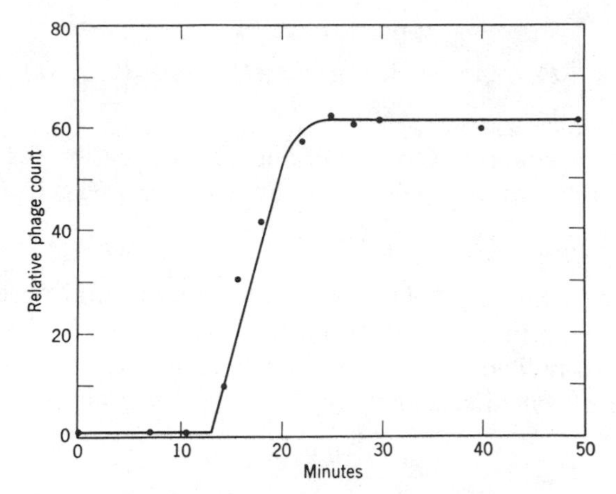

Fig. 5.1 Growth of bacteriophage T1 on *E. coli* strain B in nutrient broth at 37°C [after Luria and Darnell (1967, p. 189)]. Relative phage count is 45% of average phage yield per infected cell. Lower plateau corresponds to stages 2 and 3, and upper plateau is due to completion of lysis. (Newly liberated phages fail to meet new bacteria because of high dilution.)

The fraction of uninfected bacteria, given by Eq. (7), is also in agreement with experimental evidence. In the work reported by Luria and Darnell (1967, Fig. 3.5), the fraction of bacteria in which lysis occurs as a function of initial phage concentration is an S-shaped curve, which arrives at the plateau (=1) when initially there are 10 or more phages per bacterium. From Eq. (7), the fraction of infected bacteria is given by

$$1 - \exp\left\{-\lambda s \int_0^{t^*} x(\tau)\, d\tau\right\} \approx 1 - \exp\{-\lambda s t^* x(0)/2\} \tag{24}$$

Qualitatively, it increases from 0 to 1 as $x(0)$ changes from 0 to ∞, with the main increase taking place for small values of $x(0)$. Thus this fraction is above the one given by the S-shaped curve for small values of $x(0)$. This behavior is expected since among the infected bacteria there are some in which eventual lysis does not occur, especially when the number of attached phages is low. A discussion of this phenomenon can be found in the book by Luria and Darnell (1967, p. 48).

References

Bailey, N. T. J. (1953). The total size of a general stochastic epidemic. *Biometrika* **40**, 177.

Bailey, N. T. J. (1957). "The Mathematical Theory of Epidemics." Griffin, London.

Chiang, C. L. (1954). Competition and other interactions between species, *in* "Statistics and Mathematics in Biology" (O. Kempthorne, T. A. Bancroft, J. W. Gowen, and J. L. Lush, eds.), p. 197. Iowa State College Press, Ames.

D'Ancona, U. (1954). "The Struggle for Existence." Brill, Leiden, Netherlands.
Dietz, K. (1967). Epidemics and rumours: A survey. *J. Roy Statist. Soc. Ser. A* **130**, 505.
Gani, J. (1965). Stochastic models for bacteriophage. *J. Appl. Probability* **2**, 225.
Goel, N. S., Maitra, S. C., and Montroll, E. W. (1971). On the Volterra and other nonlinear models of interacting populations. *Rev. Modern Phys.* **43**, 231; "On the Volterra and Other Nonlinear Models of Interacting Populations." Academic Press, New York.
Gradshteyn, I. S., and Ryzhik, I. M. (1965). "Table of Integrals, Series and Products." Academic Press, New York.
Grat, J. J. (1965). Some stochastic models relating time and dosage in response curves. *Biometrics* **21**, 583.
Kendall, D. G. (1956). Deterministic and stochastic epidemics in closed populations. *Proc. Symp. Math. Statist. Probability, 3rd, Berkeley, California*, 1956, **4**, p. 149.
Kermack, W. O., and McKendrick, A. G. (1927). A contribution to the mathematical society of epidemics: I. *Proc. Roy. Soc. Ser. A* **115**, 700.
Luria, S. E., and Darnell, J. E. (1967). "General Viriology," 2nd ed. Wiley, New York.
Ohlsen, S. (1963). Further models for phage reproduction in a bacterium. *Biometrics* **19**, 441.
Richter-Dyn, N., and Goel, N. S. (1972). On the extinction of a colonizing species. *J. Theoret. Pop. Biol.* **3**, 406.
Saaty, T. L. (1961). Some stochastic processes with absorbing barriers. *J. Roy. Statist. Soc. Ser. B* **23**, 319.
Shortley, G. (1965). A stochastic model for distributions of biological response times. *Biometrics* **21**, 562.
Volterra, V. (1928). Variations and fluctuations of the number of individuals in animal species living together. *J. Cons. Perm. Int. Explor. Mer* **3**, 1; translated in R. N. Chapman "Animal Ecology." McGraw-Hill, New York.
Volterra, V. (1931). "Lecon sur le theorie Mathematique de la Lutte pour la Vie." Gauthier-Villars, Paris.
Volterra, V. (1937). Principes de biologie Mathematique. *Acta Biotheoret.* **3**, 1.
Watson, J. D. (1970). "Molecular Biology of the Gene." Benjamin, New York.
Yassky, D. (1962). A model for the kinetics of phages attachment to bacteria in suspension. *Biometrics* **18**, 185.

6

Dynamics of a Population of Interacting Species

In Chapter 4 we discussed the growth and extinction of a single-species population in the presence of some random elements affecting both the basic mechanisms of population growth and the factors determining it. In most natural systems, every species interacts with a number of other species, either through prey–predator interactions (by eating or by being eaten) or through competition for the same food. In this chapter we study two aspects of a population of interacting species.

One aspect is the mechanism behind the changes in the species diversity (which can be crudely defined as the number of species per unit area) in a particular region. The changes involve the processes of extinction† of some of the species and immigration of other species from nearby regions. A detailed insight into these two processes and how they affect the species diversity can be provided by the detailed analysis of species diversity on a set

† We have neglected the emigration of the species. As can be seen in the discussion of Section 6.1, this can be easily incorporated into the formalism.

of islands in the *same geographic region* for which a few experimental data are available. Such an analysis was carried out by MacArthur and Wilson (1963), and in the next section we present their linear model in the language of Chapter 2, introduce a nonlinear model also, and modify some of the results.

The other aspect, discussed in Section 6.2, is the dynamics of an individual species that interacts with many other species.

6.1 Species Diversity on Islands

In a set of small islands in the same region, the increase in the number of species (diversity) on a particular island takes place due to immigration of new species from a nearby big island (or continent) which will be called the "pool" of species. The number of species decreases by the extinction of some of the species, due mainly to interaction with other species and, in particular, due to competition between the species on the island for the same limited food. Because of the stochastic nature of these events, the change in the number of species occupying an island is a stochastic process which depends on the island's size (available food), location, and degree of geographic isolation. The random variable describing the process is the number of species present on the island. This number changes discretely and thus the process can be analyzed by using the mathematical analysis of Chapter 2.

Let λ_n and μ_n be the probability rates of the arrival of a new species and of the extinction of an already present species, respectively, when the number of species present on the island is n. In other words, $\lambda_n\,\Delta t + O(\Delta t)$ is the probability that in the time interval $(t, t+\Delta t)$, the number of species increases from n to $n+1$, and $\mu_n\,\Delta t + O(\Delta t)$ is the probability that it decreases from n to $n-1$. If initially $(t = 0)$ there are m species on the island, then the probability $P_{n,m}(t)$ of the island having n species at time t satisfies both the forward and backward master equations (2.0.3) and (2.0.4). These equations are to be solved subject to the initial conditions

$$P_{n,m}(0) = \delta_{n,m} \tag{1}$$

and some boundary conditions. The boundary conditions can be determined by noting that when there are no species present on the island, there is no extinction so that $\mu_0 = 0$. Further, due to immigration, $\lambda_0 \neq 0$. Also, when the number of species is equal to K, the number of species in the pool, any further immigration will not increase the number of species on the island so that $\lambda_K = 0$. Further, due to extinction, $\mu_K \neq 0$. Therefore, in the language of Chapter 2, the process is confined between the two reflecting states 0 and K. The master equations are therefore subject to the boundary conditions given in Section 2.2 for the reflecting states, and the process is subject to the analysis of that section. To be able to use this analysis, we require the explicit forms of λ_n and μ_n. We choose the following two reasonable models:

Model A. This model is defined by

$$\mu_n = \mu n, \qquad \lambda_n = \nu - \lambda n \equiv \lambda(K-n) \qquad (\nu = \lambda K \qquad \text{since} \quad \lambda_K = 0) \tag{2}$$

These forms of λ_n and μ_n are based on the following observations:

For a given island it is reasonable to assume that the probability of a new species occupying the island depends on the number of species already present. Since as more species become established on the island, fewer newly arriving individuals will belong to new species, this dependence can be taken to be monotonically decreasing with the number of species present on the island. As a first approximation, this dependence can be taken to be linear in $K-n$ with $\lambda_K = 0$. The condition $\lambda_K = 0$ requires that ν, the probability rate of species arriving at the island when no species are present on the island, is linearly proportional to the total number of species in the pool. For a pool of species with the same means of mobility (e.g., flying for birds), the proportionality constant λ is the probability rate for members of a species to arrive at the island, and is dependent only on the degree of isolation of the island. Likewise, since the area of the island is fixed, as the number of species present on the island increases, the average population size of a given species decreases. Consequently, the probability of an already existing species to become extinct can be taken to be a monotonically increasing function of the number of species already present. As a first approximation, we can assume that each species has the same probability of extinction μ, independent of the number of species present, and therefore μ_n can be taken to be linear in n with $\mu_0 = 0$.

Model B. This model is defined by

$$\mu_n = \mu n^2, \qquad \lambda_n = \lambda(K-n)^2 \tag{3}$$

and describes the process under consideration better than model A for the following reasons:

The probability of an increase in the number of species is steeper at low values of n due to the fact that more rapidly dispersing species would become established first, causing a rapid initial drop in the overall immigration rate of new species, while the later arrival of slow colonizers would drop the immigration rate probability more slowly to zero. The linear extinction rate probability (2) corresponds to the assumption that each species has the same probability of extinction μ. But when more species are present, the average size of the population of a species is smaller and thus on the average the probability of extinction for any given species is greater. Thus μ_n/n is a monotone increasing function of n.

We now analyze these two models. For model A, the forward master equation can be explicitly solved. The solution is given in Table 2.1. (See process 8. Of course, due to slightly different λ_n and μ_n, λ has to be replaced by $-\lambda$ and ν/λ by K.) For $m = 0$ (when initially there are no species present on the island), $P_{n,0}(t)$ is the binomial distribution

$$P_{n,0}(t) = \gamma^n \left(\frac{1+\gamma\sigma}{1+\gamma}\right)^K \left(\frac{1-\sigma}{1+\gamma\sigma}\right)^n \frac{K(K-1)(K-2)\cdots(K-n+1)}{n!}$$

$$= \binom{K}{n}\left(\frac{1+\gamma\sigma}{1+\gamma}\right)^{K-n}\left(\frac{\gamma-\sigma\gamma}{1+\gamma}\right)^n \tag{4}$$

where

$$\gamma \equiv \lambda/\mu, \qquad \sigma \equiv \exp(-(\lambda+\mu)t) \tag{5}$$

As $t \to \infty$, the expression for $P_{n,0}(t)$ becomes

$$P_{n,0}(\infty) = P_n^* = \binom{K}{n}\left(\frac{\lambda}{\lambda+\mu}\right)^n\left(\frac{\mu}{\lambda+\mu}\right)^{K-n} = \binom{K}{n}\left(\frac{\lambda}{\mu}\right)^n \bigg/ \left(1+\frac{\lambda}{\mu}\right)^K, \qquad 0 \leqslant n \leqslant K \tag{6}$$

which is a binomial distribution in $\lambda/(\lambda+\mu)$. Since the process is confined between two reflecting states, $P_{n,m}(\infty)$ will be independent of m (see Section 2.2C). The expression (6) can be directly obtained by substituting Eq. (2) into Eqs. (2.2.43) and (2.2.40).

Further, from Table 2.1, the explicit expressions for the average number of species $\langle n\rangle$ and its variance $\operatorname{var}(n)$ are

$$\langle n\rangle = \frac{K\lambda}{\lambda+\mu}(1-\exp(-(\lambda+\mu)t)) \tag{7a}$$

$$\operatorname{var}(n) = \frac{K\lambda}{(\lambda+\mu)^2}(\mu-(\mu-\lambda)\exp(-(\lambda+\mu)t)-\lambda\exp(-2(\lambda+\mu)t)) \tag{7b}$$

The limiting forms for $t \to \infty$ are approached with a time constant $(\lambda+\mu)^{-1}$, and are given by

$$\langle n\rangle_{t\to\infty} = K(1+\mu/\lambda)^{-1} = K\lambda(\lambda+\mu)^{-1} \tag{8a}$$

$$\operatorname{var}(n)_{t\to\infty} = K(\mu/\lambda)(1+\mu/\lambda)^{-2} = K\lambda\mu(\lambda+\mu)^{-2} \tag{8b}$$

[The latter two expressions also follow directly from Eq. (6) and the definitions of $\langle n\rangle$ and $\operatorname{var}(n)$.]

From Eq. (8) the coefficient of variation at the steady state is $(\mu/\lambda K)^{1/2}$

and by Eq. (6), the steady-state density P_n^* is maximal for $n < K/2$ if $\lambda/\mu < 1$, for $n = K/2$ if $\lambda/\mu = 1$, and for $n > K/2$ if $\lambda/\mu > 1$. Furthermore, from Eqs. (7a) and (8a), the average number of species in the steady state is $K(1+\mu/\lambda)^{-1}$ and this steady state is approached with a time constant $(\lambda+\mu)^{-1}$. As K increases, the most probable point approaches the average value since the binomial distribution approaches the normal distribution, and when $K\lambda\mu(\lambda+\mu)^{-2} \geqslant 10$, the normal distribution is already a good approximation for the steady-state distribution (6).

We now proceed to derive the corresponding results for model B. For this model we are able to obtain analytical expressions only for P_n^* and the steady-state values of $\langle n\rangle$ and $\mathrm{var}(n)$. Substituting Eq. (3) into Eqs. (2.2.43) and (2.2.40), we obtain

$$P_n^* = \binom{K}{n}^2 (\lambda/\mu)^n / Z_K \tag{9a}$$

where

$$Z_K = \sum_{n=0}^{K} \binom{K}{n}^2 (\lambda/\mu)^n = F(-K, -K; 1; \lambda/\mu) \tag{9b}$$

and F is the hypergeometric function (Abramowitz and Stegun, 1964).

By the defining equation for $\langle n\rangle$ and $\mathrm{var}\langle n\rangle$, the steady-state values of these quantities are given by

$$\langle n\rangle_{t\to\infty} = Z_K^{-1} \sum_{n=0}^{K} n \binom{K}{n}^2 \left(\frac{\lambda}{\mu}\right)^n = Z_K^{-1} \left(\frac{\lambda}{\mu}\right) \frac{\partial Z_K}{\partial(\lambda/\mu)}$$

$$\mathrm{var}_{t\to\infty}\langle n\rangle = Z_K^{-1} \sum_{n=0}^{K} n(n-1) \binom{K}{n}^2 \left(\frac{\lambda}{\mu}\right)^n = Z_K^{-1} \left(\frac{\lambda}{\mu}\right)^2 \frac{\partial^2 Z_K}{\partial(\lambda/\mu)^2}$$

Using the relation (Abramowitz and Stegun, 1964)

$$\frac{d^n}{dz^n} F(a, b, c, z) = \frac{(a)_n (b)_n}{(c)_n} F(a+n, b+n; c+n; z)$$

$$(a)_n = a(a+1)\cdots(a+n-1)$$

we can rewrite $\langle n\rangle$ and $\mathrm{var}\langle n\rangle$ as

$$\langle n\rangle = \frac{\lambda}{\mu} K^2 F\left(-K+1, -K+1; 2; \frac{\lambda}{\mu}\right) \Big/ F\left(-K, -K; 1; \frac{\lambda}{\mu}\right) \tag{10a}$$

$$\mathrm{var}\langle n\rangle = 2\left(\frac{\lambda}{\mu}\right)^2 \binom{K}{2}^2 F\left(-K+2, -K+2; 3; \frac{\lambda}{\mu}\right) \Big/ F\left(-K, -K; 1; \frac{\lambda}{\mu}\right) \tag{10b}$$

For this model, P_n^* is maximal at $K/2$ for $\lambda/\mu = 1$, but for $\lambda/\mu \neq 1$ it has its maximal value closer to $K/2$ than P_n^* of model A. This can be seen from the fact that the maximum of P_n^* in model B is attained at the same n for which P_n^* of model A with $(\lambda/\mu)^{1/2}$ is maximal, and this point is closer to $K/2$ since $(\lambda/\mu)^{1/2}$ is closer to 1.

To get some insight into the time constant for approaching the steady state, we set $\lambda = \mu$ because for this case the differential equations (2.0.8) and (2.0.13) can be analytically solved for $\langle n \rangle$ and $\mathrm{var}(n)$. Also P_n^* is simplified to

$$P_n^* = \binom{K}{n}^2 \Big/ \binom{2K}{K} \tag{11}$$

because (Gradshteyn and Ryzhik, 1965)

$$\sum_{n=0}^{K} \binom{K}{n}^2 = \binom{2K}{K} \tag{12}$$

P_n^* is thus a hypergeometric distribution (Abramowitz and Stegun, 1964).

Substituting Eq. (3) into Eqs. (2.0.8) and (2.0.13), after simple algebraic manipulations, we obtain

$$\frac{d\langle n \rangle}{dt} = \lambda \sum_{n=0}^{K} (K-n)^2 P_n - \mu \sum_{n=0}^{K} n^2 P_n = \lambda K^2 - 2\lambda K \langle n \rangle \tag{13}$$

$$\begin{aligned} \frac{d\,\mathrm{var}(n)}{dt} &= 2\lambda \sum_{n=0}^{K} (K-n)^2 (n-\langle n \rangle) P_n - 2\mu \sum_{n=0}^{K} n^2 (n-\langle n \rangle) P_n \\ &\quad + \lambda \sum_{n=0}^{K} (K-n)^2 P_n + \mu \sum_{n=0}^{K} n^2 P_n \\ &= 2\lambda(1-2K)\,\mathrm{var}(n) + \lambda K^2 + 2\lambda \langle n \rangle (\langle n \rangle - K) \end{aligned} \tag{14}$$

The solution of Eq. (13) with the initial condition $\langle n \rangle(0) = 0$ is

$$\langle n \rangle = (K/2)(1-e^{-2K\lambda t}) \underset{t\to\infty}{\to} K/2 \tag{15}$$

Substituting this solution into Eq. (14) we obtain

$$\frac{d\,\mathrm{var}(n)}{dt} = 2\lambda(1-2K)\,\mathrm{var}(n) + \frac{\lambda K^2}{2}(1+e^{-4K\lambda t}) \tag{16}$$

The solution of this equation with the initial condition $\mathrm{var}(n) = 0$ is

$$\mathrm{var}(n) = \frac{K^2}{4(2K-1)} - \frac{K^2}{4} e^{-4\lambda K t}\left(1 + \frac{2(1-K)}{2K-1} e^{2\lambda t}\right) \tag{17}$$

$$\underset{t\to\infty}{\mathrm{var}(n)} = \frac{K^2}{4(2K-1)} \tag{18}$$

Comparing Eq. (15) with Eq. (8a), with $\lambda = \mu$, we observe that the steady-state value of $\langle n \rangle$ for both models is $K/2$, while from comparison of Eqs. (8b) and (18), we conclude that the variance is smaller for model B. The approach to steady state for model B is much faster than for the linear case because, from Eq. (15), the time constant for model B is $1/2\lambda K$ as compared to $1/2\lambda$ for model A. Further, the derivative of $\langle n \rangle$ with respect to (λ/μ) at $\lambda/\mu = 1$ is greater for model B, and therefore for λ/μ close to 1, $\langle n \rangle$ in this model is farther from $K/2$ than $\langle n \rangle$ in model A.

The theory discussed above predicts a distribution of the number of species on an island when steady state is reached, as well as the rates of approach to this steady state. To test the predictions of the theory, it is necessary to have experimental data on the number of species on a group of islands of equal areas, at the same distance from the pool of species (mainland or another big island), and to compare the distribution of these numbers with the probability distributions predicted by the two models. Unfortunately, such data are not available.

One set of data which can be usefully analyzed is the data on the number of land and freshwater bird species on various islands and archipelagos of the Moluccas, Melanesia, Micronesia, and Polynesia in the Pacific Ocean. The main source of birds is New Guinea. These data are not available for various years. Therefore, we can use only the asymptotic expressions (steady state), assuming that in the era the data were taken, the steady state had already been reached. These islands do not have the same area and are not equidistant from New Guinea. Therefore we cannot use these data to test the predicted distribution directly. However, these data confirm the observation inferred from model A that on the average, far islands are more "sensitive" than nearby islands to changes in area. This follows from Eq. (8a) under the assumption that $\mu = \mu(A)$, $\lambda = \lambda(r)$, where A and r are, respectively, the area and distance of an island. From Eq. (8a) we obtain

$$\frac{\partial}{\partial r}\frac{\partial\langle n\rangle}{\partial A} = \frac{\partial}{\partial A}\frac{\partial\langle n\rangle}{\partial r} = K\frac{\partial^2\langle n\rangle}{\partial\mu\,\partial\lambda}\left(\frac{d\mu}{dA}\right)\left(\frac{d\lambda}{dr}\right) = \frac{K(\lambda-\mu)}{(\mu+\lambda)^3}\left(\frac{d\mu}{dA}\right)\left(\frac{d\lambda}{dr}\right) \tag{19}$$

Since

$$d\mu/dA < 0, \qquad d\lambda/dr < 0 \tag{20}$$

for $\lambda > \mu$, $(\partial/\partial r)(\partial\langle n\rangle/\partial A) > 0$, i.e., the changes in the number of species due to changes in area $(\partial\langle n\rangle/\partial A)$ are increasing with r, and since $(\partial\langle n\rangle/\partial A) > 0$ this increase is also an increase in magnitude. On the other hand, since $(\partial/\partial A)(\partial\langle n\rangle/\partial r) > 0$ and $\partial\langle n\rangle/\partial r < 0$, $(\partial/\partial A)(|\partial\langle n\rangle/\partial r|) < 0$, the magnitude of the change in $\langle n\rangle$ due to changes in distance decreases with the increase of area. Thus small islands are more "sensitive" to changes in distance than the larger ones.

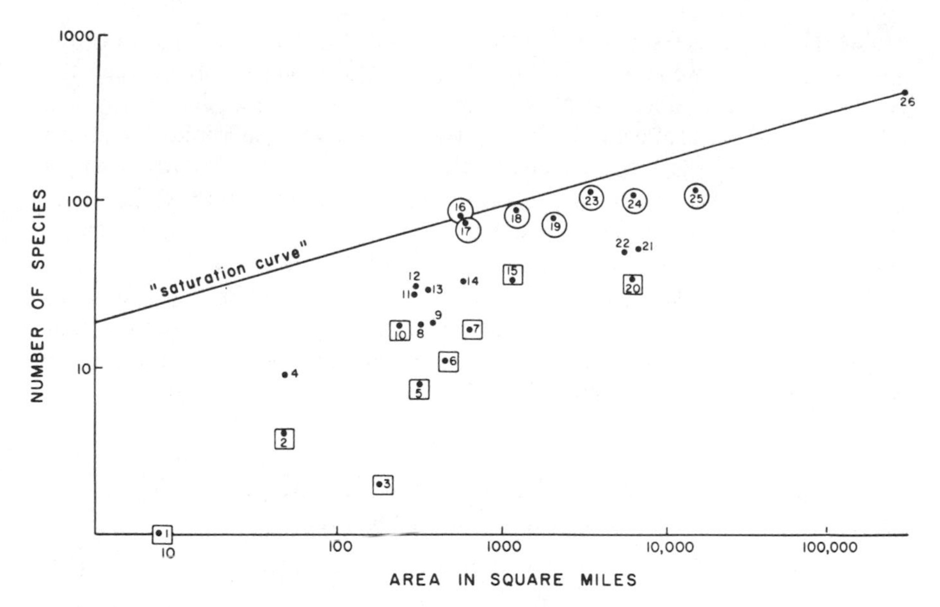

Fig. 6.1 The number of land and freshwater bird species on various islands of the Moluccas, Melanesia, Micronesia, and Polynesia versus the area of the islands. The distance from the pool of species (New Guinea) is indicated by a circle for an island which is less than 500 miles away, and by a square for an island which is more than 2000 miles away.

This predicted behavior is indicated by Fig. 6.1, where the log of the number of species is plotted versus the log of the area for the archipelagos discussed above. Closer islands are enclosed in circles and farther islands in squares. For the circled islands the increase in $\langle n \rangle$ is much smaller compared to the increase in $\langle n \rangle$ for the squared islands. For an area in the range 100–1000 square miles the far islands have a significantly lower number of species than the nearby islands, while for an area about 10,000 square miles this difference is much less pronounced.

From the same data on the number of bird species we can also find the dependence of λ [defined by Eq. (2) and discussed below] on the distance of the island from the pool, if we assume that μ, the extinction rate per species, is independent of the distance of the island and is equal for all the islands of equal area, and that model A is a good approximation. To do so we have to normalize the number of species on various islands to the same area. It is known (MacArthur and Wilson, 1967) that, in the absence of distance effects, the log of the number of species on an island versus the log of its area for a set of islands in the same vicinity is a straight line; i.e., $n = CA^{\gamma}$. As shown in Fig. 6.1, taken from MacArthur and Wilson (1963), where log number of species is plotted versus log of area for various islands of the Moluccas,

Melanesia, Micronesia, and Polynesia, due to distance effects, this is not the case. However, if we assume that the distance effect enters only through the coefficient C, then since for New Guinea (the source) and the nearby Kei Islands, the distance effect is negligible, we can take the line joining the points corresponding to these two islands as the "saturation curve" $C(0)A^{\gamma}$ which represents the predicted range of "saturation" value (maximum value on an island of a given area). The normalized number of species on a particular island is taken to be proportional to the ratio of the actual number of species and the saturation value for that area of the island, i.e., to $C(r)/C(0)$. We take this constant of proportionality to be 100 and the normalized number thus obtained is the percentage saturation. Regarding this percentage saturation n (which is a normalized number of species) as a function of distance r

$$n = f(r) \tag{21}$$

and assuming that λ_n and μ_n are linear in n [Eq. (2)], we get from Eq. (8a) by equating $f(r)$ to the average number predicted by model A

$$n = f(r) = K[1+\mu/\lambda(r)]^{-1}, \qquad K = 100\% \tag{22}$$

Thus

$$\lambda(r) = \mu f(r)(K-f(r))^{-1} \tag{23}$$

In Fig. 6.2 we have plotted $\log n$ vs. r for the islands of Fig. 6.1. As can be seen, various points lie on two distinct lines. Ignoring for the time being the upper line (with smaller slope), we conclude that $f(r)$ is an exponential function, i.e.,

$$f(r) = Ke^{-\delta r} \tag{24}$$

with $\delta \simeq 9.8\times 10^{-4}$ per mile, and from Eq. (23)

$$\lambda(r) = \mu(e^{\delta r}-1)^{-1} \tag{25}$$

which is independent of K and is the required dependence of immigration probability rate on distance.

Let us now discuss the other line with smaller slope. In our treatment we have assumed that there is only one pool of species. As can be seen from Fig. 6.3, the islands on the upper line are close to an additional source of species (South America, United States, Australia) and therefore the effective value of K increases and the number of species will be larger, consistent with the observation.

It is interesting to note that the dependence (25) of the immigration probability rate of birds on distance is not true for "islands" in continental regions. Vuilleumier (1970) studied species numbers among the birds living in 15 "islands" of paramo vegetation in the Andes of Venezuela, Colombia, and

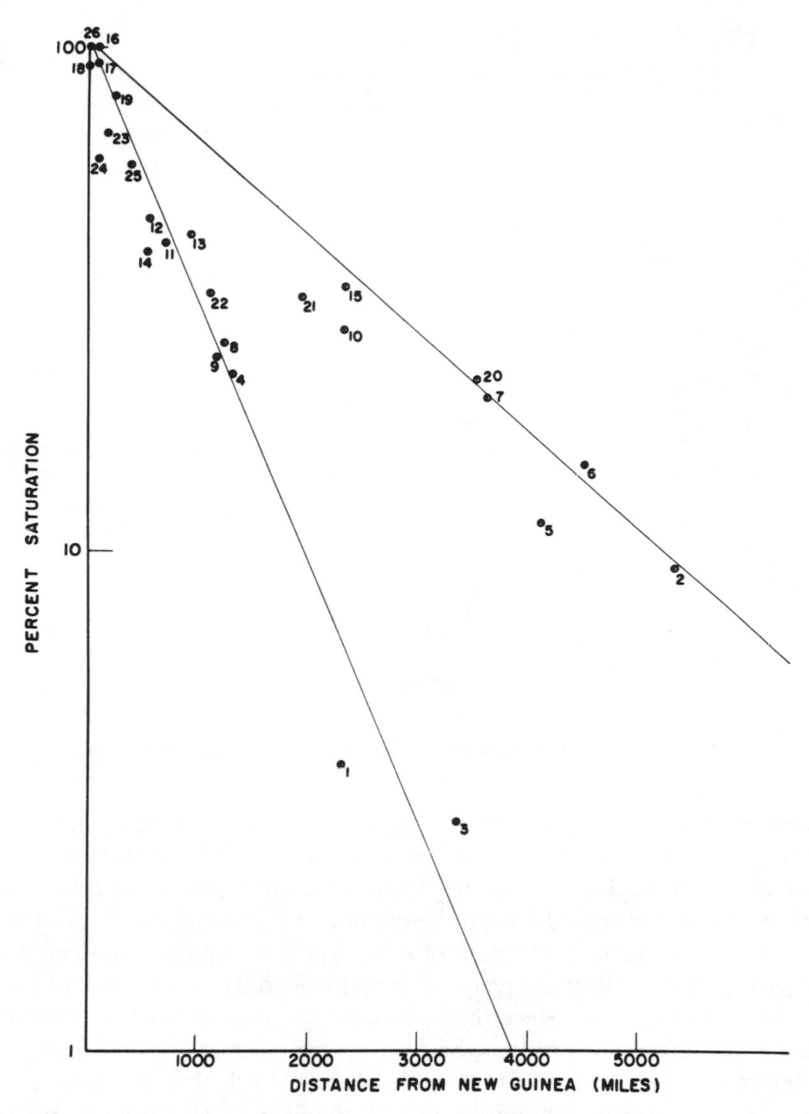

Fig. 6.2 The percentage saturation of species on the islands of Fig. 6.1 versus the distances of the islands from New Guinea.

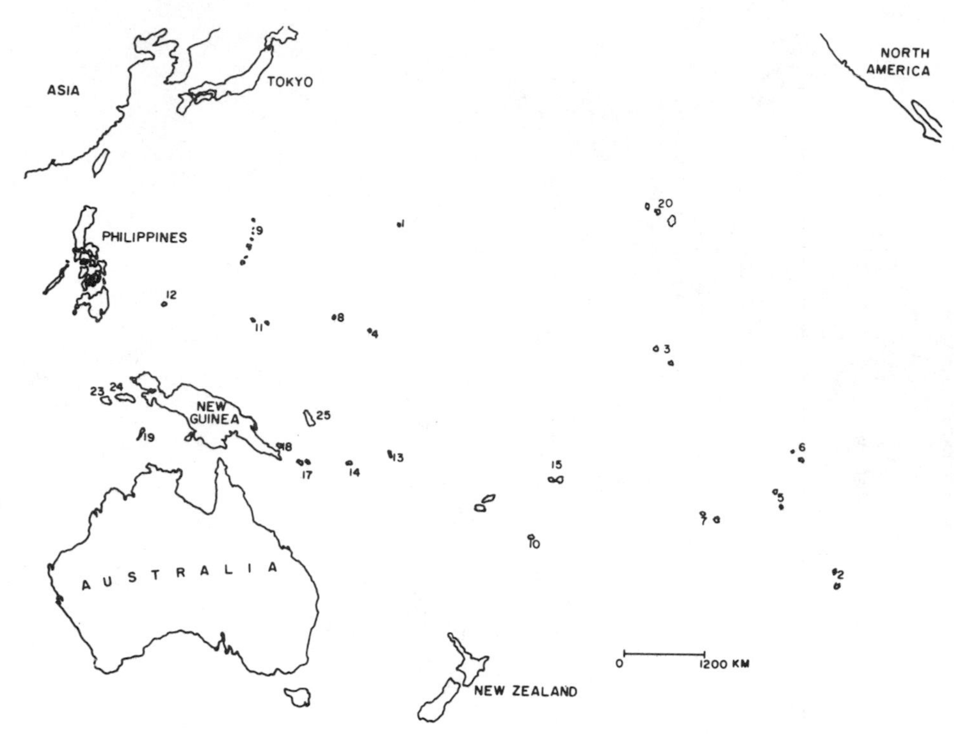

Fig. 6.3 A map of the geographic location of the islands of Fig. 6.1.

Northern Ecuador. By carrying out an analysis similar to the one presented above, one finds that there is no clean-cut correlation between the percentage saturation and the distance (though Vuilleumier suggests a linear relation). A reason for the difference between the oceanic and continental islands is that, in the former case, the probability of a propagule falling in the sea and dying before it reaches a potential recipient island is high while, for the latter case, it is considerably lower, since the islands are not isolated from each other by sea, but by different vegetation where the propagule may "rest."

Before concluding this section, we wish to remark that the assumptions $\mu = \mu(A)$, $\lambda = \lambda(r)$ are not consistent with the form of $\langle n \rangle$ given by Eq. (8a) and the experimental relation $\langle n \rangle = CA^{\gamma}$ between the area and the number of species for a given distance. It seems to us that the assumption $\mu = \mu(A)$ is the one to be corrected. The extinction rate per species, μ, can depend on distance as well as on area, because a greater supply of propagules (immigration) decreases the probability of a species becoming extinct. By this argument the dependence of μ on r is much weaker than the dependence of λ on r, so that the results derived above might still constitute a good approximation.

6.2 Population Growth of Individual Species in an Interacting Population of Species

The growth of a species in the absence of other species was investigated in 1845 by the Belgian mathematician Verhulst using the model described in Chapter 4 [Eq. (4.0.3)]. But it was not until the late 1920s that any serious analysis of a species growth in the presence of other species was carried out, though the problem was recognized. It was Volterra [(1928, 1931, 1937), an excellent English review of the original Volterra theory and its use for the interpretation of experiments involving a small number of species can be found in the work of D'Ancona (1954)] who proposed and investigated in detail an interesting model for the interaction between a number of different biological species. The two-species version of this model and its similarity to autocatalytic reactions had already been discussed independently by Lotka (1910, 1920) and was described in the preceding chapter. The Volterra model has been reviewed and analyzed thoroughly elsewhere by one of the authors and his collaborators (Goel *et al.*, 1971). In this section we describe the model briefly and analyze it by making certain reasonable assumptions about the effect of all the species on one species. We also discuss the consequences of these assumptions in light of limited experimental data.

According to Volterra, the dynamical equations describing the growth of a system of N interacting species are

$$dn_i/dt = r_i n_i + \beta_i^{-1} \sum_{j=1, j \neq i}^{N} a_{ij} n_i n_j \tag{1}$$

where n_i is the number of individuals of the ith species. The first term describes the behavior of the ith species in the absence of others: When $r_i > 0$, the ith species is postulated to grow in an exponential Malthusian manner with r_i as the "rate constant." When $r_i < 0$ and all other $n_j = 0$, the population of the ith species would die out exponentially. The quadratic terms in Eq. (1) describe the interaction of the ith species with all the other species. The jth term in the sum is proportional to the number of possible binary encounters $n_i n_j$ between members of the ith species and members of the jth species. The physical meanings of a_{ij} and β_i can be derived as follows.

Let $m_{ij} n_i n_j$ be the number of binary encounters per unit time between the ith and jth species where m_{ij} is a constant coefficient. Let q_{ij} be the number of individuals of the jth species destroyed per encounter with an individual of the ith species. (Here we have taken the ith species to be predator for the jth species.) Therefore the number of individuals of the jth species reduced per unit time is $m_{ij} q_{ij} n_i n_j$. If β_i $(i = 1, \ldots, N)$ denotes the mean weight of the individuals in the ith species, and we assume 100% transfer efficiency in each encounter, the increase in the number of individuals of the ith species per unit time is $m_{ij} q_{ij} n_i n_j (\beta_j/\beta_i)$. If we define $m_{ij} q_{ij} \beta_j = a_{ij}$, then the increase in the number of individuals of the ith species is $\beta_i^{-1} a_{ij} n_i n_j$, and the decrease in the number of individuals of the jth species is $\beta_j^{-1} a_{ij} n_i n_j$. Thus by taking $a_{ij} > 0$ if the ith species is a predator for the jth species, $a_{ij} < 0$ if vice versa, and $a_{ij} = 0$ if the ith and jth species do not interact with each other, and imposing the condition

$$a_{ji} = -a_{ij} \tag{2}$$

we can represent by the sum in Eq. (1) the interaction between the ith species with all the other species, independent of the prey–predator characteristics of the ith species.

The Volterra model can be modified by introducing other types of interactions ($a_{ij} \neq -a_{ji}$), a saturation effect in the growth of species with $r_i > 0$, and other environmental factors, in addition to the interactions between the N species. Such factors can be bacteria (if the N species are larger animals) and other parasites, plant life which varies in intensity from season to season, and unspecified migrating species, etc., which affect the population of our N species in a random way. Then our basic equations for population growth might be considered to be of the form

$$dn_i/dt = r_i n_i G(n_i/K_i^*) + n_i \left\{ U_i(t) + \beta_i^{-1} \sum_{j=1}^{N} a_{ij} n_j \right\}, \qquad a_{ii} = 0 \tag{3}$$

where $G(x)$ is a saturation-inducing term† and $U_i(t)$ represents random unspecified influences. When the number of specified related species N is large and each species interacts with a fairly large number of others, one would expect a_{jk}'s of both signs to appear in Eq. (3) for most j. Since the population of each of the species $n_1, n_2, \ldots$ varies with time, and each is influenced by random unspecified effects, the sum in Eq. (3) might also be considered to be a random function of time. The combination of $U_i(t)$ and the sum might then be defined as a random function of time, $S_i(t)$. This consideration would lead to the species being coupled in only a random way. Since only terms with index i will appear in the resulting equation, we suppress the i in the following and develop the consequences of postulating $S_i(t)$ to be a random function. Then Eq. (3) becomes

$$dx/dt = h(x) + e(x)\, i(t) \tag{4}$$

where we have taken

$$x \equiv n \tag{5a}$$

$$h(x) \equiv rnG(n/K^*) = rxG(x/K^*) \tag{5b}$$

$$e(x) \equiv n = x \tag{5c}$$

$$i(t) = S(t) \tag{5d}$$

† $G(x)$ is a saturation-inducing term [e.g., in Eq. (4.0.7)] if $G(0) = 1$, $G(1) = 0$, and $G'(x) < 0$. For $r_i < 0$ we assume $G(x) \equiv 1$ or, equivalently, $K^* = \infty$.

For a species with $r > 0$, a reasonable form of the function $G(x/K^*)$ is [see Eq. (4.0.7)],

$$G(x/K^*) = (1-(x/K^*)^{\alpha})/\alpha \tag{6}$$

where α is a parameter equal to 1 for the Verhulst model. Equation (4) is identical to Eq. (3.0.14) and if in the absence of any other information we assume $i(t)$ to have the same characteristics as $i(t)$ of Chapter 3 (see p. 37), then we can use the techniques of Chapter 3 to analyze the population growth. Carrying out the transformation (3.0.16), Eq. (4) can be written in the form of Eq. (3.0.18), i.e.,

$$dx/dt = rx[1-(x/K^*)^{\alpha}]/\alpha + mx + \sigma x F(t) = kx[1-(x/K)^{\alpha}]/\alpha + \sigma x F(t) \tag{7}$$

where $F(t)$ is defined by Eq. (3.0.16) and

$$k \equiv r + m\alpha \tag{8a}$$

$$K = K^*\{1+\alpha m/r\}^{1/\alpha} = K^*(k/r)^{1/\alpha} \tag{8b}$$

Thus the directional drift due to random factors (i.e., $\langle i(t)\rangle = m \neq 0$) has two effects: It changes the growth rate for low density population and the saturation level. From Eq. (4.0.7), the growth rate for low density populations in the absence of noise is r/α, as compared to k/α in the presence of noise. Therefore, from Eq. (8a), the low density growth rate increases by m for $m > 0$ and decreases by $|m|$ for $m < 0$, due to random factors. The saturation level changes by a factor that is a positive power of the ratio between the new and the old low density growth rates and is only well defined when $k = r+m\alpha > 0$. For the Gompertz model ($\alpha = 0$), the new saturation level is $K = K^*e^{m/r}$, and $k = r$.

The Fokker–Planck equation satisfied by the probability density $P(x\,|\,y,t)$, where y is the initial size of the population, is

$$\frac{\partial P}{\partial x} = -\frac{\partial}{\partial x}(a(x)\,P) + \frac{1}{2}\frac{\partial^2}{\partial x^2}(b(x)\,P) \tag{9}$$

where

$$a(x) \equiv kx[1-(x/K)^{\alpha}]/\alpha + \sigma^2 x/2 \tag{10a}$$

$$b(x) \equiv \sigma^2 x^2 \tag{10b}$$

As noted on page 40, this FP equation can be transformed into the FP equation of the type (3.0.27), i.e.,

$$\frac{\partial g}{\partial t} = -\frac{\partial}{\partial z}(\hat{a}(z)g) + \frac{1}{2}\frac{\partial^2 g}{\partial z^2} \tag{11}$$

with

$$dz = dx/\sigma x, \qquad z = \sigma^{-1} \ln(x/K) \tag{12a}$$

$$\hat{a}(z) = k[1-(x/K)^{\alpha}]/\sigma\alpha = k(1-e^{\alpha\sigma z})/\alpha\sigma \tag{12b}$$

$$p(x \mid y, t) = g(z \mid z_0, t)/\sigma x \tag{12c}$$

where z_0 is the initial value of z, i.e.,

$$z_0 = \sigma^{-1} \ln(y/K) \tag{12d}$$

In the following we discuss the FP equation (11) and the resulting population distribution, starting first with the *steady-state distribution*.

From the forms of $a(x)$ and $b(x)$ given by Eqs. (10), we can infer, using the classification of Section 3.1, that the singular boundary at $x = 0$ is an inaccessible natural boundary and that the singular boundary at $x = +\infty$ is an inaccessible boundary. The characteristics of the boundaries are also transferred to the boundaries $-\infty$, $+\infty$ of the state space of the variable z. Since in both boundaries there is no flow of probability from the state space, the steady-state probability density function is given by Eq. (3.1.3). For the z variable this density function is

$$g(z \mid z_0, \infty) = g_0/\pi(z) \tag{13}$$

where g_0 is a normalization constant and

$$\pi(z) = \exp\left[-2\int^{z} \hat{a}(\xi)\, d\xi\right] \tag{14}$$

Substituting for $\hat{a}(z)$ from Eq. (12b) into this equation, we get

$$\pi(z) = \exp[-\theta(\alpha\sigma z - e^{\alpha\sigma z})], \qquad \theta \equiv 2k/\sigma^2\alpha^2 \tag{15}$$

so that from Eq. (13)

$$g(z \mid z_0, \infty) = g_0 \exp[\theta(\alpha\sigma z - e^{\alpha\sigma z})], \qquad \theta \equiv 2k/\sigma^2\alpha^2 \tag{16}$$

The normalization constant g_0 is given by

$$\begin{aligned} g_0 &= \left\{\int_{-\infty}^{\infty} \exp[\theta(\alpha\sigma z - e^{\alpha\sigma z})]\, dz\right\}^{-1} \\ &= (\alpha\sigma)\,\theta^{\theta}\left\{\int_0^{\infty} u^{\theta-1} e^{-u}\, du\right\}^{-1} = (\alpha\sigma)\,\theta^{\theta}[\Gamma(\theta)]^{-1} \end{aligned} \tag{17}$$

Therefore, the final form of $g(z \mid z_0, \infty)$ is

$$g(z \mid z_0, \infty) = g(z, \infty) = \frac{(\alpha\sigma)\,\theta^{\theta}}{\Gamma(\theta)} \exp\{\theta(\alpha\sigma z - e^{\alpha\sigma z})\}, \qquad \theta \equiv 2k/\sigma^2\alpha^2 \tag{18}$$

The transformation to the population number space is obtained by substituting Eq. (18) into Eq. (12c) and using Eq. (12a). The resulting form of the probability density function is

$$P(x \mid y, \infty) = P(x, \infty) = \frac{\alpha}{x\Gamma(\theta)}[\theta(x/K)^{\alpha}]^{\theta} \exp(-\theta(x/K)^{\alpha}), \qquad \theta \equiv 2k/\sigma^2\alpha^2 \tag{19}$$

For $k > \sigma^2\alpha/2$, this density function is bell shaped with the most probable population number

$$x = K\left(1 - \frac{\sigma^2\alpha}{2k}\right)^{1/\alpha} = K^*[1 + (m - \sigma^2/2)\alpha/r]^{1/\alpha}$$

This number is less than K^*, the deterministic steady-state population size, when $m < \sigma^2/2$, and is greater than K^* when $m > \sigma^2/2$. For $0 < k < \sigma^2\alpha/2$, $P(x, \infty)$ is very high for small values of x and is monotonically decreasing to zero as $x \to \infty$.

For $\alpha = 0$, the Gompertz model [Eq. (4.0.6)], the probability density (16) becomes

$$g(z, \infty) = g_0' \exp(-kz^2)$$

where g_0' is the normalization constant which is easily found to be $(k/\pi)^{1/2}$. Therefore,

$$g(z, \infty) = (k/\pi)^{1/2} \exp(-kz^2) \tag{20a}$$

or, from Eq. (12c),

$$P(x, \infty) = (k/\pi\sigma^2)^{1/2} x^{-1} \exp\{-k[\ln(x/K)]^2/\sigma^2\} \tag{20b}$$

Since $k = r > 0$ in this case, $P(x, \infty)$ is bell shaped with maximal value at

$$x = K \exp(-\sigma^2/2r) = K^* \exp[(m - \sigma^2/2)/r]$$

For $\alpha = 1$, the Verhulst model [Eq. (4.0.3)], the probability density (18) becomes

$$g(z, \infty) = \sigma[\Gamma(2k/\sigma^2)]^{-1}((2k/\sigma^2)e^{\sigma z})^{2k/\sigma^2} \exp(-(2k/\sigma^2)e^{\sigma z}) \tag{21a}$$

or, from Eq. (12c),

$$P(x, \infty) = [\Gamma(2k/\sigma^2)]^{-1} x^{-1} (2kx/K\sigma^2)^{2k/\sigma^2} \exp(-2kx/K\sigma^2) \tag{21b}$$

The steady-state density (16) has the interesting property that if z is small [i.e., from Eq. (12a), the deviations of the population from the saturation value K are small], then $g(z, \infty)$ is independent of α and has the Gaussian form

$$g(z, \infty) = g_0' \exp(-kz^2), \qquad g_0' = \text{constant} \tag{22}$$

and is identical with the steady-state density of the Gompertz model. Equation (21b) was first derived in connection with the population of interacting species by Leigh (1969), who also derived the special form of the FP equation which is appropriate for the Verhulst model.

For later use let us calculate some averages with respect to the steady-state distribution. From Eq. (16),

$$\langle e^{\lambda z}\rangle = g_0 \int_{-\infty}^{\infty} e^{\lambda z} \exp[\theta(\alpha\sigma z - e^{\alpha\sigma z})]\, dz$$

$$= \frac{g_0}{\alpha\sigma}\frac{1}{\theta^{(\theta+\lambda/\alpha\sigma)}} \int_0^{\infty} u^{(\theta+(\lambda/\alpha\sigma)-1)} e^{-u}\, du = \frac{g_0}{\alpha\sigma}\frac{\Gamma(\theta+\lambda/\alpha\sigma)}{\theta^{(\theta+\lambda/\alpha\sigma)}}$$

Using Eq. (17), this equation reduces to

$$\langle e^{\lambda z}\rangle = \frac{\Gamma(\theta+\lambda/\alpha\sigma)}{\theta^{\lambda/\alpha\sigma}\Gamma(\theta)} \tag{23}$$

Therefore,

$$\langle x\rangle = \langle Ke^{\sigma z}\rangle = K\frac{\Gamma(\theta+1/\alpha)}{\theta^{1/\alpha}\Gamma(\theta)} \tag{24a}$$

$$\langle x^2\rangle = \langle K^2 e^{2\sigma z}\rangle = K^2\frac{\Gamma(\theta+2/\alpha)}{\theta^{2/\alpha}\Gamma(\theta)} \tag{24b}$$

$$\langle z\rangle = \frac{\partial}{\partial\lambda}\langle e^{\lambda z}\rangle\bigg|_{\lambda=0} = \{\phi(\theta)-\ln\theta\}/\alpha\sigma \tag{24c}$$

$$\langle z^2\rangle = \frac{\partial^2}{\partial\lambda^2}\langle e^{\lambda z}\rangle\bigg|_{\lambda=0} = \phi'(\theta)/(\alpha\sigma)^2 + \langle z\rangle^2 \tag{24d}$$

where ϕ is the digamma function and ϕ' is the trigamma function (Abramowitz and Stegun, 1964). For the Gompertz model these averages become

$$\langle e^{\lambda z}\rangle = \exp(\lambda^2/4k) \tag{25a}$$

$$\langle x\rangle = K\exp(\sigma^2/4k) = K^*\exp[(\sigma^2+4m)/4r] \tag{25b}$$

$$\langle x^2\rangle = K^2\exp(\sigma^2/k) = K^{*2}\exp[(\sigma^2+2m)/r] \tag{25c}$$

$$\langle z\rangle = 0 \tag{25d}$$

$$\langle z^2\rangle = 1/2k \tag{25e}$$

We now proceed to solve the FP equation and obtain the time-dependent distribution. To solve the FP equation (11), we consider the eigenvalue equation (3.0.37), which in view of Eqs. (3.0.38), (11), and (12b) is

$$d^2\psi/dz^2 + [E - k\{k(\sigma\alpha)^{-2}(1-e^{\alpha\sigma z})^2 - e^{\alpha\sigma z}\}]\psi = 0 \tag{26a}$$

and which is subject to the boundary conditions (the boundaries are inaccessible ones)

$$\psi(z \to \pm\infty) = 0 \tag{26b}$$

If we introduce new quantities E', z', ξ, and c, defined by

$$kE' = E - k^2(\sigma\alpha)^{-2}, \qquad E = k(E' + k(\sigma\alpha)^{-2}) \tag{27a}$$

$$\xi = \sigma(z - z^*), \qquad \exp(\alpha\sigma z^*) = 1 + (\sigma\alpha)^2/2k \tag{27b}$$

$$c = k(\sigma\alpha)^{-2}[1 + (\sigma\alpha)^2/2k]^2 \tag{27c}$$

then Eqs. (26) become

$$\left(\frac{\sigma^2}{k}\right)\frac{d^2\psi}{d\xi^2} + \{E' - c(e^{2\alpha\xi} - 2e^{\alpha\xi})\}\psi = 0 \tag{28a}$$

$$\psi(\xi \to \pm\infty) = 0 \tag{28b}$$

Equation (28a) is just the Schrödinger equation for a diatomic molecule with a Morse potential (Morse, 1929) when the reduced mass is taken to be $\frac{1}{2}$ and $\hbar^2$ is identified with σ^2/k. Generally ξ is replaced by $-\xi$ in studying diatomic molecules, although mathematically this difference is of no importance. Before we give the solution for the general case, let us analyze the Gompertz case ($\alpha = 0$), which is easier to do. For this case, Eq. (26a) becomes

$$d^2\psi/dz^2 + (E + k - k^2 z^2)\psi = 0 \tag{29}$$

which is the eigenvalue equation for the OU process discussed in Appendix G [see Eq. (G. 4)], with r replaced by k. Therefore the probability density $g(z \mid z_0, t)$ is given by Eq. (G. 10), which can be rewritten, by using the identity (G. 12), in the following form:

$$g(z \mid z_0, t) = \left\{\frac{k}{\pi(1 - e^{-2kt})}\right\}^{1/2} \exp\{-k(z - z_0 e^{-kt})^2/(1 - e^{-2kt})\} \tag{30a}$$

and transforming back to the variable x [Eqs. (12)], we get

$$P(x \mid y, t) = \left\{\frac{K}{\pi\sigma^2 x^2(1 - e^{-2kt})}\right\}^{1/2} \exp\left\{-\frac{k[\ln u]^2}{\sigma^2(1 - e^{-2kt})}\right\},$$

$$u = \frac{x}{K}\left(\frac{K}{y}\right)^{e^{-kt}} \tag{30b}$$

The first few moments of (x/K) are easily found from the probability density (30a), when x is known to have the value y at $t = 0$. The calculation of

$$\langle (x/K)^{2\lambda} \rangle = \langle \exp(2\lambda\sigma z) \rangle \tag{31}$$

with Eq. (30a) as the probability density function, leads to easily carried out Gaussian integrals. One finds that (Goel *et al.*, 1971)

$$\langle (x/K)^{2\lambda} \rangle = \exp\{\lambda[2\sigma z_0 e^{-kt} + \lambda(\sigma^2/k)(1-e^{-2kt})]\} \tag{32}$$

which implies that

$$\begin{aligned} \langle x/K \rangle &= \exp\{\sigma z_0 e^{-kt} + (\sigma^2/4k)(1-e^{-2kt})\} \\ &= (y/K)^{\exp(-kt)} \exp\{(\sigma^2/4k)(1-e^{-2kt})\} \end{aligned}$$

or

$$\langle x/K^* \rangle = (y/K^*)^{\exp(-kt)} \exp\{(\sigma^2/4k)(1-e^{-2kt}) + (m/r)(1-e^{-kt})\} \tag{33}$$

while

$$\langle (x-\langle x \rangle)^2 \rangle / (\langle x \rangle)^2 = -1 + \exp\{(\sigma^2/2k)(1-e^{-2kt})\} \tag{34}$$

Comparing Eqs. (33) and (34) with the corresponding expressions in the absence of noise, derived from the deterministic equation (4.0.6), i.e.,

$$x/K^* = (y/K^*)^{\exp(-rt)} \tag{35}$$

with zero coefficient of variation, we conclude that the average size of the population in the presence of noise is above (below) the population size in the absence of noise for all $t > 0$ when $m > -(\sigma/2)^2$ ($m < -\sigma^2/2$). When $-\sigma^2/2 < m < -(\sigma/2)^2$, $\langle x \rangle$ is above the deterministic value only for $0 < t < r^{-1} \ln[\sigma^2/(4|m| - \sigma^2)]$, and is below it thereafter.

We now come back to the general case ($\alpha > 0$), and discuss the results in three regimes.

a. *Population far from saturation.* When the population is far from saturation, we can let $K \to \infty$, so that in the FP equation (11), $\hat{a}(z) \to k/\alpha\sigma$, and the FP equation for the process is

$$\frac{\partial g}{\partial t} = -\frac{\partial}{\partial z}\left(\frac{k}{\alpha\sigma} g\right) + \frac{1}{2}\frac{\partial^2 g}{\partial z^2} \tag{36}$$

with

$$P(x \mid y, t) = g(z \mid z_0, t)/\sigma x, \qquad z = \sigma^{-1} \ln x \tag{37}$$

Equation (36) is the FP equation for a Wiener process, and from Table 3.4

$$g(z \mid z_0, t) = (2\pi t)^{-1/2} \exp\{-(z - z_0 - kt/\alpha\sigma)^2/2t\} \tag{38}$$

Therefore, for Eq. (37), the probability that x lies between x and $x + dx$ at time t is

$$dx\, P(x \mid y, t) = \frac{\exp\{-(\ln[(x/y)\, e^{-kt/\alpha}])^2/2\sigma^2 t\}}{x(2\pi t\sigma^2)^{1/2}}\, dx, \qquad 0 < x < \infty \tag{39}$$

The first moment and the coefficient of variation of this distribution are

$$\langle x \rangle = y \exp\{(k/\alpha + \sigma^2/2)\,t\} \tag{40a}$$

$$\langle (x - \langle x \rangle)^2 \rangle / \langle x \rangle^2 = -1 + \exp(\sigma^2 t) \tag{40b}$$

b. *Population fluctuating around the average steady-state value.* For this case $x/\langle x \rangle \simeq 1$, where $\langle x \rangle$ is given by Eq. (24a). If in addition we work in the regime $\sigma^2\alpha^2/k \to 0$ (small intensity of noise), i.e., $\theta \to \infty$, then from Eq. (24a), $\langle x \rangle = K$ for all α. In this regime, population fluctuates around its saturation value and $x/K \simeq 1$, i.e., z is very small, so that in the FP equation (11), $\hat{a}(z) \to -kz$, and the FP equation for the process is

$$\frac{\partial g}{\partial t} = \frac{\partial}{\partial z}(kzg) + \frac{1}{2}\frac{\partial^2 g}{\partial z^2} \tag{41a}$$

with

$$P(x \,|\, y, t) = g(z \,|\, z_0, t)/\sigma x, \qquad z = \sigma^{-1} \ln(x/K) \tag{41b}$$

Equation (41a) is the FP equation for an OU process, with the eigenvalue equation

$$d^2\psi/dz^2 + (E + k - k^2 z^2)\psi = 0 \tag{42}$$

the same as for the Gompertz model ($\alpha = 0$). Therefore $g(z \,|\, z_0, t)$ is given by Eq. (30a), with $\langle x/K \rangle$ and coefficient of variation given by Eqs. (33) and (34).

c. *Population in regimes other than a and b.* In these regimes we have to solve Eq. (28a) with the boundary conditions (28b). Since at the boundary $\xi \to -\infty$, $c(e^{2\alpha\xi} - 2e^{\alpha\xi}) \to 0$, from the remarks made in Section 3.0 (p. 42), the eigenvalue problem will involve both discrete and continuous spectra. The discrete region has been worked out in connection with the Schrödinger equation for the Morse potential by Trischka and Salwen (1959). The energy levels for the Morse potential have been found to be

$$E_n' = -c\{1 - (\hbar^2\alpha^2/c)^{1/2}(n + \tfrac{1}{2})\}^2, \qquad 0 \leqslant n \leqslant c^{1/2}/\hbar\alpha - \tfrac{1}{2} \tag{43}$$

In terms of our parameters (which are related to $\hbar$), Eq. (43) is transformed into the equation

$$\tfrac{1}{2}k[E' + k(\sigma\alpha)^{-2}] = nk(1 - n\sigma^2\alpha^2 k^{-1}/2), \qquad n = 0, 1, 2, \ldots, [k/\sigma^2\alpha^2] \tag{44}$$

where $[x]$ is the integral part of the number x, i.e., the largest integer less than or equal to x. Using Eq. (27a), Eq. (44) becomes

$$\tfrac{1}{2}E = nk[1 - n\sigma^2\alpha^2 k^{-1}/2] \tag{45}$$

These values of E characterize the relaxation time of the population size to its steady state. The eigenfunctions are given by Trischka and Salwen (1959) if we rewrite Eq. (28a) as

$$\frac{d^2\psi}{d\xi^2} + \left(\frac{k}{\sigma^2}\right)\{(E'+c) - c(e^{\alpha\xi}-1)^2\}\,\psi = 0 \tag{46}$$

The normalized eigenfunctions are

$$\psi_n(\xi) = M_n u^{1/2(q-2n-1)} e^{-u/2} F_n(u) \tag{47}$$

where

$$u = q\exp\{\alpha(\xi-\xi_0)\}, \qquad q = 2k(\sigma\alpha)^{-2}[1+(\sigma\alpha)^2/2k] \tag{48a}$$

$$\tfrac{1}{2}(q-2n-1) = k(\sigma\alpha)^{-2} - n \tag{48b}$$

$$F_n(u) = \sum_{i=1}^{n}\binom{n}{i}\frac{(-u)^i}{(q-2n)_i}, \qquad \binom{n}{i} = \frac{n!}{i!(n-i)!} \tag{48c}$$

$$M_n^{\ 2} = \frac{1}{n!}\frac{\alpha(q-2n)_n}{\Gamma(q-2n-1)} \tag{48d}$$

with $(a)_n$ defined by

$$(a)_n = \begin{cases} 1, & \text{if } n = 0 \\ a(a+1)\cdots(a+n-1), & \text{if } n > 1 \end{cases} \tag{49}$$

Knowing the eigenfunction and using Eqs. (27b), (3.0.36), and (3.0.41), the contribution of the discrete part of the eigenspectrum to $g(z/z_0, t)$ can be calculated. The contribution of the continuous part of the spectrum has not yet been calculated analytically. For some remarks and approximate form for the case $\alpha = 1$, we refer the reader to Goel *et al.* (1971).

We conclude this section by describing and analyzing a few experimental data which throw some light on the validity of the theory described in this section. The theory gives the time-dependent probability density function and various moments of the size of a population of a species in the presence of other species, provided its growth law in the absence of other species is known. To be able to test the theory rigorously, in particular the validity of the probability density function, one needs an ensemble of systems of interacting populations and the data on the population growth of a species in each system of the ensemble. Such an ensemble is not available in natural systems and thus a rigorous test of the theory is not possible. A less ambitious and rigorous but feasible verification of the theory is provided if we make the following assumptions:

(a) the population distribution is very close to the steady-state distribution, and

(b) the averages over this distribution are equal to the long time averages.

The former assumption is somewhat justified in a natural system because the biological systems have had enough time to reach the steady state, or to be close to it. The latter assumption, known as the ergodic theorem (Goel *et al.*, 1971), is hard to justify rigorously and is commonly made in the analysis of many complex systems. With these assumptions, as we will now show, the theory implies certain relations between certain moments of the size of a species in a population of interacting species, which can be experimentally verified. For the purpose of illustration, let us assume that each of the N species grows according to the Verhulst law of growth [Eq. (4.0.3)]. Therefore from Eqs. (24a) and (24b) for each of any two species of the set N

$$\langle x_i \rangle = K_i, \qquad i = 1,2 \tag{50a}$$

$$\langle x_i^2 \rangle = K_i^2 (1+\theta_i^{-1}), \tag{50b}$$

and from Eqs. (24c), (24d), and (19),

$$\langle v_i \rangle = \{\phi(\theta_i) - \ln \theta_i\}, \qquad v_i \equiv \ln(x_i/K_i) \tag{50c}$$

$$\langle v_i^2 \rangle = \phi'(\theta_i) + \langle v_i \rangle^2 \tag{50d}$$

$$\langle v_1 v_2 \rangle = \langle v_1 \rangle \langle v_2 \rangle \tag{50e}$$

Measuring the sizes of the two species at many different times, we can use Eq. (50a) to calculate K_1 and K_2, and then use Eq. (50c) to calculate θ_1 and θ_2. The remaining three equations (50b), (50d), and (50e) can be used to check whether these relations, predicted by thoery, hold among the various computed experimental averages. Thus, Eqs. (50) can be used to calculate v_i^2 for both species, and these results can be checked against the corresponding experimental averages. Further verification is provided by the relation (50e) and the relations

$$\langle (x_1 - K_1)^2 \rangle = \langle x_1 \rangle^2 / \theta_1 \tag{51a}$$

$$\langle (x_2 - K_2)^2 \rangle = \langle x_2 \rangle^2 / \theta_2 \tag{51b}$$

which follows from Eqs. (50a) and (50b).

Unfortunately the experimental data on the long-range fluctuations of two species in the same system of interacting species are not extensive. The only data which have been extensively studied and which are surrounded by controversy regarding the causes of the oscillations are on Canadian lynx and snowshoe hare, obtained by Hudson's Bay Company for the years 1847–1903. In Fig. 6.4 changes in the abundance of the two species are given. These two species lived in a wild environment and were subjected to random changes in the environment. By taking the time averages of the data [as read from the

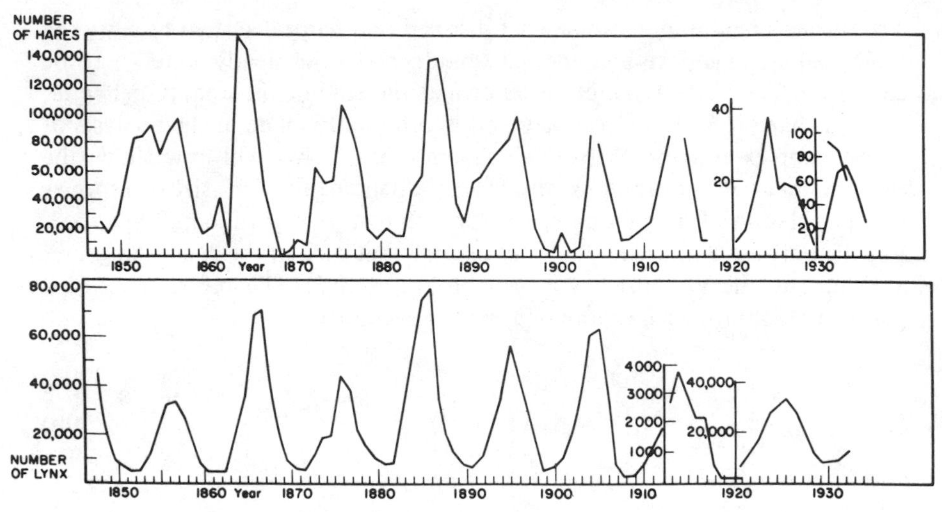

Fig. 6.4 Abundance of varying hare and lynx in winter, plotted over the end of the year in which the winter began, i.e., after the season of biological reproduction [modified from MacLulich (1937)].

graphs by Leigh (1969)], we find that (index 1 is for lynx and 2 is for hare)

$$\langle x_1 \rangle = K_1 = 4.89 \times 10^4, \qquad \langle x_2 \rangle = K_2 = 2.34 \times 10^4$$

$$\langle v_1 \rangle = -0.201, \qquad \langle v_2 \rangle = -0.121, \qquad \langle v_1 v_2 \rangle = 0.11$$

$$\langle (x_1 - K_1)^2 \rangle = 1.54 \times 10^9, \qquad \langle (x_2 - K_2)^2 \rangle = 3.6 \times 10^8$$

$$\theta_1 = 2.64, \qquad \theta_2 = 4.3$$

From the "experimental averages" given above the value of the right-hand side of Eq. (51a) is 0.9×10^9 as compared to the value 1.54×10^9 for the left-hand side. The corresponding numbers for Eq. (51b) are 1.27×10^8 and 3.61×10^8, respectively. Further, by using the experimental values for θ_1, θ_2, $\langle v_1 \rangle$, and $\langle v_2 \rangle$, the averages $\langle {v_1}^2 \rangle$ and $\langle {v_2}^2 \rangle$ can be calculated from Eq. (50d). These values turn to be 0.50 and 0.28, respectively, as compared to their experimental values of 0.24 and 0.16. Also, from Eq. (50e) the calculated value of $\langle v_1 v_2 \rangle$ is 0.0243, as compared to the experimental value of 0.11.

The agreement between theory and experimental data is not very encouraging. The disagreement may be due to one or more of many factors, e.g., inaccurate number of catches (Hudson's Bay Company caught only what they could), different causes of oscillations (mainly due to the prey–predator type of interactions between hare and lynx, hare being the primary food of lynx, rather than the interactions with other species), a wrongly guessed form of the saturation effect, and so on. Our main purpose in presenting and analyzing these data was to illustrate a method by which the theory can be tested.

References

Abramowitz, M., and Stegun, I. A., eds. (1964). "Handbook of Mathematical Functions." Nat. Bur. Stand., Washington, D.C.

D'Ancona, V. (1954). "The Struggle for Existence." Brill, Leiden, Netherlands.

Goel, N. S., Maitra, S. C., and Montroll, E. W. (1971). On the Volterra and other nonlinear models of interacting populations. *Rev. Modern Phys.* **43**, 231; "On the Volterra and Other Nonlinear Models of Interacting Populations." Academic Press, New York.

Gradshteyn, I. S., and Ryzhik, I. M. (1965). "Tables of Integrals, Series and Products." Academic Press, New York.

Leigh, E. G., Jr. (1969). The ecological role of Volterra's equations, *in* "Some Mathematical Problems in Biology" (M. Gerstenhaber, ed.). Amer. Math. Soc., Providence, Rhode Island.

Lotka, A. J. (1910). Contribution to the theory of periodic reactions. *J. Phys. Chem.* **14**, 271.

Lotka, A. J. (1920). Analytical note on certain rhythmic relations in organic systems. *Proc. Nat. Acad. Sci. U.S.A.* **6**, 410.

MacArthur, R. H., and Wilson, E. O. (1963). An equilibrium theory of insular zoogeography. *Evolution* **17**, 373.

MacArthur, R. H., and Wilson, E. O. (1967). "The Theory of Island Biogeography." Princeton Univ. Press, Princeton, New Jersey.

MacLulich, D. A. (1937). "Fluctuations in the Numbers of the Varying Hare." Univ. of Toronto Press, Toronto.

Morse, P. M. (1929). Diatomic molecules according to the wave mechanics. II. Vibrational levels. *Phys. Rev.* **34**, 57.

Trischka, J., and Salwen, H. (1959). Dipole moment function of diatomic molecules. *J. Chem. Phys.* **31**, 218.

Volterra, V. (1928). Variations and fluctuations of the number of individuals in animal species living together. *J. Cons. Perm. Int. Explor. Mer* **3**, 1; translated in R. N. Chapman "Animal Ecology." McGraw-Hill, New York.

Volterra, V. (1931). "Lecon sur la Theorie Mathematique de la Lutte pour le Vie." Gauthier-Villars, Paris.

Volterra, V. (1937). Principes de biologie mathematique. *Acta Biotheoret.* **3**, 1.

Vuilleumier, F. (1970). Insular biogeography in continental regions. I. The northern Andes of South America. *Amer. Natur.* **109**, 373.

7

Population Genetics

In the preceding three chapters, we modeled the dynamics of a population with interactions either within the population (Chapter 4) or with populations of other species (Chapters 5 and 6). The individuals constituting a population were regarded as identical, and their number characterized the state of the population. To investigate the population more precisely, one has to consider the structure of the population at a somewhat microscopic scale by taking into account the differences among the individuals of the same species. The two main structural features of the population are its age distribution and genetic consitution. In this chapter we attempt to study the genetic consitution and changes in it as the population evolves, by using the mathematical analysis of stochastic processes presented in this monograph. The study of this subject started with Mendel and has since been carried out very extensively. There exists a large amount of literature on the subject and, in this book, we can only discuss and review the most important factors underlying genetic processes and demonstrate how the general theory of Chapter 3 leads to a quantitative understanding of these processes. A list of references is given at the end of the chapter where a more detailed treatment of this subject can be found. We will assume that the reader is familiar with the basic Mendelian genetics. If he is not, he can easily familiarize himself by reading an elementary monograph on gentics, e.g., that by Barish (1965).

In a single species, the number of different possible genotypes† far exceeds the number of individuals, since the individual genotype is determined by a large number of genes, many of which have more than one possible form (allele). Therefore, it is impossible to study the number of individuals in each genotype group. Instead, the investigation is carried out by considering the number of genes of the same allele in the population.

The "population" of genes of the same allele is well characterized by the number of genes, since all such genes are identical and reproduce their own kind except rarely when mutation occurs. In our simplified treatment we consider one particular locus‡ for which there are two possible alleles in the population. The relative proportion of an allele in the population changes due to two types of factors. The first type consists of factors that have a directional effect on the population and are assumed to occur at a constant rate. The three main factors in this category, which were termed "systematic evolutionary pressures" by Wright (1949), are

1. mutation—errors in the reproduction process in which the newly formed allele is different from the original allele;
2. migration—exchange of individuals with other populations of different genetic structure; and
3. selection—differences in the numbers of progeny among the various genotypes, due to differences in survival, mating, and fertility.

The second type of factors are the stochastic factors:

1. random fluctuations in systematic evolutionary pressures, and especially in selection intensities, due to changes in environmental conditions; and
2. random sampling of gametes§ in a finite population.

The finiteness of the population introduces a stochastic element into the process when random mating of gametes is assumed (i.e., each new gene has the same probability to be a descendant of any one of the genes in the previous generation), although the average proportions of genes remain constant. To see this we consider a population of N genes, j of which are of type A and $N-j$ of type a. The probability P that a randomly chosen allele is of type A

† Genotype: the genetic makeup of an individual determined by the totality of genetic factors.

‡ Locus: the location of a gene on a chromosome.

§ Gamete: the germ cell carrying one set of chromosomes from the parent to the offspring.

is j/N, and therefore the probability of having i alleles of type A in the next generation is given by

$$\text{prob}[n_A = i] = \binom{N}{i}\left(\frac{j}{N}\right)^i\left(1-\frac{j}{N}\right)^{N-i} \tag{1}$$

Let us take the unit of time as one generation, i.e., measure the time in the number of generations. Let $q_A(t)$ be the proportion of allele A in the tth generation. The average proportion $\langle q_A(t+1)\rangle$ of allele A in the $(t+1)$th generation, when it is given that $q_A(t) = j/N$ in the tth generation, is

$$\langle q_A(t+1)\rangle = \sum_{i=1}^{N} i\binom{N}{i}\left(\frac{j}{N}\right)^i (1-j/N)^{N-i} = j/N = q_A(t) \qquad (2)^\dagger$$

which is the given proportion in the tth generation. Thus on the average there is no change in the gene proportions due to random mating of gametes in a finite population, though the process is described by a random variable $q_A(t)$.

The numbers n_A, n_a of the alleles A and a in the population become stochastic variables when either one or both of the stochastic factors described above are taken into account. The resulting process is a bivariate Markov process (Feller, 1951) which is beyond the treatment of this monograph. Only under the additional assumption of constant population size ($dN/dt = 0$) does the process reduce to a univariate Markov process. This assumption is quite reasonable if changes in genetic constitution are investigated in a population that has already reached its ecological steady state. This assumption has been made extensively in the literature (Fisher, 1930; Wright, 1942; Kimura, 1964) and is also made in the following.

To obtain the stochastic equations which characterize the changes in gene proportions in a finite population, we first formulate the deterministic equations describing the systematic evolutionary pressures, and then use the method of Section 3.0 to get the corresponding stochastic diffusion equations by assuming certain properties of the stochastic factors.

7.1 Genetic Changes under Systematic Evolutionary Pressures—Deterministic Equations

In this section we derive the deterministic equations for the changes in gene proportions from generation to generation due to various systematic evolutionary processes which are assumed to occur at a constant rate. We start with changes due to mutations.

† Equation (2) implies that the set $\{q_A(t)\}_{t=1}^{\infty}$ for the process under consideration is a martingale, i.e., $\langle q_A(t)\rangle = q_A(0)$ for all t. As $t \to \infty$, q_A becomes either 1 or 0 so that $q_A(0) = 0 \cdot \text{prob}[q_A(\infty) = 0] + 1 \cdot \text{prob}[q_A(\infty) = 1]$; i.e., the probability of allele A getting fixed is its initial proportion.

1. *Mutations.* Let u_A and u_a be the mutation rates per generation of allele A to and from its allele a, respectively, and $q(t)$ the proportion of allele A in the population. The change in this proportion due to mutations is described by the equation

$$q(t+1) - q(t) = -u_A q(t) + [1-q(t)]\, u_a = -(u_A+u_a)\, q(t) + u_a \quad (1)$$

since by mutations $u_A q(t) N$ A genes are lost and $u_a(1-q)N$ A genes are gained where N is the total number of genes.

2. *Migration.* Let the population exchange a certain fraction r of its individuals per generation with another larger population, characterized by the proportion q' of allele A. The changes in the proportion q of allele A are described by the equation

$$q(t+1) - q(t) = r(q'-q) \quad (2)$$

since from one generation to the next $rq'N$ A genes are added to the population while rqN A genes are lost.

Comparing Eq. (2) with Eq. (1), we note that both are of the same form except that rq' and r in Eq. (2) are replaced, respectively, by u_a and u_A+u_a in Eq. (1). Therefore, by changing u_A into $r(1-q')$ and u_a into rq', we can transform any solution of Eq. (1) into a solution of Eq. (2).

3. *Genotype selection.* The pressure of selection operates through the different properties of the genotypes, which are of three different types with respect to the locus under consideration: AA, Aa, and aa. In the simplified approach we take, we neglect the cooperative effects of several loci on the ability of the genotype to survive and reproduce. We consider the differences in selection due to only that one locus whose allele proportions are studied. Assuming random mating of gametes as the mating system in the population, the proportions of the genotypes AA, Aa, and aa determined by the gene proportions are q^2, $2qp$, and p^2, respectively, where $p = 1-q$.

To introduce selection we introduce three quantities, w_{AA}, w_{Aa}, and w_{aa}: $2w_{AA}$ is the number of individuals contributed by one parent of type AA to the next generation, with a similar meaning for $2w_{Aa}$ and $2w_{aa}$ with respect to parents of types Aa and aa. Since each progeny is a descendant of two parents, each of w_{AA}, w_{Aa}, and w_{aa} is a growth rate (birth rates minus death rates) per generation of the corresponding genotype. For a population consisting of N genes, n_A of which are of type A, the equation describing the number of A genes in the next generation is

$$n_A(t+1) = [q(t)]^2 \frac{N}{2} 2w_{AA} + [q(t)][1-q(t)] \frac{N}{2} \cdot 2w_{Aa} \quad (3)$$

where, in writing this equation, we have incorporated the facts that the total population size is $N/2$ (in diploids†), and that each parent contributes to the next generation one gene per one progeny. An equation similar to Eq. (3) describes the change in n_a. Adding this equation to Eq. (3), we get the equation describing the change in the overall number of genes,

$$N(t+1) = n_A(t+1) + n_a(t+1)$$
$$= [q(t)]^2 N w_{AA} + 2[q(t)][1-q(t)]\, N w_{Aa} + [1-q(t)]^2 N w_{aa} \tag{4}$$

Since by definition

$$q(t+1) = n_A(t+1)/N(t+1)$$

from Eqs. (3) and (4) we get

$$q(t+1) = \frac{sq(t)\{q(t)+h[1-q(t)]\}+w_{aa}\,q(t)}{sq(t)\{q(t)+2h[1-q(t)]\}+w_{aa}} \tag{5a}$$

where

$$s \equiv w_{AA} - w_{aa}, \qquad sh = w_{Aa} - w_{aa} \tag{5b}$$

are the selective coefficients. With the average fitness $\bar{w}(t)$ at the tth generation given by

$$\bar{w}(t) = [q(t)]^2 w_{AA} + 2q(t)[1-q(t)]\, w_{Aa} + [1-q(t)]^2 w_{aa}$$
$$= w_{aa} + sq(t)\{q(t) + 2h[1-q(t)]\} \tag{6}$$

Eq. (5a) can be written as

$$q(t+1) - q(t) = sq(t)[1-q(t)][h+q(t)(1-2h)]/\bar{w}(t) \equiv s\phi[q(t)] \tag{7}$$

which is the equation describing the change in the proportion of gene A due to selection. From this equation we can write particular equations for different types of selection.

(a) *No dominance.* This type of selection occurs when the properties of the heterozygote *Aa* are the average of the properties of the homozygotes *AA* and *aa*, i.e.,

$$w_{Aa} = \tfrac{1}{2}(w_{AA}+w_{aa}) \qquad \text{or} \qquad h = \tfrac{1}{2} \tag{8}$$

In this case, Eq. (7) becomes

$$q(t+1) - q(t) = \frac{sq(t)[1-q(t)]}{w_{aa}+sq(t)} \tag{9}$$

† Diploid: an individual carrying two sets of chromosomes (in haploids and in gametes a cell contains only one set of chromosomes).

(b) *Dominance.* This type of selection occurs when the properties of heterozygote *Aa* are the same as the properties of the homozygote *AA*, i.e.,

$$w_{Aa} = w_{AA} \qquad \text{or} \qquad h = 1 \tag{10}$$

so that Eq. (7) becomes

$$q(t+1) - q(t) = sq(t)[1-q(t)]^2/\bar{w}(t) \tag{11}$$

Equation (11) with $s > 0$ describes the case of dominant† homozygote favored by selection, and with $s < 0$ the case of recessive‡ homozygote favored by selection.

(c) *Overdominance.* In this case the heterozygote *Aa* has some selective advantages over the two homozygotes *AA* and *aa*. From the defining equations (5b), this case is therefore characterized by $h > 1$ if $s > 0$ and $h < 0$ if $s < 0$. A more convenient set of selective coefficients is defined by

$$\sigma_1 = w_{Aa} - w_{AA} = s(h-1) > 0 \tag{12a}$$

$$\sigma_2 = w_{Aa} - w_{aa} = sh > 0 \tag{12b}$$

so that σ_1 and σ_2 are the selective coefficients of heterozygote *Aa* relative to the homozygotes *AA* and *aa*, respectively. With these new selective coefficients, Eq. (7) takes the form

$$q(t+1) - q(t) = q(t)[1-q(t)][\sigma_2 - q(t)(\sigma_1+\sigma_2)]/\bar{w}(t) \tag{13}$$

It may be noted that overdominance explains the existence of genic variability in natural populations in the absence of mutation or migration, since in the deterministic approach selection of type (a) or (b) results in the fixation of the advantageous gene, while in case of overdominance the two alleles *A* and *a* coexist with steady-state proportions $\sigma_2(\sigma_1+\sigma_2)^{-1}$ and $\sigma_1(\sigma_1+\sigma_2)^{-1}$, respectively.

4. *Genic selection.* When selection takes place at the gametic level, or in populations of haploids,§ one has to consider the selective properties of two disjoint populations, one with allele *A* and the other with allele *a*, with the growth rates w_A and w_a, respectively. The equations analogous to Eqs. (3) and (4) are

$$n_A(t+1) = n_A(t)\,w_A \tag{14}$$

$$N(t+1) = n_A(t)\,w_A + n_a(t)\,w_a = w_a N(t) + n_A(t)(w_A - w_a) \tag{15}$$

† Dominant: a factor that determines the property of the heterozygote.

‡ Recessive: a factor that fails to determine the property of the heterozygote.

§ See footnote on page 148.

so that

$$q(t+1) - q(t) = \frac{n_A(t+1)}{N(t+1)} - q(t) = s\frac{q(t)[1-q(t)]}{\bar{w}(t)} \tag{16}$$

where

$$s = w_A - w_a, \qquad \bar{w} = (1-q)\,w_a + qw_A = w_a + sq \tag{17}$$

Each of the deterministic equations derived above takes into account one of the various systematic evolutionary pressures. When two or more pressures operate, they occur at different stages (migration occurs among adults, mutation at the stage of reproduction, and selection at both stages). When the effects of the different pressures can be approximated by a sequence of effects, then $q(t+1)$ can be derived by a number of steps similar to those given by Eqs. (1), (2), (7), and (16). For example, when mutation and genotype selection operate and the latter is due mainly to differences in viability, then if $q(t)$ stands for the A gene proportion among gametes of generation t, the proportion $\tilde{q}$ among the adults of generation t is given by Eq. (7), i.e., $\tilde{q} = q(t)+s\phi[q(t)]$, and $q(t+1)$ is therefore given by Eq. (1) with $q(t)$ replaced by $\tilde{q}$. Thus

$$q(t+1) = q(t) + u_a - (u_A+u_a)\,q(t) + s\phi[q(t)] - (u_a+u_A)\,s\phi[q(t)] \tag{18}$$

If the mutation rates and the selective coefficient are small, so that nonlinear terms in these quantities can be neglected, then the right-hand side of Eq. (18) is the sum of the right-hand sides of Eqs. (1) and (7), and the effects of the two types of pressures are additive. This last statement is true for any two pressures which operate very slowly, i.e., cause small changes per generation.

Having discussed the deterministic equations for the changes in gene proportion under various evolutionary pressures, we introduce and discuss in the next section the stochastic factors due to random sampling of gametes in a finite population.

7.2 Genetic Changes in a Finite Population—Stochastic Equations

In this section we take into account the finiteness of the population and the stochastic elements which are introduced by it to the process of genetic changes. When the change in gene proportions is due to random sampling of gametes, the proportion of A genes, $q(t+1)$, in the $(t+1)$th generation, when the corresponding quantity $q(t)$ for the tth generation equals j, is a random variable with the binomial distribution given in Eq. (7.0.1). From this distribution, the first few conditional moments of $\Delta q \equiv q(t+1)-q(t)$, given that

$q(t) = x$, can easily be found to be

$$\langle \Delta q \rangle = 0 \tag{1a}$$

$$\langle (\Delta q)^2 \rangle = N^{-1}x(1-x) \tag{1b}$$

$$\langle (\Delta q)^3 \rangle = N^{-2}x(1-x)(1-2x) \tag{1c}$$

$$\langle (\Delta q)^4 \rangle = N^{-2}3x^2(1-x)^2 + N^{-3}x(1-x)(1-6x(1-x)) \tag{1d}$$

with higher moments of the form

$$\langle (\Delta q)^n \rangle = N^{-2}x(1-x)\sum_{i=0}^{n-2} N^{-i}v_{n,i}(x) = O(N^{-1}), \qquad n \geqslant 3 \tag{1e}$$

where $v_{n,i}(x)$ is a polynomial in x of degree $n-2$.

In natural populations, the size of the population is finite but can be quite large, and the changes in the gene proportions occur slowly and are smaller in size compared to the size of the population. Therefore $q(t)$ can be approximated by $x(t)$ a continuous function of time and, as was noted in the beginning of this chapter, $x(t)$ becomes a univariate Markov process in case the population is of constant size. In this continuous approximation, according to Eqs. (1), the moments defined by Eqs. (3.0.13) are

$$M_1(x) = 0 \tag{2a}$$

$$M_2(x) = N^{-1}x(1-x) \tag{2b}$$

$$M_n(x) = O(N^{-1}), \qquad n \geqslant 3 \tag{2c}$$

where we may recall that we are measuring time in units of generations. From Eqs. (2), up to first order in N^{-1}, all the moments for $n \geqslant 3$ vanish, and the differential equations satisfied by the probability density of this Markov process are the FP equation (3.0.2) and the backward equation (3.0.3), with $a(x) = M_1(x)$ and $b(x) = M_2(x)$ where M_1 and M_2 are given by Eqs. (2a) and (2b), respectively, i.e.,

$$\frac{\partial P(x|y,t)}{\partial t} = \frac{1}{2}\frac{\partial^2}{\partial x^2}\left[\frac{x(1-x)}{N}P(x|y,t)\right] \tag{3}$$

$$\frac{\partial P(x|y,t)}{\partial t} = \frac{1}{2}\frac{y(1-y)}{N}\frac{\partial^2 P(x|y,t)}{\partial y^2} \tag{4}$$

These equations describe the evolution of allele A in the absence of systematic evolutionary pressures. The effects of these evolutionary pressures can be incorporated using the deterministic equations of the preceding section, which were derived with the underlying assumption that the pressures occur at a constant rate. We now derive the corresponding stochastic equations, for a finite population.

1. *Mutation.* In this process we assume that each gamete has the same probability of being a descendant of any one of the parents of the previous generation, and that in each generation a fraction u_A of the A genes mutate to gene a and a fraction u_a of a genes mutate back to gene A. By the first assumption, if no mutation occurs, the probability of having i A genes in the next generation is

$$\binom{N}{i} x^i (1-x)^{N-i} \tag{5}$$

where x is the proportion of A genes in the present generation. In the presence of mutation at constant rate, this probability corresponds to the event that in the next generation the number of A genes is

$$n_A(t+1) = i + u_a(N-i) - u_A i$$

i.e.,

$$\operatorname{prob}[n_A(t+1) = i(1-u_A) + (N-i)u_a \,|\, q(t) = x] = \binom{N}{i} x^i (1-x)^{N-i} \tag{6}$$

Since by definition $q(t+1) \equiv n_A(t+1)/N$, the first two moments of $\Delta q = q(t+1) - x$ are given by

$$\langle \Delta q \rangle = \sum_{i=0}^{N} \left[\frac{i}{N}(1-u_A-u_a) + u_a - x \right] \binom{N}{i} x^i (1-x)^{N-i}$$

$$= u_a - x(u_A+u_a) \tag{7}$$

$$\langle (\Delta q)^2 \rangle = N^{-1} x(1-x) + u_a{}^2 + (u_A+u_a)$$
$$\times \left[(u_A+u_a)(x^2 + N^{-1}x(1-x)) - 2u_a x - 2N^{-1}x(1-x) \right] \tag{8}$$

Comparing Eq. (7) with (7.1.1), we conclude that on the average the changes in gene proportions are identical with the deterministic changes. Since, as was noted before, these changes occur very slowly, we assume u_A and u_a to be of the order of $1/N$, i.e.,

$$u_A = u_1/N + O(1/N), \qquad u_a = u_2/N + O(1/N) \tag{9}$$

With this assumption, the first two moments defined by Eqs. (7) and (8), up to first order in N^{-1}, are given by

$$\langle \Delta q \rangle = \langle q(t+1) - x \rangle = N^{-1}[u_2 - x(u_1+u_2)] + O(N^{-1}) \tag{10a}$$

$$\langle (\Delta q)^2 \rangle = \langle [q_A(t+1) - x]^2 \rangle = N^{-1} x(1-x) + O(N^{-1}) \tag{10b}$$

It can be easily shown that all the higher moments are of the order of N^{-2}.

Therefore, the probability density describing the process of random mating of gametes in a finite population under constant mutation rates, up to first order in N^{-1}, satisfies the equations

$$\frac{\partial P(x \mid y,t)}{\partial t} = (N)^{-1} \frac{\partial}{\partial x} \{[(u_1+u_2)x - u_2]\, P(x \mid y,t)\} + (2N)^{-1} \frac{\partial^2}{\partial x^2} [x(1-x)\, P(x \mid y,t)] \tag{11}$$

$$\frac{\partial P(x \mid y,t)}{\partial t} = (N)^{-1} [u_2 - (u_1+u_2)y] \frac{\partial}{\partial y} P(x \mid y,t) + (2N)^{-1} y(1-y) \frac{\partial^2}{\partial y^2} P(x \mid y,t) \tag{12}$$

2. *Migration.* As was noted in Section 7.1, the deterministic equations describing the genetic changes in the presence of migration are the same as those in the presence of mutation except that u_a is replaced by rq' and u_A by $r(1-q')$. Therefore, from Eqs. (11) and (12), the probability density in the presence of migration satisfies the diffusion equations (3.0.2) and (3.0.3), up to first order in N^{-1}, with

$$a(x) = N^{-1} r_1 (q'-x), \qquad b(x) = N^{-1} x(1-x) \tag{13a}$$

where

$$r = r_1 N^{-1} + O(N^{-1}) \tag{13b}$$

3. *Selection.* To incorporate selection pressures into the model of finite population undergoing random sampling of gametes, we divide the transition from generation to generation into two stages. In the first stage gametes are produced with each gamete having equal probability of being a descendant of any of the parents of the previous generation, and therefore the gene proportions are distributed according to the binomial law. In the second stage, the selection pressure operates with constant rates according to the deterministic equations (7.1.7) or (7.1.16). Thus, the probability distribution of the gene proportion in the next generation is given by

$$\text{prob}\left[q(t+1) = \frac{i}{N} + \frac{s}{\overline{w}(t)} \frac{i}{N}\left(1 - \frac{i}{N}\right)\left(\frac{i}{N}(1-2h) + h\right) \middle| q(t) = x\right] = \binom{N}{i} x^i (1-x)^{N-i} \tag{14}$$

for the genotype selection, while for the genic selection

$$\text{prob}\left[q(t+1) = \frac{i}{N} + \frac{s}{\overline{w}(t)} \frac{i}{N}\left(1 - \frac{i}{N}\right) \middle| q(t) = x\right] = \binom{N}{i} x^i (1-x)^{N-i} \tag{15}$$

where $\bar{w}(t)$ is defined by Eqs. (7.1.6) and (7.1.17), respectively, with $q(t)$ replaced by i/N. Let us assume small but not negligible selective differences (to maintain slow changes) and take

$$s = s_1/N + O(N^{-1}), \qquad h \ll N, \qquad |s_1| \gg N^{-1} \tag{16}$$

With this assumption, according to Eq. (14), the first two conditional moments of $\Delta q = q(t+1) - x$ are

$$\langle \Delta q \rangle = \langle q(t+1) - x \rangle = N^{-1} s_1 x(1-x)[x(1-2h) + h] + O(N^{-1}) \tag{17}$$

$$\langle (\Delta q)^2 \rangle = \langle [q(t+1) - x]^2 \rangle = N^{-1} x(1-x) + O(N^{-1}) \tag{18}$$

with all the higher moments of the order of N^{-2}. Similarly, from Eq. (15), we get

$$\langle q(t+1) - x \rangle = N^{-1} s_1 x(1-x) + O(1/N) \tag{19}$$

with second moment given by Eq. (18) and all higher moments of the order of N^{-2}.

Therefore, up to order N^{-1}, the coefficients of the diffusion equations, satisfied by the probability density in case of a genotype selection, are

$$a(x) = N^{-1} s_1 x(1-x)[x(1-2h) + h], \qquad b(x) = N^{-1} x(1-x) \tag{20}$$

while, in case of genic selection,

$$a(x) = N^{-1} s_1 x(1-x), \qquad b(x) = N^{-1} x(1-x) \tag{21}$$

This concludes the derivation of the diffusion equations describing the genetic changes in a finite population when evolutionary pressures occur at constant rates. Similar diffusion equations are found when the evolutionary pressures occur *randomly* with constant low probabilities.

To illustrate this observation we consider in detail the mutation pressure, and assume that in any replication of an A gene there is a probability U_A of mutation into an a gene, and similarly, with probability U_a an a gene replicates into an A gene. Therefore, if at generation t a fraction q of the genes is of type A, at generation $t+1$ the probability of a randomly chosen gene to be of type A is

$$\tilde{q} = q(1 - U_A) + (1-q) U_a \tag{22}$$

and the probability of having i genes of type A becomes

$$\binom{N}{i} (\tilde{q})^i (1-\tilde{q})^{N-i} \tag{23}$$

which is again the binomial distribution with average $\tilde{q}$ and variance

$N^{-1}\tilde{q}(1-\tilde{q})$. Thus

$$\langle q(t+1)-q(t)\,|\,q(t)=q\rangle = \tilde{q}-q = -U_A q + U_a(1-q) \tag{24}$$

$$\begin{aligned}\langle [q(t+1)-q(t)]^2\,|\,q(t)=q\rangle &= \langle (q(t+1)-\tilde{q})^2 \\ &\quad + 2(q(t+1)-\tilde{q})(\tilde{q}-q)+(\tilde{q}-q)^2\rangle \\ &= N^{-1}\tilde{q}(1-\tilde{q}) + [U_a-(U_a+U_A)q]^2\end{aligned} \tag{25}$$

with higher moments of $q(t+1)-\tilde{q}$ of the order of N^{-2}. Now if we assume that

$$U_A = U_1 N^{-1} + O(N^{-1}), \qquad U_a = U_2 N^{-1} + O(N^{-1}) \tag{26}$$

then in the continuous approximation, up to terms of order N^{-1}, the coefficients of the FP equations for the probability density of the process are

$$a(x) = N^{-1}[U_2-(U_2+U_1)x] \tag{27a}$$

$$b(x) = N^{-1}x(1-x) \tag{27b}$$

which have the same form as $a(x)$ and $b(x)$ given by Eq. (10). The case of migration can be handled similarly by introducing R, the probability that an individual in the population is replaced by migration before reproduction. Then $\tilde{q}$ of Eq. (22) is to be replaced by

$$\hat{q} = q + R(q'-q), \qquad R = R_1 N^{-1} + O(N^{-1}) \tag{28}$$

where q' is the A gene proportion in the immigrating population.

In case of genotype selection we assume that the probabilities for a given gamete to be a descendant of a parent of type AA, Aa, aa are in the ratio $W_1 : W_2 : 1$ due to differences in fertilities and viabilities, where $W_1 = 1+S_1N^{-1}+O(N^{-1})$ and $W_2 = 1+HS_1N^{-1}+O(N^{-1})$. In this case $\tilde{q}$ of Eq. (22) is to be replaced by

$$q^* = [q^2W_1 + q(1-q)W_2]/[q^2W_1 + 2q(1-q)W_2 + (1-q)^2] \tag{29}$$

and analysis similar to that carried out above will yield $a(x)$ and $b(x)$ of the same form as in Eq. (20) but with s_1 and h replaced by S_1 and H, respectively. The case of genic selection is handled by assuming that the probability of a newly formed gamete to be a descendant of an A gene is W times its probability of being a descendant of an a gene, with $W = 1+S_1N^{-1}+O(N^{-1})$. $\tilde{q}$ of Eq. (22) is then replaced by

$$\bar{q} = qW[qW+(1-q)]^{-1} \tag{30}$$

and the coefficients $a(x)$ and $b(x)$ turn out to be of the form given in Eq. (21), but with s_1 replaced by S_1.

Another approach by which the randomness of the evolutionary pressures can be incorporated, regards the rates of the various pressures as random variables which change in time according to some distribution. These rates are then related to the gene proportions by the deterministic equations of Section 7.1, from which the first two moments of $q(t+1)-q(t)$ are derived and substituted as $a(x)$ and $b(x)$ in the FP equation characterizing the process. In this approach the population is assumed to be infinite and the random elements are introduced only through the randomness in the rates of the pressures. Analysis of this approach in the case of random changes in the intensity of genic selection can be found in the work of Crow and Kimura

In the following we will investigate genetic changes in a finite population in the presence of pressures that occur at constant rates, but the same results, with a different interpretation of the pressures' rates, describe also a finite population in the presence of pressures that occur randomly with constant low probabilities.

It should be noted that in all the diffusion equations derived above $b(x) = N^{-1}x(1-x)$ and $a(x)$ is the same as the deterministic change per generation. Further, the variables x and y are confined to the interval (0, 1) by the singular boundaries 0, 1 (see the discussion in Section 3.1). Thus no boundary condition can be imposed on $P(x\,|\,y,t)$.

In the next section we analyze the diffusion equations given above by using the methods of Chapter 3. In particular we address ourselves to the following questions.

(a) What is the ultimate fate of the population, i.e., the steady-state probability density $P(x\,|\,y,\infty)$?

(b) At what rate does the process approach the steady state?

(c) What is the probability of fixation of a mutant gene with or without selective advantage?

(d) What is the average time for the fixation of a gene?

We start with the simplest diffusion equations (3) and (4) which correspond to genetic changes in a randomly mating finite population in the absence of any systematic evolutionary pressure.

7.3 Genetic Changes in a Finite Population with Random Mating and No Systematic Evolutionary Pressures

The equations characterizing the genetic changes for this case are (7.2.3) and (7.2.4). Since $b(x)$ vanishes at $x=0$ and $x=1$, these boundaries are singular, and their properties are determined by the functions $h_1(x)$ and $h_2(x)$ defined by Eqs. (3.1.1). For $a(x)=0$, $b(x)=N^{-1}x(1-x)$, $\pi(x)=1$, and

$$h_1(x) = N \ln x - N \ln(1-x) + \text{constant}$$
$$h_2(x) = N(1-x)^{-1} + Nx_0[x(1-x)]^{-1}$$

Thus $h_1(x)$ is integrable near both boundaries while $h_2(x)$ is not. According to Table 3.3 both boundaries are exit boundaries (acting as absorbing boundaries), in agreement with the biological situation discussed in this section, where one of the alleles can become fixed in the population while the other is lost, and there is no mechanism (mutation or immigration) by which the lost allele can be introduced back.

The solution of Eq. (7.2.3), given in Table 3.4, is

$$\begin{aligned} P(x\,|\,y,t) &= \sum_{i=1}^{\infty} y(1-y)\,i(i+1)(2i+1)\,F(1-i,i+2;2;y)\,F(1-i,i+2;2;x) \\ &\quad \times \exp\{-[i(i+1)/2N]\,t\} \\ &= \sum_{i=1}^{\infty} \frac{(2i+1)(1-\rho^2)}{i(i+1)} T_{i-1}^{1}(\rho)\,T_{i-1}^{1}(z)\exp\{-[i(i+1)/2N]\,t\} \end{aligned} \tag{1}$$

where

$$\rho = 1 - 2y, \qquad z = 1 - 2x, \tag{2a}$$

$T_n{}^1(x)$ are the Gegenbauer polynomials

$$T_0{}^1(x) = 1, \qquad T_1{}^1(x) = 3x, \qquad T_2{}^1(x) = (3/2)(5x^2-1), \quad \ldots \tag{2b}$$

and $F(a,b;c;x)$ is the hypergeometric function (Abramowitz and Stegun, 1964). Letting $t \to \infty$ in Eq. (1), we get $P(x\,|\,y,\infty) = 0$, which implies that the fixation of one of the two genes is inevitable. The asymptotic approach to the steady-state distribution depends on the first few terms of the sum in Eq. (1), which are

$$P(x\,|\,y,t) = 6y(1-y)e^{-t/N} + 30y(1-2y)(1-2x)e^{-3t/N} + \cdots \tag{3}$$

Thus for large t the probability density decays exponentially at a rate $1/N$, and takes a value which depends only on y and is independent of x. This value is maximal at $y = \frac{1}{2}$ (equal proportions of two alleles) and approaches zero as y approaches either of the two boundaries ($y = 0$ and 1). The same dependence on y characterizes the probability of still having the two alleles after a long time, since this probability is given by

$$\int_0^1 P(x\,|\,y,t)\,dx = 6y(1-y)e^{-t/N} + \cdots \tag{4}$$

The probability density of the first passage time to fixation of allele A (i.e., the random variable x taking the value 1 for the first time), according to Eqs. (3.2.57) and (1), is given by

$$F(1\,|\,y,t) = -(2N)^{-1}\frac{\partial}{\partial x}[x(1-x)\,P(x\,|\,y,t)]_{x=1} = (2N)^{-1}P(1\,|\,y,t)$$

$$= (2N)^{-1}\sum_{i=1}^{\infty} y(1-y)\,i(i+1)(2i+1)\,F(1-i,i+2;2;y)$$

$$\times \exp\{-[i(i+1)/2N]\,t\} \tag{5}$$

The probability density of the first passage time to fixation of allele a can be similarly calculated and is given by Eq. (5) with y replaced by $1-y$.

Since one of the two alleles is bound to disappear, the probability of fixation of each of the genes is of central interest, together with the average time for fixation in each case, and the average time of coexistence of the two alleles in the population. To derive these quantities, we note that the boundaries act as absorbing boundaries because once x takes the value 0 or 1, it can not return to the open interval (0, 1). Therefore, we can use the results of Section 3.2B. Thus, from Table 3.7, the probability of fixation of gene A when its initial value is y is given by

$$R(1\,|\,y) = \int_0^y \pi(\xi)\,d\xi \Big/ \int_0^1 \pi(\xi)\,d\xi \tag{6a}$$

where

$$\pi(\xi) = \exp\left(-2\int^{\xi}\frac{a(x)}{b(x)}\,dx\right) \tag{6b}$$

For the process under consideration $a(x) = 0$, and therefore $\pi(x) = 1$ and

$$R(1\,|\,y) = y \tag{6c}$$

the initial proportion. This result, derived from the continuous approximation, agrees with the result found by the finite Markov chain corresponding to randomly mating finite population, as was presented in the footnote on page 146.

The average number of generations (average time) for fixation of the gene A, when a priori it is known that fixation of A is bound to occur, can be obtained by using Table 3.7 and Eq. (3.2.33):

$$M_1{}^*(1\,|\,y) = \frac{M_1(1\,|\,y)}{R(1\,|\,y)} = -\frac{2N}{y}\int_0^y d\eta\int_0^\eta (1-\xi)^{-1}\,d\xi + 2N\int_0^1 d\eta\int_0^\eta (1-\xi)^{-1}\,d\xi$$

$$= 2Ny^{-1}(1-y)\,|\ln(1-y)| = 2N(1-y)[1+y/2+y^2/3+\cdots] \tag{7}$$

Thus the average number of generations for fixation of gene A is linear in the size of the population and decreases monotonically with y, from $2N$ at $y = 0$ to 0 at $y = 1$. For small values of y, the average time is linear in the distance from the initial proportion y to the fixation proportion 1, while for y close to 1, small changes in y change the average time drastically.

By symmetry the average number of generations for fixation of allele a, starting from proportion $1-y$, when fixation of allele a is a priori known to occur, is

$$M_1{}^*(0\,|\,y) = \frac{M_1(0\,|\,y)}{R(0\,|\,y)} = 2Ny(1-y)^{-1}\,|\ln y| \tag{8}$$

From Eqs. (6)–(8), the average number of generations in which the two alleles coexist in the population is

$$M(y) = M_1(0\,|\,y) + M_1(1\,|\,y) = 2N[(1-y)\,|\ln(1-y)| + y\,|\ln y|] \tag{9}$$

Thus $M(y)$ is symmetric with respect to $y = \frac{1}{2}$ where it attains its maximal value and drops very fast to zero as y approaches one of the boundaries.

7.4 Genetic Changes in a Finite Population with Random Mating and Mutation or Migration

In this section we analyze the process of genetic changes in a finite population in the presence of mutation pressure. Such a process is described by the diffusion equations (7.2.11) and (7.2.12). As pointed out in Section 7.2, if instead of mutation there is migration pressure, the diffusion equations are the same as Eqs. (7.2.11) and (7.2.12) except that u_a is replaced by rq' and u_A by $r(1-q')$, where q' is the proportion of allele A in the other larger population. Thus by making these changes in the results derived in this section for mutation pressure, one can obtain the corresponding results in the presence of migration. We give these results at the end of this section.

In discussing mutation pressure, we distinguish between the two cases of mutation, irreversible and reversible. In the first case, mutation occurs only in one direction (one of the mutation rates, u_a or u_A, is zero), which implies that the ultimate fate of the population is the fixation of that gene which reproduces without errors. The boundary corresponding to fixation is an exit (absorbing) boundary while the other is a regular-reflecting or an entrance boundary, due to mutation which occurs at a constant nonzero rate. In the second case where both mutation rates are positive, no fixation is possible. The boundaries are regular-reflecting or entrance boundaries, and the distribution among the unfixed states tends to a steady-state distribution.

We first discuss the case of reversible mutation and then derive the expression for the case of irreversible mutation by taking the limit $u_a \to 0$.

1. *Reversible mutation.* In the case of reversible mutation, $u_a > 0$, $u_A > 0$,

and the solution to the FP equation (7.2.11), from Table 3.4, is

$$P(x \mid y, t) = \sum_{i=0}^{\infty} \frac{[2(u_1+u_2)+2i-1]\,\Gamma(2u_1+i)\,\Gamma[2(u_1+u_2)+i-1]}{i!\,[\Gamma(2u_1)]^2\,\Gamma(2u_2+i)} \cdot x^{2u_2-1}(1-x)^{2u_1-1} \cdot \tilde{F}_i(1-x)\,\tilde{F}_i(1-y)\exp(-\alpha_i t) \tag{1}$$

where

$$\tilde{F}_i(x) \equiv F[2(u_1+u_2)+i-1, -i; 2u_1; x],$$
$$\alpha_i \equiv i[2(u_1+u_2)+i-1]/2N \tag{2}$$

and $F(a,b;c;z)$ is the hypergeometric function (Abramowitz and Stegun, 1964). $F_i(x)$ is a polynomial in x since the second argument of the hypergeometric function is a negative integer.

The probability of having the two types of alleles at time t is given by the integral of $P(x \mid y,t)$ over $(0,1)$. Using the following integration formula (Gradshteyn and Ryzhik, 1965):

$$\int_0^1 x^{\gamma-1}(1-x)^{\rho-1}F(\alpha,\beta;\gamma;x)\,dx = \frac{\Gamma(\gamma)\,\Gamma(\rho)\,\Gamma(\gamma+\rho-\alpha-\beta)}{\Gamma(\gamma+\rho-\alpha)\,\Gamma(\gamma+\rho-\beta)} \tag{3}$$

we find that the integral of the terms in the summation in Eq. (1) vanishes except for $i = 0$, for which the value is 1. Therefore,

$$\text{prob}[0 < x < 1, t] = 1 \tag{4}$$

Thus in the process under consideration, there is a conservation of probability in $(0,1)$ and no flow from the interval is possible. For $u_1 < \frac{1}{2}$ ($u_2 < \frac{1}{2}$) the boundary $x = 1$ ($x = 0$) is a regular-reflecting boundary, while for $u_1 > \frac{1}{2}$ ($u_2 > \frac{1}{2}$) the boundary $x = 1$ ($x = 0$) is an entrance boundary. In both cases the probability current vanishes at the boundaries and $\lim_{t\to\infty} P(x \mid y,t)$ is nonzero and independent of the initial proportion,

$$\lim_{t\to\infty} P(x \mid y,t) \equiv P(x) = \frac{\Gamma[2(u_1+u_2)]}{\Gamma(2u_1)\,\Gamma(2u_2)}\,x^{2u_2-1}(1-x)^{2u_1-1} \tag{5}$$

The rate of approach to the steady-state distribution is given by the term corresponding to $i = 1$ in the sum in Eq. (1), i.e., by

$$[2(u_1+u_2)+1]\frac{u_1}{u_2}P(x)\left[\frac{u_1}{u_2}x-(1-x)\right]\left[\frac{u_1}{u_2}y-(1-y)\right]\exp(-N^{-1}(u_1+u_2)t) \tag{6}$$

The steady-state distribution is approached at the rate $N^{-1}(u_1+u_2)$, compared to N^{-1} [see Eq. (7.3.3)] in the absence of mutation. The density (5) is the beta density for which the integral over $(0,1)$ is unity, and which is extremal at the point (see Table 3.9).

$$x_0 = (u_2 - \tfrac{1}{2})(u_2+u_1-1)^{-1} \tag{7}$$

The location of x_0 [which can be outside (0, 1)] and the type of the extremal value $P(x_0)$ depend on the relation between the parameters u_1 and u_2. When $u_1 > \frac{1}{2}$ and $u_2 < \frac{1}{2}$, x_0 is outside (0, 1) and $P(x)$ is monotonically decreasing from very large values near $x = 0$ to 0 at $x = 1$. Similarly, for $u_1 < \frac{1}{2}$ and $u_2 > \frac{1}{2}$, $P(x)$ is monotonically increasing from zero to very large values near $x = 1$. For u_1 and u_2 greater than $\frac{1}{2}$, $P(x_0)$ is the maximal value of $P(x)$, which is a bell-shaped unimodel density tending to zero as x approaches the boundaries. x_0 is located below $\frac{1}{2}$ if $u_2 < u_1$ and above $\frac{1}{2}$ if $u_1 < u_2$. For u_1 and u_2 less than $\frac{1}{2}$, $P(x)$ is minimal at x_0 and increases without limit as x approaches the boundaries (U shaped).

Noting that $u_1 \sim Nu_A$ ($u_2 \sim Nu_a$) is the number of mutant a genes (A genes) produced per generation when most of the population consists of A genes (a genes), we conclude that if a mutant gene appears only once in several ($\gg 2$) generations, one of the two alleles will be practically fixed due to the random sampling of gametes, and the rarely produced mutant genes will disappear shortly after being formed. Similarly, when mutation occurs rarely in one direction whereas in the reverse direction it occurs quite frequently, the gene that mutates rarely will become practically fixed. It is only when mutations in both directions are frequent that the coexistence of both alleles is of high probability, with proportions around x_0 and $1-x_0$ where $x_0 > \frac{1}{2}$ when $u_a > u_A$ and $x_0 < \frac{1}{2}$ when $u_a < u_A$. In case $u_1 = u_2 = \frac{1}{2}$, $P(x)$ is independent of x and all unfixed states are equally probable.

The average gene proportion in the steady state is

$$\langle x \rangle_\infty = \int_0^1 xP(x)\,dx = u_2/(u_2+u_1) \tag{8}$$

which differs from x_0 whenever $u_2 \neq u_1$, but is near it when $u_1 \gg 1$ or $u_2 \gg 1$.

When mutation occurs frequently in one direction and rarely in the opposite direction, more insight can be acquired by investigating the limiting case of irreversible mutation.

2. *Irreversible mutation.* Taking $u_2 = 0$, $u_1 > 0$ in Eq. (1), and noting that the term corresponding to $i = 0$ vanishes in the limit $u_2 \to 0$, we get for the probability density the expression

$$\begin{aligned} P(x\,|\,y,t) &= \sum_{i=1}^{\infty} \frac{(2u_1+2i-1)(2u_1+i-1)[\Gamma(2u_1+i-1)]^2}{(i-1)!\,i!\,[\Gamma(2u_1)]^2} x^{-1}(1-x)^{2u_1-1} \\ &\quad \cdot \tilde{F}_i(1-x)\tilde{F}_i(1-y)\exp(-\alpha_i t) \qquad (9) \\ &= \sum_{i=0}^{\infty} \frac{(2u_1+2i+1)(2u_1+i)[\Gamma(2u_1+i)]^2}{(i-1)!\,i!\,[\Gamma(2u_1)]^2} (1-x)^{2u_1-1} \\ &\quad \cdot yF_i^*(1-x)\,F_i^*(1-y)\exp(-\alpha_i^* t) \end{aligned}$$

where

$$F_i^*(x) = F(2u_1+i, -i; 2u_1; x), \qquad \alpha_i^* = (i+1)(2u_1+i)/2N \tag{10}$$

From Eq. (9), as t increases, the probability of having both alleles in the population decreases. For large t, $P(x|y,t)$ decays to zero according to

$$P(x|y,t) \sim 2u_1(2u_1+1)(1-x)^{2u_1-1} y \exp(-(u_1/N)t) \tag{11}$$

The rate of decay at each unfixed state, $0 < x < 1$, is $u_1 N^{-1}$ compared to N^{-1} in the absence of mutation. For $u_1 > \frac{1}{2}$ and for large t, $P(x|y,t)$ decreases as x increases, while for $u_1 < \frac{1}{2}$, $P(x|y,t)$ increases from a small value near $x = 0$ to very large values near $x = 1$ with a drastic increase near $x = 1$. The high value of $P(x|y,t)$ near $x = 1$ for small u_1 reflects the possibility that by random sampling of gametes, the population consists mostly of A genes for a long time when $u_1 \sim Nu_A$, the number of mutant a genes, is less than 1 per two generations. However, the boundary $x = 0$ is an exit boundary, and fixation of the allele a is inevitable.

The probability of having both types of alleles after a long time t is

$$\int_0^1 P(x|y,t) = (2u_1+1) y \exp(-(u_1/N)t) + \cdots \tag{12}$$

This probability is linear in the distance from the initial proportion y to the ultimate proportion 0, and decays to zero at a rate u_1/N.

Since near $x = 1$, $P(x|y,t)$, given by Eq. (9), behaves as $c(1-x)^{2u_1-1}$ $[F_i^*(1) = 1]$, where c is a function of y but is independent of x, the probability current at $x = 1$ according to Eq. (3.0.52) becomes

$$\begin{aligned} \lim_{x\to 1} J(x,t) &= N^{-1} c \lim_{x\to 1} \left\{ -u_1 x(1-x)^{2u_1-1} - \frac{1}{2}\frac{\partial}{\partial x}[x(1-x)^{2u_1}] \right\} \\ &= N^{-1} c \lim_{x\to 1} \{ -u_1(1-x)^{2u_1-1} + u_1(1-x)^{2u_1-1} \} = 0 \end{aligned} \tag{13}$$

With no flow through the boundary $x = 1$, all the loss of probability in $(0, 1)$ is due to flow through the boundary at $x = 0$. Therefore, the probability density of the time for fixation of allele a is given by

$$\begin{aligned} F(0|y,t) &= -\frac{\partial}{\partial t}\int_0^1 P(x|y,t)\,dx \\ &= (2N)^{-1} y \sum_{i=0}^{\infty} (-1)^i \frac{\Gamma(2u_1+i+1)(2u_1+2i+1)}{\Gamma(2u_1)\,i!} \\ &\quad \cdot F(2u_1+i+1, -i; 2u_1; 1-y) \exp(-\alpha_i^* t) \end{aligned} \tag{14}$$

where we have used formula (3) in integrating $P(x|y,t)$. The average time for the fixation of allele a can be derived by substituting the $a(x)$ and $b(x)$ of the present process into Eq. (3.2.29) which applies for processes confined between a reflecting boundary at $x = B$ $[J(B,t) = 0]$ and an absorbing boundary at $x = A$. With

$$a(x) = -N^{-1}u_1 x, \qquad b(x) = N^{-1}x(1-x) \tag{15a}$$

$$\pi(x) = \exp\left\{-2\int^x d\xi\, a(\xi)/b(\xi)\right\} = (1-x)^{-2u_1} \tag{15b}$$

the average time for the fixation of allele a is given by

$$\begin{aligned} M_1(0\,|\,y) &= 2\int_0^y dy\, \pi(y) \int_\eta^1 [b(\xi)\pi(\xi)]^{-1}\, d\xi \\ &= 2N\int_0^y d\eta(1-\eta)^{-2u_1} \int_\eta^1 \xi^{-1}(1-\xi)^{-1+2u_1}\, d\xi \\ &= (N/u_1)\int_0^y d\eta\; F(2u_1, 1; 2u_1+1; 1-\eta) \end{aligned} \tag{16}$$

Expanding the integrand in powers of $1-\eta$ and integrating, we obtain

$$M_1(0\,|\,y) = 2N\sum_{i=0}^{\infty}(i+1)^{-1}[1-(1-y)^{i+1}](2u_1+i)^{-1} \tag{17}$$

Thus the average time for the fixation of allele a is proportional to the size of the population and decreases as u_1 increases. We can bound $M_1(0\,|\,y)$ by

$$M_1(0\,|\,y) \leqslant 2N\left\{\frac{1}{2u_1} + \sum_{i=1}^{\infty}[(i+1)i]^{-1}\right\} = 2N\{(2u_1)^{-1}+1\} \tag{18}$$

Thus for $u_1 > 1$, $M_1(0\,|\,y)$ is less than $3N$.

Before closing the discussion of mutation pressure, we note that within the continuous approximation, irreversible mutation applies not only when there is no mutation in one direction but also when the mutation rate in one direction is of the order of N^{-2}.

As noted in the beginning of this section, the results derived for a population under mutation pressure can be easily transferred into results for a population under migration pressure by replacing u_2 by $r_1 q_A'$ and u_1 by $r_1(1-q_A')$, where

$$r_1 = rN + O(1) \ll N \tag{19}$$

is the number of genes exchanged per generation. The case of immigration from populations which are homogeneous with respect to the discussed locus, corresponds to irreversible mutation. When more than one gene per two generations of each of the two alleles immigrates $(r_1 q') > \frac{1}{2}$, $r_1(1-q') > \frac{1}{2}$, then the two alleles coexist in the population. On the other hand, when $r_1 q' < \frac{1}{2}, r_1(1-q') < \frac{1}{2}$, practical fixation of one of the two genes is inevitable and the population becomes homogeneous. An isolated population which starts to exchange individuals with a population in which one of the alleles is already fixed, is going to lose the other allele if more than one gene per two generations is exchanged ($r_1 > \frac{1}{2}$). In the presence of both mutation and migration pressures, the results depend on $u_2 + r_1 q'$ and $u_1 + r_1(1-q')$; thus a high proportion of alleles of one type in the arriving population can compensate for a low mutation rate to this allele or a high mutation rate from this allele.

7.5 Genetic Changes in a Finite Population with Random Mating and Selection

In a finite population with selection pressure, the fate of the population is fixation of one of the two alleles, which is irreversible. By Eqs. (7.2.20) and (7.2.21) the random process of genetic changes is confined between two (absorbing) exit boundaries. The quantities that give insight into the process are the probabilities of fixation of each of the alleles, the average time for this fixation, and the average time during which the two alleles coexist. We now discuss these quantities and their dependence on the selection coefficients and the initial state and their relationship to the corresponding quantities in the absence of selection ($s = 0, h = 1$). The latter provides insight into the effects of the different types of selections. We start with the simple case of genic selection.

a. *Genic selection.* In this process (see p. 154) $P(x \mid y, t)$ obeys the FP equation (3.0.2) with

$$a(x) = N^{-1} s_1 x(1-x), \qquad b(x) = N^{-1} x(1-x) \tag{1}$$

where $s_1 \sim Ns + O(1)$ is the excess in the number of progeny in a population of A genes as compared to a population of a genes. According to the assumptions made in Section 7.2 [Eq. (7.2.16)]

$$N^{-1} \ll s_1 \ll N \tag{2}$$

The probability of the process reaching the boundary $x = 1$, i.e., the probability of fixation of allele A, when its initial proportion is y, is given by (see Table 3.7)

$$R(1 \mid y) = \int_0^y \pi(\eta)\, d\eta \Big/ \int_0^1 \pi(\eta)\, d\eta \tag{3a}$$

where

$$\pi(\eta) = \exp\left[-\int^\eta \frac{2a(\xi)}{b(\xi)}\, d\xi \right] = \exp(-2 s_1 \eta) \tag{3b}$$

i.e., by

$$R(1\,|\,y) = [1-\exp(-2s_1 y)]/[1-\exp(-2s_1)] \tag{4}$$

In particular for $y = 1/N$

$$R(1\,|\,N^{-1}) \simeq \frac{2s}{1-\exp(-2s_1)} + O(N^{-1}) \tag{5}$$

which is the probability of fixation of an individual mutant gene that might appear in the population. For $\exp(-2s_1) \ll 1$ $(s_1 > 10)$, this probability is equal to $2s$ as compared to $1/N$ in the absence of selection [see Eq. (7.3.6c)]. Thus the effect of genic selection is to increase the probability of fixation of a favorable mutant gene by a factor of $2sN \simeq 2s_1$.

For arbitrary y, $R(1\,|\,y)$ increases from 0 to 1 as y increases from 0 to 1. Since for $\exp(-2s_1) < 10^{-8}$ $(s_1 > 10)$

$$\frac{dR(1\,|\,y)}{dy} \simeq 2s_1 \exp(-2s_1 y) \tag{6}$$

the main increase in $R(1\,|\,y)$ takes place for small values of y, and the derivative becomes practically zero when y is greater than some critical value y_c. In Fig. 7.1, $R(1\,|\,y)$ is plotted as a function of y for a variety of values of s_1. Note that for any initial frequency of allele A which is greater than the critical value y_c,

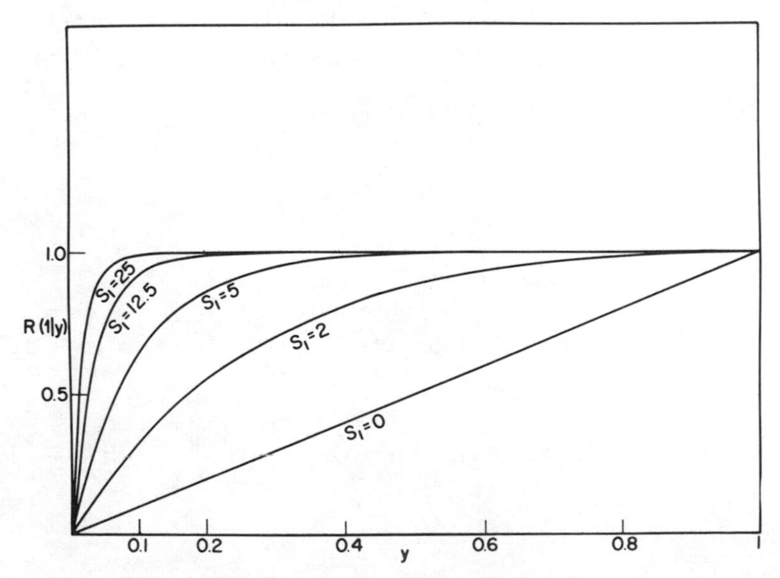

Fig. 7.1 The probability $R(1\,|\,y)$ of fixation of an advantageous allele versus its initial proportion y in the case of genic selection. The curve $s_1 = 0$ corresponds to a neutral allele.

the fixation of A is practically a certain event. The existence of this critical value guarantees the stability of allele A in the population relative to any disadvantageous mutants which might appear due to errors in reproduction in a low proportion (less than $1-y_c$). On the other hand, the same existence of critical value implies that the fixation of a mutant advantageous gene is guaranteed only if it appears in a fraction $y > y_c$. Such an event cannot be realized by a random mutation ($y \sim N^{-1}$) but might be realized either by mutations caused by unnatural environmental factors, e.g., radiation, or by a large migration from a population in which the advantageous gene is already fixed.

The average time for the fixation of allele A is

$$M_1(1\,|\,y) = 2N\left\{R(1\,|\,y)\int_0^1 d\eta \exp(-2s_1\eta)r(\eta) - \int_0^y d\eta \exp(-2s_1\eta)r(\eta)\right\} \tag{7a}$$

with

$$r(\eta) = \int^{\eta} \frac{\exp(2s_1 x)-1}{x(1-x)(1-\exp(-2s_1))}\,dx \tag{7b}$$

(See Table 3.7 and expressions (1) and (3)). $M_1(1\,|\,y)$ depends linearly on N as was the case in the absence of selection (see p. 158). These expressions cannot be simplified because an analytical integration is not possible. However, for the interesting case of $\exp(-2s_1) \ll 1$ and $y > y_c$, some simple bounds on $M_1(1\,|\,y)$ can be obtained. For this case, $R(1\,|\,y) \simeq 1$, $M_1{}^*(1\,|\,y)$, the conditional average time for fixation, equals $M_1(1\,|\,y)$, and Eqs. (7) become

$$M_1{}^*(1\,|\,y) = 2N\int_y^1 d\eta \exp(-2s_1\eta)r(\eta) \tag{8}$$

with

$$r(\eta) = \int_y^{\eta} \frac{\exp(2s_1 x)}{x(1-x)}\,dx \tag{9}$$

Since $x(1-x) \leqslant \frac{1}{4}$, $r(\eta)$ is bounded by

$$4\int_y^{\eta} \exp(2s_1 x)\,dx \leqslant r(\eta) \leqslant \exp(2s_1\eta)\int_y^{\eta}[x(1-x)]^{-1}\,dx \tag{10a}$$

or

$$\frac{2}{s_1}(\exp(2s_1\eta)-\exp(2s_1 y)) \leqslant r(\eta) \leqslant \exp(2s_1\eta)\left[\ln\left(\frac{\eta}{1-\eta}\right) - \ln\left(\frac{y}{1-y}\right)\right] \tag{10b}$$

Therefore from Eq. (8), $M_1{}^*(1\,|\,y)$ is bounded by

$$\frac{2N}{s_1{}^2}[\exp(-2s_1(1-y)) - 1 + 2s_1(1-y)] \leqslant M_1{}^*(1\,|\,y) \leqslant 2N|\ln y| \tag{11a}$$

For y close to 1, the lower limit is

$$4N(1-y)^2[1-(2/3)s_1(1-y)] \leqslant M_1^*(1|y) \tag{11b}$$

For completeness, we may point out that the FP equation satisfied by $P(x|y,t)$ can be solved by using the methods of Chapter 3, which involve solving an eigenvalue problem. The eigenfunctions are oblate spheroidal functions (Kimura, 1955; Crow and Kimura, 1956). However, there is no analytical expression for the coefficients in the eigenfunction expansion nor for the eigenvalues, and they have to be evaluated numerically. Crow and Kimura (1970) expand the eigenvalue with smallest absolute value in a power series in s_1 which converges for small values of s_1. This eigenvalue determines the rate of decay of $P(x|y,t)$ to its steady-state value 0 for large values of t, i.e., when one of the two alleles is about to be fixed.

We now consider the case of genotype selection, and assume that allele A is the advantageous allele.

b. *Genotype selection.* In this case the coefficients in the FP equation are

$$a(x) = N^{-1}s_1 x(1-x)[h+(1-2h)x], \qquad b(x) = N^{-1}x(1-x) \tag{12}$$

with s_1 and h as defined in Section 7.1. The function $\pi(\eta)$, defined by the first part of Eq. (3b), is

$$\pi(\eta) = \exp\left\{-2s_1\int_0^\eta [h+(1-2h)x]\,dx\right\} = \exp\{-s_1\eta[2h+(1-2h)\eta]\} \tag{13}$$

so that from Eq. (3a), the probability of fixation of allele A, with initial proportion y, is

$$R(1|y) = \frac{\int_0^y \exp\{-s_1 x[2h+(1-2h)x]\}\,dx}{\int_0^1 \exp\{-s_1 x[2h+(1-2h)x]\}\,dx} \tag{14}$$

or

$$R(1|y) = \frac{\int_0^y \exp\{-s_1(1-2h)[x+h/(1-2h)]^2\}\,dx}{\int_0^1 \exp\{-s_1(1-2h)[x+h/(1-2h)]^2\}\,dx}, \qquad h \neq \tfrac{1}{2} \tag{15}$$

These expressions for $R(1|y)$ simplify for various special values of h which determine the type of selection.

(i) $h = \frac{1}{2}$, *no dominance.* Equation (14) becomes

$$R(1|y) = \frac{1-\exp(-s_1 y)}{1-\exp(-s_1)} \tag{16}$$

which is the same expression as for the genic selection [Eq. (4)] but with s_1

replacing $2s_1$. Thus the behavior of probability of fixation is the same as for the genic selection, including the existence of critical initial proportion y_c of allele A.

(ii) $h = 0$, *A completely recessive.* The probability of fixation of allele A is

$$R(1 \mid y) = \frac{\int_0^y \exp(-s_1 x^2)\, dx}{\int_0^1 \exp(-s_1 x^2)\, dx} = \frac{\operatorname{erf}(s_1^{1/2} y)}{\operatorname{erf}(s_1^{1/2})} \tag{17}$$

where $\operatorname{erf}(x)$ stands for the error function (Abramowitz and Stegun, 1964). $\operatorname{Erf}(x)$ increases monotonically with x, with the main increase taking place in the interval $0 < x < 2$ [$\operatorname{erf}(2) = 0.995, \operatorname{erf}(\infty) = 1.0$]. For large x, $\operatorname{erf}(x)$ has the asymptotic form

$$\operatorname{erf}(x) = 1 - (\pi x^2)^{-1/2} \exp(-x^2)\left[1 - \frac{1}{2x^2} + \frac{3}{4x^4} - \cdots\right] \tag{18}$$

Thus for large enough s_1 and $y > 0.5$, Eq. (17) becomes

$$R(1 \mid y) = \frac{1 - (\pi s_1)^{-1/2} y^{-1} \exp(-s_1 y^2)}{1 - (\pi s_1)^{-1/2} \exp(-s_1)} \tag{19}$$

and

$$\frac{d}{dy} R(1 \mid y) = \frac{\exp(-s_1 y^2)}{\operatorname{erf}(s_1^{1/2})} = \frac{\exp(-s_1 y^2)}{1 - (\pi s_1)^{-1/2} \exp(-s_1)} \simeq \exp(-s_1 y^2) \tag{20}$$

When $\exp(-s_1) < 10^{-8}$ ($s_1 \geqslant 20$), Eq. (20) implies the existence of a critical frequency y_r such that for all $y > y_r$ the increase in $R(1 \mid y)$ is negligible and $R(1 \mid y)$ is practically 1. For example, if $s_1 = 25$, $dR(1 \mid y)/dy < 10^{-6}$ for $y^2 > 16/25$, i.e., for $y > 0.8$, while if $s_1 = 100$, $dR(1 \mid y)/dy < 10^{-6}$ for $y^2 > 16/100$, i.e., for $y > 0.4$. These critical values y_r can be compared with the corresponding values in case of no dominance ($h = \frac{1}{2}$). From Eq. (16) $dR(1 \mid y_c)/dy \simeq s_1 \exp(-s_1 y_c) < 10^{-6}$, so that the corresponding critical values y_c are 0.5 and 0.1 for $s_1 = 25$ and 100, respectively. In general y_c is of the order of the square of y_r, and therefore $y_r > y_c$. From the discussion above we conclude that a recessive advantageous gene is also stable in a population but its range of stability is smaller than in the case of no dominance. To get an order of magnitude estimation of the fixation probability, we consider the case of $s_1 > 20$ and small values of y. Equation (17) becomes (Abramowitz and Stegun, 1964)

$$R(1 \mid y) = \operatorname{erf}(s_1^{1/2} y) = \left(\frac{2}{\pi}\right)^{1/2} \exp(-s_1 y^2) \sum_{n=0}^{\infty} \frac{2^n}{1 \cdot 3 \cdot 5 \cdots (2n+1)} (s_1^{1/2} y)^{2n+1} \tag{21}$$

Thus the probability of one mutant advantageous gene ($y = 1/N$) to become fixed in the population, when $s_1^{1/2}N^{-1} \ll 1$, is given by

$$R(1 \mid N^{-1}) = \left(\frac{2}{\pi}\right)^{1/2} \exp(-s_1 N^{-2}) s_1^{1/2} N^{-1} \simeq \left(\frac{2}{\pi}\right)^{1/2} \frac{s_1^{1/2}}{N} \simeq \left(\frac{2s}{\pi N}\right)^{1/2} \tag{22}$$

Since $s_1 > 1$, this probability is smaller than the corresponding probability in case of no dominance, which is given by $s_1 N^{-1}$ [Eq. (16)]. This result can be explained by noting that the heterozygote in the no dominance case has some advantage over the *aa* homozygotes, a property that increases the chance of spreading the *A* genes in the population, while in the recessive case the heterozygote has the same properties as the *aa* homozygote.

(iii) $h = 1$, *A completely dominant.* From Eq. (15), the probability of fixation of allele *A* is

$$R(1 \mid y) = \frac{\int_0^y \exp(s_1(x-1)^2)\,dx}{\int_0^1 \exp(s_1(x-1)^2)\,dx} = \frac{\int_{1-y}^1 \exp(s_1 x^2)\,dx}{\int_0^1 \exp(s_1 x^2)\,dx} \tag{23}$$

and

$$\frac{dR(1 \mid y)}{dy} = \frac{\exp(s_1(1-y)^2)}{\int_0^1 \exp(s_1 x^2)\,dx} = \frac{s_1^{1/2}\exp(s_1(1-y)^2)}{\int_0^{s_1^{1/2}} \exp(x^2)\,dx} \tag{24}$$

Using the function (Abramowitz and Stegun, 1964, p. 319)

$$g(x) = x\exp(-x^2)\int_0^x \exp(\xi^2)\,d\xi = x^2 - \tfrac{2}{3}x^4 + \cdots \tag{25}$$

which increases from 0 to 0.6 as x increases from 0 to 2 and then decreases to 0.5 as x increases from 2 to ∞, Eq. (24) can be rewritten in the form

$$\frac{dR(1 \mid y)}{dy} = \frac{s_1 \exp\{s_1[(1-y)^2 - 1]\}}{g(s_1^{1/2})} \simeq 2s_1 \exp\{-s_1[1 - (1-y)^2]\} \tag{26}$$

As in the previous case, when $s_1 > 20$ there exists a critical proportion y_d such that for all $y > y_d$, $dR(1 \mid y)/dy$ is practically zero and fixation is a certain event [since $R(1 \mid y) = 1$]. Comparing Eq. (26) with Eq. (6), with $2s_1$ replaced by s_1 (the no dominance case), we find that

$$(1-y_d)^2 = 1 - y_c \quad \text{or} \quad y_d = 1 - (1-y_c)^{1/2} < y_c \tag{27}$$

which implies that the range of stability of allele *A* is greater when *A* is dominant as compared to the no dominance case.

When the initial proportion is small, using the function $g(x)$ of Eq. (25), we get

$$R(1\,|\,y) = 1 - \frac{\exp\{-s_1[1-(1-y)^2]\}}{1-y}\,\frac{g(s_1^{1/2}(1-y))}{g(s_1^{1/2})}$$

$$\simeq 1 - \frac{\exp(-2s_1 y)}{1-y} \simeq (2s_1-1)\,y \tag{28}$$

Therefore, for an individual mutant allele the probability of fixation is

$$R(1\,|\,N^{-1}) = (2s_1-1)/N \tag{29}$$

which is almost twice the corresponding probability for the case of no dominance, when $s_1 > 10$.

From the analysis above we conclude that a dominant advantageous gene has better chances to become fixed when its initial proportion is small, than any advantageous gene for which the heterozygote has intermediate properties ($0 < h < 1$), and once it is fixed it is more stable than genes with $0 < h < 1$.

(iv) $h > 1$ ($s_1 > 0$), *overdominance.* For this case, from Eq. (15),

$$R(1\,|\,y) = \frac{\int_0^y \exp\{N(\sigma_1+\sigma_2)[x-\sigma_2/(\sigma_1+\sigma_2)]^2\}\,dx}{\int_0^1 \exp\{N(\sigma_1+\sigma_2)[x-\sigma_2/(\sigma_1+\sigma_2)]^2\}\,dx} \tag{30}$$

and

$$\frac{dR(1\,|\,y)}{dy} = \frac{\exp\{c_1(y-x_0)^2\}}{\int_0^1 \exp\{c_1(x-x_0)^2\}\,dx} \tag{31}$$

where σ_1 and σ_2 are the selection coefficients

$$\sigma_2 = N^{-1}s_1 h, \qquad \sigma_1 = N^{-1}s_1(h-1)$$

x_0 is the steady-state proportion of allele A in the deterministic approach, and

$$c_1 = N(\sigma_1+\sigma_2) = s_1(2h-1), \qquad N^{-1} \ll c_1 \ll N$$

The derivative $dR(1\,|\,y)/dy$ is smallest when $y \simeq x_0$ and increases as y approaches either of the two boundaries 0 and 1. Therefore, the dependence of the probability of fixation on the initial proportion is weakest for proportions in the neighborhood of x_0. Furthermore, whenever

$$c_1 x_0{}^2 = N\sigma_2{}^2/(\sigma_1+\sigma_2) > 20$$

there exists a critical distance $\Delta = |y-x_0|$ such that for all y in the interval $(x_0-\Delta, x_0+\Delta)$, $dR(1\,|\,y)/dy$ is practically zero, while near the boundaries this derivative is large. In Fig. 7.2 we have plotted $R(1\,|\,y)$ as a function of y for the case of overdominance; for comparison we have also plotted $R(1\,|\,y)$ for the cases of no dominance, allele A completely recessive, and allele A completely dominant. An estimate of Δ can be made by rewriting Eq. (31) in terms

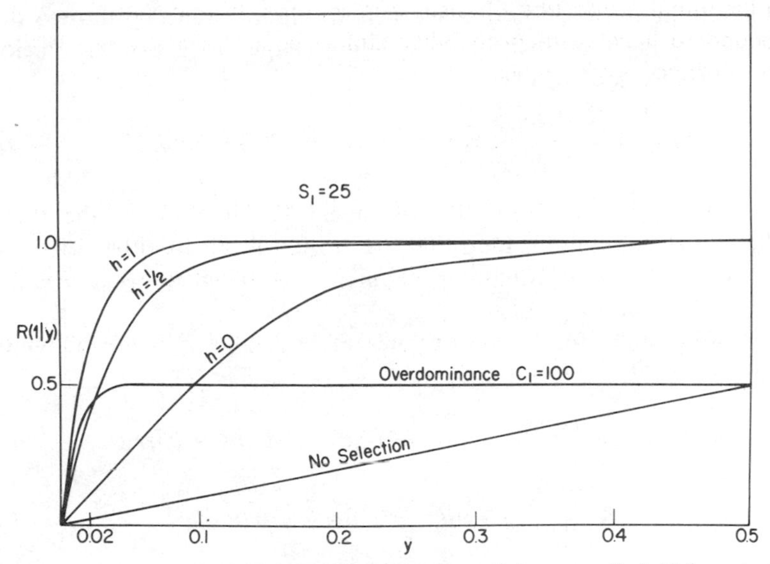

Fig. 7.2 The probability $R(1\,|\,y)$ of fixation of an allele versus its initial proportion y in the case of genotype selection of various types: $h=1$—advantageous allele completely dominant; $h=1/2$—advantageous allele with no dominance; $h=0$—advantageous allele completely recessive.

of the function $g(x)$ defined by Eq. (25),

$$\frac{dR(1\,|\,y)}{dy} = \frac{c_1 x_0(1-x_0)\exp[-c_1\{x_0{}^2-(y-x_0)^2\}]}{(1-x_0)g(c_1^{1/2}x_0)+x_0 g(c_1^{1/2}(1-x_0))} \tag{32}$$

For $s_1 = \sigma_1 - \sigma_2 = O(1)$, i.e., when there is no significant difference between the two types of homozygotes, $x_0 \simeq \frac{1}{2}$, and Eq. (30) becomes

$$R(1\,|\,y) = \tfrac{1}{2}\left[1 - \frac{\exp(-c_1\{1-(1-2y)^2\}/4)g(|(1-2y)|\,c_1^{1/2}/2)}{(1-2y)g(c_1^{1/2}/2)}\right] \tag{33}$$

$$= 1 - R(1|1-y)$$

Comparing Eq. (33) with (28) we conclude that if $y \leqslant \frac{1}{2}$ and c_1 in Eq. (33) is four times s_1 of Eq. (28), then

$$R(1\,|\,y) = \tfrac{1}{2}R_{\text{d}}(1\,|\,y/2) \tag{34}$$

where R_{d} stands for the probability of fixation when A is dominant. Therefore, the critical distance Δ in the present case of overdominance is related to the critical proportion y_{d} for the case of dominance by

$$\Delta = (1-y_{\text{d}})/2 \tag{35}$$

When the initial proportion of allele A is within a distance at most Δ of x_0, it is bound to have (with probability almost equal to 1) any proportion x within a distance Δ of x_0 since

$$R(x\,|\,y) = \frac{R(1\,|\,y)}{R(1\,|\,x)} \simeq 1, \qquad y, x \quad \text{in} \quad (x_0-\Delta, x_0+\Delta) \tag{36}$$

This implies that the proportion of allele A will stay in the interval $(x_0-\Delta, x_0+\Delta)$ for quite a long time before it will go to either of the two boundaries, which are equally probable to be reached since $R(1\,|\,x) \simeq R(1\,|\,\frac{1}{2}) = \frac{1}{2}$ for $|x-x_0| < \Delta$.

The average time for the coexistence of both alleles in the population, according to Table 3.7, is given by

$$M_1(y) = 2N\left\{R(1\,|\,y)\int_0^1 d\eta \exp(-2hs_1\eta(1-\eta))r(\eta) - \int_0^y d\eta \exp(-2hs_1\eta(1-\eta))r(\eta)\, d\eta\right\} \tag{37a}$$

where

$$r(\eta) = \int^\eta \frac{\exp(2hs_1\xi(1-\xi))}{\xi(1-\xi)}\, d\xi \tag{37b}$$

The expressions for the average times for fixation or loss of the alleles can be written in terms of quadratures, but the integration cannot be carried out analytically, and therefore we do not discuss these quantities. The study of their dependence on the selective coefficients and initial proportions can be accomplished via numerical computations.

Finally we point out that as in the case of genic selection, the FP equation for the genotype selection can also be formally solved by solving the eigenvalue problem (Kimura, 1957). The eigenfunctions can be expressed as a sum of Gegenbauer polynomials with coefficients which for small s_1 are power series in s_1. These coefficients and the eigenvalues, once again, can only be calculated numerically. Only for the case of overdominance, where selective coefficients are fairly large, has an asymptotic form of the eigenvalue with smallest absolute value been analytically worked out (Miller, 1962).

So far in the discussion of selection we assumed that initially either both alleles exist in the population or one of them is introduced by a rare event of mutation (or migration). But the same mutation can happen more than once, and if the rate of mutation is not negligible in both directions, fixation cannot occur (the boundaries act as entrance boundaries) and a steady-state distribution is reached after a long time. The way this steady state is approached can be found by considering the eigenvalues of the corresponding FP equation. Since no analytic expressions for the eigenvalues are known, in the following section we discuss only the steady-state distribution. By comparing it with the steady-state distribution in the absence of selection, the effects of selection can be studied.

7.6 Steady-State Distributions in a Finite Population with Random Mating, Mutation, Migration, and Selection

In the presence of mutation (migration) pressure, the boundaries 0 and 1 act as reflecting (entrance) boundaries and a steady-state distribution is reached. To derive this distribution, we assume that $a(x)$ of the FP equation is the sum of the $a(x)$ corresponding to the various types of operating pressures, while $b(x)$, which is due to random sampling of gametes in a finite population, is the same as in the case of one operating pressure. These assumptions are justified (p. 149) when all pressure rates are small compared to the size of the population and if in the continuous approximation we retain terms only up to order N^{-1}. Therefore, in the presence of mutation and genic selection, the coefficients of the FP equation are

$$a(x) = N^{-1}[-u_1 x + u_2(1-x) + s_1 x(1-x)], \qquad b(x) = N^{-1}x(1-x) \tag{1}$$

while in the case of mutation and genotype selection,

$$a(x) = N^{-1}[-u_1 x + u_2(1-x) + s_1 x(1-x)[h + (1-2h)x]],$$
$$b(x) = N^{-1}x(1-x) \tag{2}$$

If, in addition, migration pressure is operating, u_1 and u_2 have to be replaced by $u_1 + r_1(1-q')$ and $u_2 + r_1 q'$ (see p. 147). To simplify the discussion, we restrict it to the case of mutation without migration, but the results can easily be transformed to include migration as well.

For the case of *genic selection*, from Eq. (1) and Table 3.9,

$$1/\pi(x) = \exp\{2u_1 \ln(1-x) + 2u_2 \ln x + 2s_1 x\} = (1-x)^{2u_1} x^{2u_2} \exp(2s_1 x) \tag{3}$$

and the steady-state distribution is of the form

$$P(x) = P_0 (1-x)^{2u_1 - 1} x^{2u_2 - 1} \exp(2s_1 x) \tag{4}$$

where

$$P_0^{-1} = \int_0^1 (1-x)^{2u_1-1} x^{2u_2-1} \exp(2s_1 x)\, dx \tag{5}$$

Comparing $P(x)$ of Eq. (4) with the steady-state distribution in the absence of selection pressure ($s_1 = 0$), we note that, as is to be expected, genic selection increases (decreases) the probability of higher proportions of the A gene when $s_1 > 0$ ($s_1 < 0$).

When $u_1 < \frac{1}{2}$, $u_2 < \frac{1}{2}$ (small number of mutations in the population), the most probable states are near the boundaries as in the absence of selection, although in the present case more probability is concentrated near $x = 1$ ($x = 0$) when $s_1 > 0$ ($s_1 < 0$). When both $u_1 > \frac{1}{2}$ and $u_2 > \frac{1}{2}$, $P(x)$ has its maximal value for the proportion (see Table 3.9)

$$x_0 = \frac{s_1 - \alpha - \beta + ((\alpha + \beta - s_1)^2 + 4s_1\beta)^{1/2}}{2s_1} \tag{6a}$$

where

$$\alpha = u_1 - \tfrac{1}{2} > 0, \qquad \beta = u_2 - \tfrac{1}{2} > 0 \tag{6b}$$

The proportion x_0 is in $(0, 1)$ for all values of s_1, positive or negative, and is monotonically increasing with s_1; for $s_1 > 0$ ($s_1 < 0$) it is to the right (to the left) of the maximal point corresponding to $s_1 = 0$ (no selection). For $|s_1| \gg \alpha, \beta$

$$x_0 = 1 - \alpha/s_1, \qquad s_1 > 0 \tag{7a}$$

$$x_0 = \beta/|s_1|, \qquad s_1 < 0 \tag{7b}$$

In the deterministic model corresponding to infinite population under the pressures of mutation and genic selection, the gene proportion at steady state is determined by equating $a(x)$ to zero, i.e.,

$$x_\infty = \frac{s_1 - u_1 - u_2 + ((u_1 + u_2 - s_1)^2 + 4s_1 u_2)^{1/2}}{2s_1} \tag{7c}$$

which is the same expression as (6a) but with α and β replaced by u_1 and u_2, respectively. For u_1 and u_2 much greater than 1, $u_1 \sim \alpha$, $u_2 \sim \beta$, and the most probable point is close to the deterministic steady-state value.

For the case of *genotype selection*, from Eq. (2) and Table 3.9,

$$1/\pi(x) = (1-x)^{2u_1} x^{2u_2} \exp(s_1 x^2 + hs_1 x(1-x)) \tag{8}$$

and the steady-state distribution is of the form

$$P(x) = P_0(1-x)^{2u_1-1} x^{2u_2-1} \exp(s_1 x^2 + 2hs_1 x(1-x)) \tag{9}$$

where

$$P_0^{-1} = \int_0^1 dx(1-x)^{2u_1-1} x^{2u_2-1} \exp(s_1 x^2 + 2hs_1 x(1-x)) \tag{10}$$

Comparing the steady-state distribution for the case $s_1 > 0$ with the steady-state distribution in the absence of selection pressure ($s_1 = 0$), we conclude that:

(a) For $0 \leqslant h \leqslant 1$ (the heterozygote has intermediate properties), since $s_1[x^2+2hx(1-x)]$ is positive and monotonically increasing with x, larger values of x are more probable than for the case of no selection ($s_1 = 0$). The case $h = \frac{1}{2}$ (no dominance) is identical with genic selection, with s_1 replaced by $s_1/2$.

(b) For $h > 1$ (overdominance), since $s_1[x^2+2hx(1-x)]$ is positive with maximal value at $\tilde{x} = h/(2h-1)$, proportions in the neighborhood of $\tilde{x}$ have higher probabilities as compared to the case of no selection. The proportion $\tilde{x}$ is the steady-state proportion in a deterministic model with no mutation pressure ($u_1 = u_2 = 0$), and is the most probable proportion in case $u_1 = u_2 = \frac{1}{2}$. For these mutation rates and no selection there is a uniform distribution over (0, 1) [Eq. (7.4.5)]. By changing u_1 and u_2 continuously we conclude that for $u_1 < \frac{1}{2}$, $u_2 < \frac{1}{2}$ there is a peak of $P(x)$ inside (0, 1), although in case $s_1 = 0$, $P(x)$ is U shaped. Thus small mutation rates and overdominance can maintain two alleles in the population.

(c) For $h < 0$ (disadvantageous heterozygote), $\exp\{s_1 x[(1-2h)x+2h]\}$ is less than 1 for all $0 < x < \gamma$ and is greater than 1 thereafter, where $\gamma = 2|h|/(2|h|+1)$. Thus the probability of proportions above (below) $x = \gamma$ is increased (decreased) as compared to the case with $s_1 = 0$. For $u_1 = u_2 = \frac{1}{2}$, $P(x)$ decreases from 1 to a minimal value at $x = \gamma/2 < \frac{1}{2}$ and then increases monotonically with x.

References

Abramowitz, M., and Stegun, I. A., eds. (1964). "Handbook of Mathematical Functions." Nat. Bur. Stand., Washington, D. C.

Barish, N. (1965). "The Gene Concept." Van Nostrand-Reinhold, Princeton, New Jersey.

Crow, J. F., and Kimura, M. (1956). Some genetic problems in natural populations. *Proc. Symp. Math. Statist. Probability, 3rd, Berkeley, California*, p. 1.

Crow, J. F., and Kimura, M. (1970). "An Introduction to Population Genetics Theory." Harper, New York.

Feller, W. (1951). Diffusion processes in genetics. *Proc. Symp. Math. Statist. Probability, 2nd, Berkeley, California*, p. 227.

Fisher, R. A. (1930). "The Genetical Theory of Natural Selection." Oxford Univ. Press, London and New York.

Gradshteyn, I. S., and Ryzhik, I. M. (1965). "Tables of Integrals, Series and Products." Academic Press, New York.

Kimura, M. (1955). Stochastic processes and distribution of gene frequencies under natural selection. *Cold Spring Harbor Symp. Quant. Biol.* **20**, 33.

Kimura, M. (1957). Some problems of stochastic processes in genetics. *Ann. Math. Statist.* **28**, 882.

References

Kimura, M. (1964). Diffusion models in population genetics. *J. Appl. Probability* **1**, 177.
Miller, G. F. (1962). The evaluation of eigenvalues of a differential equation arising in a problem in genetics. *Proc. Cambridge Philos. Soc.* **58**, 588.
Wright, S. (1942). Statistical genetics and evolution (Gibbs lecture). *Bull. Amer. Math. Soc.* **48**, 223.
Wright, S. (1949). Adaptation and selection, *in* "Genetics, Paleontology and Evolution" (G. L. Jepson, G. G. Simpson, and E. Mayer, eds.), p. 365. Princeton Univ. Press, Princeton, New Jersey.

Additional References

Crow, J. F., and Morton, N. E. (1955). Measurement of gene frequency drift in small populations. *Evolution* **9**, 202.
Karlin, S. (1968). Rates of approach to homozygosity for finite stochastic models with variable population size. *Amer. Natur.* **102**, 443.
Karlin, S., and McGregor, J. (1968). Rates and probabilities of fixation for two locus random mating finite populations without selection. *Genetics* **58**, 141.
Kimura, M. (1970). The length of time required for a selectively neutral mutant to reach fixation through random frequency drift in a finite population. *Genet. Res.* **15**, 131.
Kimura, M., and Ohta, T. (1969). The average number of generations until fixation of a mutant gene in a finite population. *Genetics* **61**, 763.
Kojima, K., ed. (1970). "Mathematical Topics in Population Genetics." Springer-Verlag, Berlin and New York.
Maruyama, T. (1970). On the fixation probability of mutant genes in a subdivided population. *Genet. Res.* **15**, 221.
Moran, P. A. P. (1962). "The Statistical Processes of Evolutionary Theory." Oxford Univ. Press (Clarendon), London and New York.
Moran, P. M. P. (1963). Some general results on random walks, with genetic applications. *J. Austral. Math. Soc.* **3**, 468.

8

Firing of a Neuron

In Chapter 6 we studied a population of interacting species by making a primitive statistical model. This system has the characteristics that each element influences many other elements through competition or predation. Further, there is no detailed knowledge about the interaction between various elements and the state of the system at any one time. Even if all the information were available, it is computationally too difficult to solve the dynamical equations exactly because of the large number of variables involved. In this chapter we study another system which has similar characteristics. This is the nervous system, which consists of a large number of elements known as neurons. The number of neurons that interact with each other depends on the organism and on the specific part of the nervous system, but for a moderately complex organism, this number is in the hundreds of thousands and millions. As in Chapter 6, where by making reasonable assumptions about the effect of other species on a single species we were able to study the growth of a single species, in this chapter by making reasonable assumptions about the effects of all the neurons on a single neuron, we can study the behavior of a single

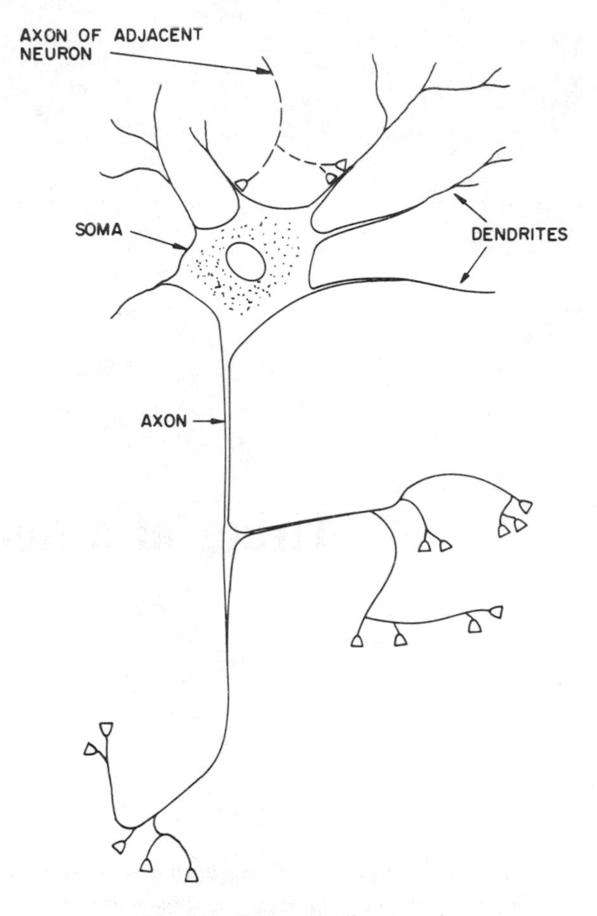

Fig. 8.1 Schematic diagram of a neuron.

neuron. Let us first digress to a brief review of the properties and functions of neurons.

The neuron is a biological cell consisting of three different parts: the soma or the cell body; the axon, an elongated part of the cell; and the dendrites which form a complicated receptive network with many branches converging on the soma. A sketch of a neuron, showing these three parts, is given in Fig. 8.1 where we have also shown the connections of these parts with a neighboring neuron. The impulses are initiated in the soma which is also the integrator of impulses arriving at the neuron's dendrites. The impulses propagate down the axon and are transferred by the axon's branches into dendrites of other neurons. These impulses are electrochemical in nature, and here we give a simplified description of the way they are initiated, propagated, and integrated. For more details, we refer to an excellent but somewhat outdated primer on neurophysiology by Stevens (1966).

The entire surface of the neuron is bounded by an electrochemically sensitive membrane, which is selectively permeable to ions, and whose permeabilities change with the potential across it. In the absence of an input the potential across the membrane is ~70 mV, with the inside negative due to a higher concentration of K^+ and lower concentration of Na^+ on the inside relative to the outside. These differences in ionic concentrations are kept by an active metabolism. The potential difference of 70 mV across the membrane is termed the resting potential, and the membrane is said to be polarized. At this potential the membrane is more permeable to K^+ than to Na^+. As the membrane is depolarized (the inside becomes less negative) the permeabilities of the membrane to both the ions Na^+ and K^+ increase but the increase in permeability to Na^+ is much faster than to K^+. When a certain threshold value (55–60 mV, inside negative) is reached, the flow of Na^+ ions into the cell depolarizes the membrane, the inside becomes +30 mV relative to the outside within 1 msec or less, and the neuron is said to be firing. As the permeability to K^+ exceeds the permeability to Na^+, the membrane potential is pulled back toward its resting value, and with the potential the permeabilities return to their previous smaller values. The rapid increase in potential across the membrane is the spike or impulse generated by the neuron, and is termed action potential. The impulse is initiated at the axon hillock—the junction of soma and axon—where the threshold has the smallest value. The depolarization at the axon hillock induces a current flow and depolarization in the neighboring segments of the membrane, and action potential is initiated there. In this fashion the action potential spreads down the axon and into the axon branches, with a constant velocity typical of the axon, and with no attenuation. After the occurrence of the action potential, no other action potential can be evoked in a so-called absolute refractory period of about 2 msec, and during the next several milliseconds (known as the relative refractory period), the threshold decays from very high values to its stationary value, approximately in an exponential fashion.

The branches of the axon are connected to different dendrites of other neurons by synapses. Each branch ends in a synaptic knob, a minute enlargement of the branch, where there is a gap of about 200 Å (known as the synaptic cleft) between the membrane of the knob and that of the cell on which it sits. The action potential propagating into the synaptic knob is stopped there and stimulates the release of a certain chemical, transmitter substance, into the synaptic cleft. The transmitter diffuses across the cleft (in 0.5–1.0 msec) and induces hyperpolarization or depolarization at the postsynaptic membrane by selective changes of the permeabilities of this membrane to K^+ and Na^+. The dendrite potential rises up during ~1 msec and then decays down to its equilibrium value at a rate which depends on the functional and structural properties of the receiving neuron. This potential, called the postsynaptic potential (PSP), spreads from the dendrite into the soma and changes the potential across the membrane. The effect of various PSPs arriving in the same time at the soma is approximately additive, but the effect of various PSPs arriving in different times is not additive, since the membrane potential is decaying to its resting value as long as it is below the threshold level.

In the absence of detailed information on the connections between neurons, it can be assumed that a neuron responds to stochastic stimulations. This assumption of random input is reasonable for neurons in the central nervous system which have a large number of dendritic connections such as the Purkinjee cells and perhaps the so-called basket and stellate cells (Eccles, 1969). With this as the basic assumption, one can make stochastic models of neural behavior which are capable of predicting input–output behavior. The stochastic models can either be discrete or continuous.

In the discrete model, the states are the levels of the membrane potential, which is transferred to a higher or to a lower level by the arrival of an excitatory or an inhibitory impulse, respectively. When the membrane potential reaches a certain state K the neuron fires and the process is then reset. Thus the state K can be regarded as an absorbing state in this model, and the mathematical analysis of Chapter 2 (in particular of Section 2.2D) can be applied. In the continuous model we replace the assumption of finite discrete jumps in the potential level by continuous gradual changes in this potential as long as it is below some value B, corresponding to the threshold potential. Therefore, B is the absorbing boundary of the continuous process, and by using the techniques of Chapter 3 (in particular of Section 3.3D) this model can be analyzed. In this chapter we discuss both types of models, discrete models in Section 8.1 and continuous models in Section 8.2. In the continuous model, we also discuss the case of a threshold, $B = B(t)$, which is changing with time, to incorporate the refractory period after firing.

8.1 Discrete Models

The firing of a neuron, using discrete stochastic models, has been studied for almost two decades. A list of references can be found in Goel *et al.* (1972) (see also ten-Hoopen, 1966) where we propose and study two models. These models have as much physiological basis as is mathematically feasible. In this section we describe these models and then analyze them by using the techniques of Chapter 2. The results will be the same as those obtained in our paper, but the procedure used here (i.e., of Chapter 2) is considerably simpler.

The first model (*model A*) we study has the following assumptions as its basis:

(a) The heights of the excitatory and inhibitory input impulses are the same. Therefore the membrane potential is assumed to be changing in discrete steps, with fixed spacings between the possible potential levels. For convenience let us number these levels $-\infty, \ldots, -3, -2, -1, 0, 1, 2, \ldots, K$, with the 0 level corresponding to the resting potential and level K to the threshold potential.

(b) The threshold for firing is independent of time; i.e., we exclude the refractory behavior of the model neuron. Inclusion of this behavior makes the analysis very difficult to carry out analytically.

(c) After firing, the membrane potential returns to the resting value.

(d) The subthreshold membrane potential decays spontaneously toward the resting potential in an exponential fashion with time constant τ.

(e) The model neuron receives stochastic excitatory and inhibitory inputs which are Poisson distributed in time with f_e (f_i) as the average frequency of the excitatory (inhibitory) input. In other words, the probability that the neuron receives an excitatory (inhibitory) input in the time interval $(t, t+\Delta t)$ is $f_e \, \Delta t$ ($f_i \, \Delta t$), and the probability that two pulses arrive in this time interval is $0 \, (\Delta t)$.

(f) The effect of each impulse on that part of the membrane where the spike is initiated, is additive.

The second model (*model B*) we study has assumptions (a)–(d) and (f) as its basis, but assumption (e) is modified. In this model we assume that the probability per unit time to increase the membrane potential by one unit due to excitatory impulse, when this potential is above its resting value, is proportional to the membrane potential. Thus this model is relevant for those cases in which the probability of a neuron receiving excitatory inputs increases as the membrane potential approaches the threshold.

In both models, the process is restricted between an absorbing state K and $-\infty$. Therefore we can analyze the firing of a neuron, using the mathematical techniques and analysis given in Section 2.2D, if the expressions for the transition probabilities per unit time λ_n and μ_n are given. In the context of the firing of a neuron, λ_n and μ_n are the probabilities per unit time of the membrane potential changing from level n to level $n+1$ and level $n-1$, respectively.

Assumption (d), based on experimental data, can be incorporated into a discrete probabilistic model by a set of λ_n and μ_n for which the average membrane potential, in the absence of any input, is given by

$$\langle n \rangle = \sum_{n=-\infty}^{K} n P_{n,m}(t) = m e^{-t/\tau} \tag{1}$$

where m is an initial subthreshold level (positive or negative) of the membrane potential, and $P_{n,m}(t)$ is the probability function of the membrane potential being at the level n at time t. Although $\langle n \rangle$ decays exponentially, the decay to the resting level (0) in the discrete models is in discrete steps, the height of the steps being the distance between the levels. At first glance, the decay in steps may appear to be unrealistic, but this is not so, especially if the spacings between levels are small and if one recalls that the fundamental process responsible for the decay, the transport of ions across the neuronal membrane, is also a discrete process. From Eq. (2.0.8), the transition probabilities per unit time which will lead to Eq. (1) are

$$\lambda_n^{(0)} = \begin{cases} 0 & \text{for} \quad n \geqslant 0 \\ -n/\tau & \text{for} \quad n < 0 \end{cases} \tag{2a}$$

$$\mu_n^{(0)} = \begin{cases} n/\tau & \text{for} \quad n > 0 \\ 0 & \text{for} \quad n \leqslant 0 \end{cases} \tag{2b}$$

The final expression for λ_n and μ_n in the presence of inputs can be obtained by including assumption (e). Since this assumption is different for the two models, we divide the analysis into two parts.

Model A. In view of assumption (e) and Eqs. (2), λ_n and μ_n, in the presence of inputs, are given by

$$\lambda_n = \begin{cases} f_e & \text{for} \quad 0 \leqslant n \leqslant K-1 \\ 0 & \text{for} \quad n \geqslant K \\ f_e - n/\tau & \text{for} \quad n < 0 \end{cases} \tag{3a}$$

$$\mu_n = \begin{cases} f_i + n/\tau & \text{for} \quad 0 \leqslant n \leqslant K-1 \\ 0 & \text{for} \quad n \geqslant K \\ f_i & \text{for} \quad n < 0 \end{cases} \tag{3b}$$

Since we have assumed that after each firing the membrane potential returns to the resting value, the initial level of the membrane potential is 0. From Table 2.9, the average of the first passage time to reach state K, i.e., the average time $\langle t \rangle$ for firing, is given by

$$\langle t \rangle \equiv M_{K,0} = \sum_{i=0}^{K-1} M_{i+1,i} \tag{4}$$

where

$$M_{i+1,i} = \frac{1}{\lambda_i} + \frac{\mu_i}{\lambda_i \lambda_{i-1}} + \frac{\mu_i \mu_{i-1}}{\lambda_i \lambda_{i-1} \lambda_{i-2}} + \cdots, \qquad i \leqslant K-1 \tag{5}$$

For the transition probabilities (3a) and (3b), with

$$a \equiv f_e \tau, \qquad b \equiv f_i \tau \tag{6}$$

Eq. (5) becomes

$$\frac{1}{\tau} M_{-i+1,-i} = \frac{1}{a+i} + \frac{b}{(a+i)(a+i+1)} + \frac{b^2}{(a+i)(a+i+1)(a+i+2)} + \cdots$$

$$= \frac{1}{a+i} F(1; a+i+1; b), \qquad i \geqslant 0 \tag{7}$$

and

$$\frac{1}{\tau} M_{i+1,i} = \frac{1}{a} + \frac{b+i}{a^2} + \frac{(b+i)(b+i-1)}{a^3} + \cdots + \frac{(b+i)(b+i-1)\cdots(b+2)}{a^{i+1}}$$

$$+ \frac{(b+i)\cdots(b+1)}{a^i} M_{1,0}/\tau$$

$$= \frac{1}{a} \sum_{k=0}^{i-1} \frac{(b+i-k+1)_k}{a^k} + \frac{(b+1)_i}{a^i} M_{1,0}/\tau, \qquad i \geqslant 1 \tag{8}$$

where F is the standard confluent hypergeometric function (Abramowitz and Stegun, 1964, Chapter 13) and

$$(a)_n = a(a+1)\cdots(a+n-1), \qquad n > 0$$

$$(a)_0 = 1$$

Substituting for $M_{1,0}$ from Eq. (7) into Eq. (8) and using Eq. (4), we obtain

$$M_{K,0}/\tau = \frac{F(1; a+1; b)}{a\Gamma(b+1)} \sum_{j=0}^{K-1} \frac{\Gamma(b+j+1)}{a^j} + \frac{1}{a} \sum_{j=1}^{K-1} \frac{\Gamma(b+j+1)}{a^j} \sum_{i=1}^{j} \frac{a^i}{\Gamma(b+i+1)} \tag{9}$$

By the formula

$$\sum_{n=j}^{N} \frac{\Gamma(b+n+l)}{l!\Gamma(b+n)} = \frac{\Gamma(b+N+l+1)}{(l+1)!\Gamma(b+N)} - \frac{\Gamma(b+j+l)}{(l+1)!\Gamma(b+j-1)}$$

Eq. (9) reduces to

$$\langle t \rangle/\tau = \frac{F(1; a+1; b)-1}{a\Gamma(b+1)} \sum_{j=0}^{K-1} \frac{\Gamma(b+j+1)}{a^j} + \Gamma(b+N+2) \sum_{j=1}^{K} \frac{a^{-j}}{j\Gamma(b+N+2-j)}$$

$$- \frac{b}{\Gamma(b+1)} \sum_{j=1}^{K} \frac{a^{-j}}{j} \Gamma(b+j) \tag{10}$$

which is the required expression for the average time for firing. To calculate the variance, we again use Table 2.9 according to which the variance σ is

$$\sigma^2 \equiv V_{K,0} = \sum_{i=0}^{K-1}\left[2\sum_{n=0}^{\infty}\Pi_{i-n,i}(M_{i-n,i-n-1})^2 + (M_{i+1,i})^2\right] \tag{11}$$

where

$$\Pi_{i,j} \equiv \frac{\mu_i\mu_{i+1}\cdots\mu_j}{\lambda_i\lambda_{i+1}\cdots\lambda_j}, \qquad i \leqslant j, \qquad \Pi_{i,i-1} = 1 \tag{12}$$

The complicated expression (11) for the variance is not simplified by the substitution of the expressions for λ_n and μ_n given in Eqs. (3). One can use Eq. (11) to compute numerically the values of $V_{K,0}$ for various sets of parameters.

In Figs. 8.2 and 8.3 we have given the so-called input–output relations [graph between dimensionless output frequency $\tau/\langle t\rangle$ versus relative input

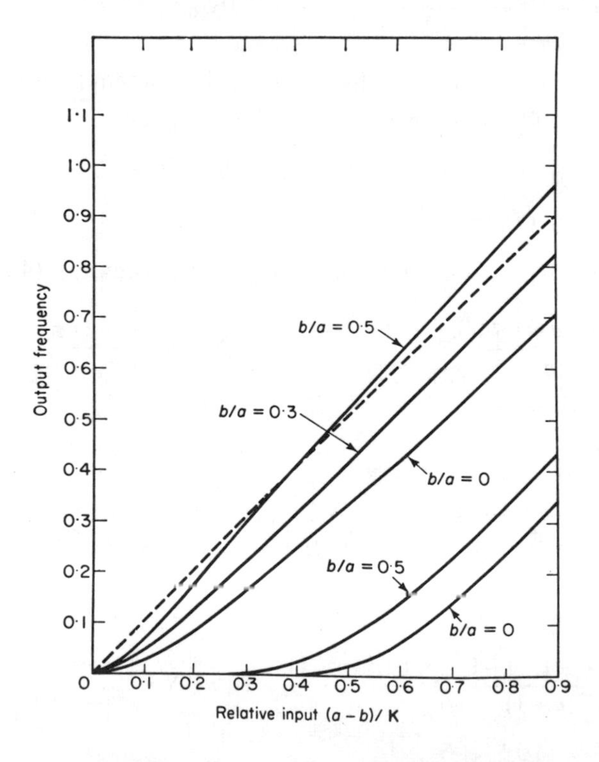

Fig. 8.2 Input–output relations for model A displaying the effect of inhibitory input for thresholds of 2 and 20 units. Dimensionless output frequency $\tau/\langle t\rangle$ is plotted versus relative input $(a-b)/K$. Left-hand series of curves are for $K=2$ and right-hand series for $K=20$. Dashed line corresponds to Eq. (8.1.16) in the text.

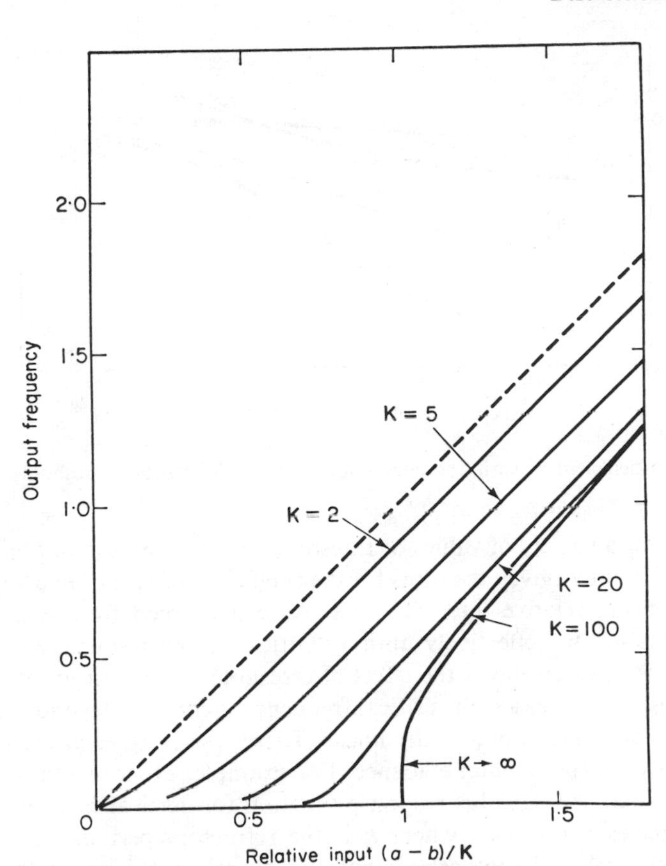

Fig. 8.3 Model A input–output relations for various thresholds. For all cases shown $b/a = 0.2$.

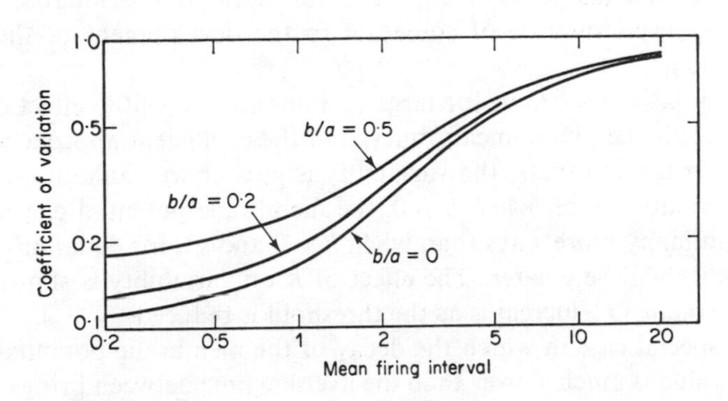

Fig. 8.4 Coefficient of variation curves for model A with $K = 100$ and for various levels of inhibition.

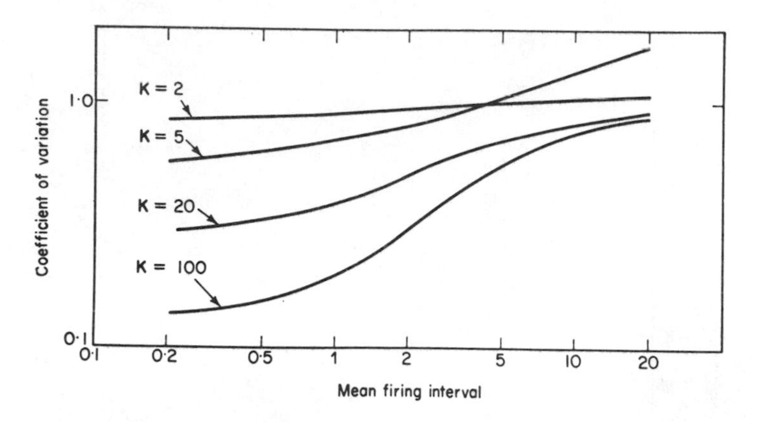

Fig. 8.5 Coefficient of variation curves for model A for various thresholds, $b/a = 0.2$.

$(a-b)/K]$ for a variety of values of thresholds K and the ratio of b/a. In Figs. 8.4 and 8.5 we have given the plots for the coefficient of variation $\sigma/\langle t \rangle$ versus the mean firing relative interval $\langle t \rangle/\tau$. As can be seen from Fig. 8.2, the neuron does not fire effectively until a "critical" input frequency is reached. From Fig. 8.3, which shows the effect of threshold on the output, we see that as the threshold increases, the critical frequency increases. Beyond the critical frequency, the curve is practically linear. These results agree at least qualitatively with the experimental evidence. For example, experimental data seem to indicate that the stimulus response function for cortical neurons is linear for frequencies below $1/t_r$, where t_r is the refractory period. Furthermore, Mountcastle (1967) hypothesizes that the overall behavior of the central nervous system (CNS) is linear even though afferent (sensory) fibers which feed input about the external environment to the CNS are generally nonlinear transforming systems. This linear behavior both at the microscopic and macroscopic levels may be of some aid to the development of theories of brain function.

From Fig. 8.4 we see that, for large K, inhibition has little effect on $\sigma/\langle t \rangle$ for small inputs, i.e., large mean times, and the coefficient approaches unity. However, for larger inputs, the variability is greater with inhibition present This is reasonable since, when $b \neq 0$, the membrane potential can reach the threshold in many more ways than when $b = 0$ and so, for equal mean times, the variability will be greater. The effect of K on variability is shown in Fig. 8.5. In general, $\sigma/\langle t \rangle$ increases as the threshold is reduced.

For the special case in which the decay of the membrane potential toward its resting value is much slower than the average time between firings, one can assume $\tau \to \infty$. Within this assumption one can not only calculate $\langle t \rangle$ and σ but also obtain an analytical expression for the probability density of firing time, $F_{K,0}(t)$. For $\tau \to \infty$, from Eq. (3),

$$\lambda_n = \begin{cases} f_e, & n \leqslant K-1 \\ 0, & n \geqslant K \end{cases} \tag{13a}$$

$$\mu_n = \begin{cases} f_i, & n \leqslant K-1 \\ 0, & n \geqslant K \end{cases} \tag{13b}$$

From Eq. (2.2.13)

$$F_{K,0}(t) = \lambda_{K-1} P_{K-1,0}(t) \tag{14a}$$

and from Table 2.2 (row 0, with $v = f_e$, $\rho = f_i$, $m = 0$, $u = K$),

$$P_{K-1,0} = \left(\frac{f_e}{f_i}\right)^{(K-1)/2} \exp(-(f_e+f_i)t)[I_{K-1} - I_{K+1}] \tag{14b}$$

where I_n stands for $I_n(2(f_e f_i)^{1/2} t)$. Using the standard identity (Abramowitz and Stegun, 1964),

$$I_{K-1}(z) - I_{K+1}(z) = \frac{2K}{z} I_K(z)$$

we get from Eqs. (14a) and (14b)

$$F_{K,0}(t) = K\left(\frac{f_e}{f_i}\right)^{K/2} t^{-1} \exp(-(f_e+f_i)t) I_K(2(f_e f_i)^{1/2} t) \tag{15}$$

The first moment and variance are given by

$$\langle t \rangle = \int_0^t t F_{K,0}(t)\, dt = \frac{K}{(f_e - f_i)} \tag{16}$$

$$\sigma^2 = \langle t^2 \rangle - \langle t \rangle^2 = \int_0^t t^2 F_{K,0}\, dt - \langle t \rangle^2 = \frac{K(f_e+f_i)}{(f_e-f_i)^3} \tag{17}$$

As a check of our expression (10) for $\langle t \rangle$, we note that as $\tau \to \infty$, since (Abramowitz and Stegun, 1964)

$$\frac{\Gamma(z+\alpha)}{\Gamma(z+\beta)} = z^{\alpha-\beta}\left[1 + \frac{1}{2z}(\alpha-\beta)(\alpha+\beta-1) + O\left(\frac{1}{z}\right)\right], \qquad \text{large } z \tag{18}$$

$$\begin{aligned} F(1; a+1; b) &\simeq 1 + \frac{b}{a} + \frac{b^2}{a^2} + \cdots, \qquad \text{large } a, b \\ &= (1 - b/a)^{-1} \end{aligned} \tag{19}$$

the leading term of the right-hand side of Eq. (10) is $K/a(1-b/a)\tau = K/(f_e - f_i)\tau$.

When τ is not infinite but very large, the expression (10) for $\langle t \rangle$ still simplifies. If we keep the next order term in Eq. (18), after some algebraic manipulations, the expression is

$$\langle t \rangle = \langle t \rangle_{\tau=\infty} \left[1 + \frac{\langle t \rangle_{\tau=\infty}}{2\tau} - \frac{1+b/a}{2a(1-b/a)^2} \right] + O\left(\frac{1}{\tau}\right) \tag{20}$$

Thus the first-order correction is proportional to $(\langle t \rangle_{\tau=\infty})/\tau$.

Since $\tau \to \infty$ is a good assumption when the average interval between firings is much smaller than τ, we expect the limiting value of the coefficient of variation for $\langle t \rangle \ll \tau$ to be merely the ratio of expression (16) and (17) for $\tau \to \infty$, i.e.,

$$\sigma/\langle t \rangle = (a+b)^{1/2}/K(a-b)^{1/2} \tag{21}$$

This parameter can be greater than unity in the case $\langle t \rangle \ll \tau$, as well as for larger values of $\langle t \rangle$ as shown in Fig. 8.5, $K = 2$ and $K = 5$.

Having discussed model A and its consequences, we now discuss model B.

Model B. In this model, since the assumptions (a)–(d) and (f) of model A are retained, and since according to assumption (e) the mean frequency of excitatory input increases linearly as the membrane potential approaches the threshold level, the transition probabilities are given by

$$\lambda_n = \begin{cases} (n+1)f_e & \text{for } 0 \leqslant n \leqslant K-1 \\ 0 & \text{for } n \geqslant K \\ f_e - n/\tau & \text{for } n < 0 \end{cases} \tag{22}$$

$$\mu_n = \begin{cases} f_i + n/\tau & \text{for } 0 \leqslant n \leqslant K-1 \\ 0 & \text{for } n \geqslant K \\ f_i & \text{for } n < 0 \end{cases} \tag{23}$$

The above λ_n and μ_n for $n \leqslant 0$ are the same as in model A, and therefore from Eq. (5), $M_{i+1,i}$ $(i \leqslant 0)$ is also the same and is given by Eq. (7), while from Eqs. (5), (22), and (23), for $i > 0$, it is given by

$$M_{i+1,i} = \frac{\Gamma(b+i+1)}{(i+1)!\,a^i} \frac{M_{1,0}}{\Gamma(b+1)} + \sum_{l=1}^{i} \frac{l!\,a^{l-1}}{(b+l+1)}, \qquad i \geqslant 0 \tag{24}$$

Substituting for $M_{i+1,i}$ in Eq. (4), after simple algebraic manipulations, we obtain

$$\langle t \rangle/\tau = \frac{F(1, a+1, b-1)}{(a+1)\Gamma(b+1)} \sum_{j=0}^{K-1} \frac{\Gamma(b+j+1)}{(j+1)!\,a^j} + \frac{1}{a} \sum_{i=0}^{K-1} a^{-i} \sum_{j=i}^{K-1} \frac{\Gamma(b+j+1)(j-i)!}{(j+1)!\,\Gamma(b+j+1-i)} \tag{25}$$

To calculate σ, we use Eq. (11) with λ_n and μ_n given by Eqs. (22) and (23), and $M_{i+1,i}$ by Eq. (24).

We have compared the results of models A and B in Figs. 8.6 and 8.7. In Fig. 8.6 we have plotted the ratio of the output frequency of model B to that of model A as a function of input. For small thresholds, the two models predict nearly the same mean first passage time independent of inhibition. In fact, for $K = 1$, the models are identical. However, as K increases, the models diverge, as is to be expected. The membrane potential of model B cascades toward the threshold when it is above the resting level and so the neuron fires even for small input levels, contrary to model A where the neuron does not fire effectively until a "critical" input frequency is reached. The effect of increasing K in model B is quite different from that in model A. For the same relative input, $(a-b)/K$, the mean time to fire decreases as K becomes larger because the probability of receiving excitatory input increases linearly with levels above the resting state.

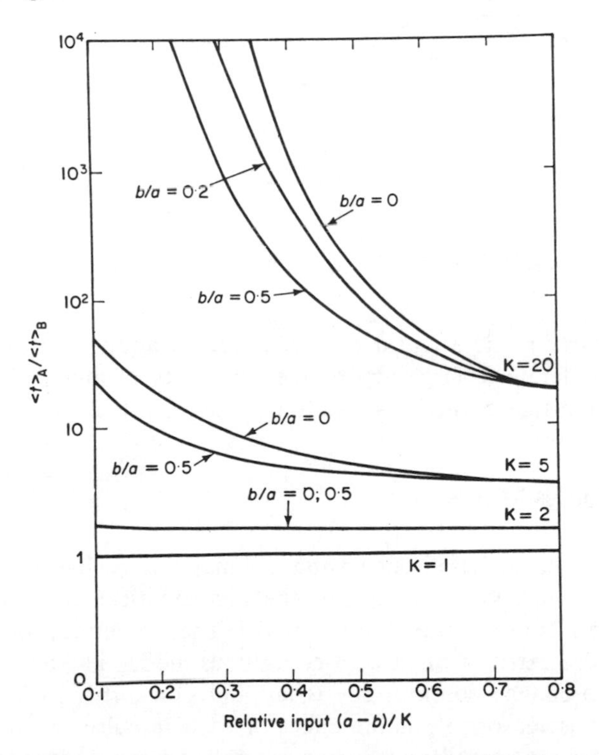

Fig. 8.6 Direct comparison of models A and B. The ratio of output frequencies, or $\langle t_A \rangle / \langle t_B \rangle$, is given versus input. For small threshold or large input, inhibition does not affect this ratio.

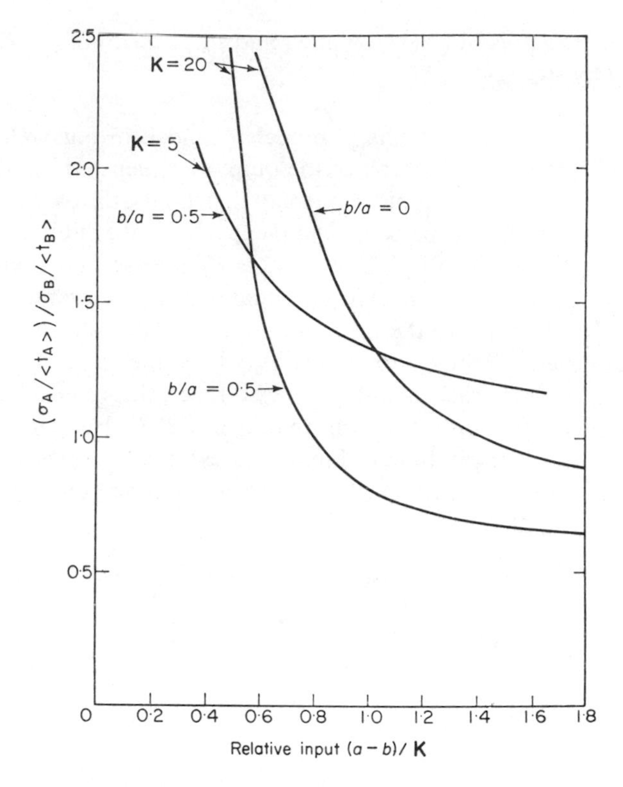

Fig. 8.7 Ratio of $\sigma/\langle t\rangle$ for model A to that for model B.

The remarks made about $\sigma/\langle t\rangle$ for model A apply also for model B. In fact, for moderate to large inputs, the variability of model B is not greatly different from that of model A as can be seen from Fig. 8.7.

8.2 Continuous Models

In the discrete models discussed above, a major defect has been the exclusion of refractory behavior, because of the mathematical complications which ensue when it is incorporated in the model. These complications can be overcome to some extent if one uses a continuous model, i.e., assume membrane potential to change continuously (Clay and Goel, 1973; Clay, 1972). This assumption is reasonable if the effect of each impulse on that part of the membrane where the spike is initiated, is small compared to the total excitatory effect needed to fire the neuron, i.e., to the threshold potential.

Let $x(t)$ denote the membrane (somatic) potential at time t, and $B(t)$ the threshold potential, with $t = 0$ as the time of the last firing of the neuron. In the continuous model we assume that the membrane potential satisfies the differential equation

$$dx/dt = h(x) + e(x)i(t) \tag{1}$$

where $i(t)$ is the input signal to the neuron under consideration, $e(x)$ describes the effect of the input signal, and $h(x)$ is the function describing the change (decay) of the membrane potential in the absence of the input signal. Both $h(x)$ and $e(x)$ are assumed to be differentiable functions. Since the input signal $i(t)$ comes from the many neurons that are connected to the particular neuron under consideration, it depends on the firing of other neurons and the connections between neurons. In the absence of any detailed knowledge about both of these factors, we assume $i(t)$ to have the same characteristics as $i(t)$ of Chapter 3 (see p. 37). With this assumption, for a given form of $e(x)$ and $h(x)$, we can use the techniques of Chapter 3 to analyze the firing of neurons. Since we assume that in the absence of input the potential drops to the resting value (set to be 0) at a rate which is proportional to the deviation from this value, $h(x)$ in Eq. (1) takes the form

$$h(x) = -x/\tau. \tag{2}$$

where τ is the rate constant of the decay process. We further assume that the change in potential due to arriving input is proportional to the input and is independent of the present value of the potential. Thus Eq. (1) becomes

$$dx/dt = -x/\tau + i(t) \tag{3}$$

where $i(t)$ is measured in voltage units. Equation (3) is equivalent to [see Eqs. (3.0.16), (3.0.18), and (3.0.19)]

$$\frac{dx}{dt} = \left(-\frac{x}{\tau} + m\right) + \sigma F(t) \tag{4}$$

where

$$F(t) = [i(t) - m]/\sigma \tag{5}$$

satisfies Eqs. (3.0.17); i.e., it is a white noise. This assumption is valid if the fluctuations in the input functions occur much faster than the changes in the somatic potential. Comparing Eq. (4) with Eq. (3.0.18), we find that the probability density $P(x|y,t)$ of somatic potential having the value x at time t, given that it was equal to y at $t = 0$, satisfies the Fokker–Planck equation

$$\frac{\partial P}{\partial t} = -\frac{\partial}{\partial x}\left(\left(m - \frac{x}{\tau}\right)P\right) + \frac{\sigma^2}{2}\frac{\partial^2 P}{\partial x^2} \tag{6a}$$

and the backward diffusion equation

$$\frac{\partial P}{\partial t} = \left(m - \frac{y}{\tau}\right)\frac{\partial P}{\partial y} + \frac{\sigma^2}{2}\frac{\partial^2 P}{\partial y^2} \tag{6b}$$

In an actual nervous system, the input impulses do not arrive continuously, and the diffusion equations (6) can also be derived by considering all possible discrete transitions and taking a limit. If we assume that the neuron receives excitatory and inhibitory inputs which are Poisson distributed in time with f_e (f_i) as the average frequency of the excitatory (inhibitory) input, and e (i) as the amplitude of an excitatory (inhibitory) input, then in the continuous approximation [in view of Eq. (3.0.1), where $e = i = 1$],

$$m = f_e e - f_i i \tag{7a}$$

$$\sigma^2 = f_e e^2 + f_i i^2 \tag{7b}$$

Further discussion of these relations is given by Capocelli and Ricciardi (1971). The variable x is restricted to the region $-\infty < x \leqslant B(t)$, with $B(t)$, the threshold membrane potential, as the absorbing boundary. Equation (6a) is subjected to the initial and boundary conditions

$$P(x \mid y, 0) = \delta(y - x) \tag{8a}$$

$$P(B(t) \mid y, t) = 0 \tag{8b}$$

$$P(-\infty \mid y, t) = 0 \tag{8c}$$

The initial condition (8a) reflects the fact that no changes can occur in a zero time interval; the second condition reflects the existence of a threshold potential $B(t)$ such that when $x(t)$ takes the value $B(t)$ the process is absorbed; and the last condition is a reflection of the fact that in finite time, only finite changes can take place.

We divide our analysis in two parts, one in which we take threshold to be time independent (constant threshold), and the other in which it is time dependent (variable threshold)

Constant threshold. Here we assume that the threshold potential is immediately reset to the constant value $B(t) \equiv B$ after each firing. Since the process is restricted to the region $-\infty < x \leqslant B$, with B an absorbing boundary, this case can be analyzed by using the methods of Section 3.2, and in particular of Section 3.2D.

For the special case in which the decay of the potential toward its resting value is much slower than the time between firings, we can take $\tau \to \infty$ in Eq. (6a) to get the equation

$$\frac{\partial P}{\partial t} = -m\frac{\partial P}{\partial x} + \frac{\sigma^2}{2}\frac{\partial^2 P}{\partial x^2} \tag{9}$$

The process described by this equation is the Wiener process. For this process, from Table 3.4,

$$P(x\,|\,y,t) = \rho\left(\frac{x-y-mt}{\sigma(t^{1/2})}\right) - \rho\left(\frac{x+y-2B-mt}{\sigma(t^{1/2})}\right)\exp\{2m(B-y)/\sigma^2\} \tag{10}$$

where $\rho(z/\sigma)$ is the normal distribution

$$\rho\left(\frac{z}{\sigma}\right) = \frac{1}{(2\pi\sigma^2)^{1/2}}\exp\left(-\frac{z^2}{2\sigma^2}\right) \tag{11}$$

The first passage time density can be calculated by using Eq. (3.2.24), and is given by

$$F(B\,|\,y,t) = \frac{B-y}{t}\,\rho\left(\frac{B-y-mt}{\sigma(t^{1/2})}\right) \tag{12}$$

Since for the process under consideration $a(x) = m$, $b(x) = \sigma^2$, from Table 3.11,

$$\pi(x) = \exp(-2mx/\sigma^2) \tag{13}$$

and the probability of membrane potential taking the value z before firing (i.e., the variable taking the value z before getting absorbed) is

$$R(z\,|\,y) = \frac{\exp(-2mB/\sigma^2)-\exp(-2my/\sigma^2)}{\exp(-2mB/\sigma^2)-\exp(-2mz/\sigma^2)} \qquad \text{for} \quad z < y < B \tag{14}$$

Therefore, the probability of neuron ever firing (i.e., the probability of the variable taking the value B without drifting to $-\infty$) is

$$R(B\,|\,y) = 1 - R(-\infty\,|\,y) = \begin{cases} 1, & m > 0 \\ \exp(-2(B-y)|m|/\sigma^2), & m < 0 \end{cases} \tag{15}$$

Thus the neuron will definitely fire when $m > 0$. For this case, from Table 3.11, the first moment and variance of firing time are

$$M_1(B\,|\,y) = (B-y)/m, \qquad m > 0 \tag{16a}$$

$$V(B\,|\,y) = M_2(B\,|\,y) - \{M_1(B\,|\,y)\}^2 = (B-y)\,\sigma^2/m^3, \qquad m > 0 \tag{16b}$$

Thus the average frequency of firing, $1/M_1$, is directly proportional to the average input m and inversely proportional to the difference between threshold potential and the reset potential.

Let us now consider the more realistic case where τ is finite and $m > 0$. Making the substitution

$$\xi = x/\tau - m, \qquad \xi_0 = y/\tau - m \tag{17}$$

the FP equation (6a) becomes

$$\frac{\partial P^*}{\partial t} = \frac{1}{\tau}\frac{\partial}{\partial \xi}(\xi P^*) + \frac{\sigma^2}{2\tau^2}\frac{\partial^2 P^*}{\partial \xi^2} \tag{18}$$

where

$$P^*(\xi \mid \xi_0, t)\, d\xi = P(x \mid y, t)\, dx \tag{19a}$$

and the absorbing boundary is

$$B^* = B/\tau - m \tag{19b}$$

Equation (18) is the FP equation of the OU process discussed in Appendix G. From Table 3.11

$$\pi(\xi) = \exp(\tau\xi^2/\sigma^2) \tag{20}$$

and the probability of firing is

$$R(B^* \mid \xi_0) = 1 - R(-\infty \mid \xi_0) = \int_{-\infty}^{\xi_0} \exp(\tau\xi^2/\sigma^2)\, d\xi \bigg/ \int_{-\infty}^{B^*} \exp(\tau\xi^2/\sigma^2)\, d\xi = 1 \tag{21}$$

i.e., the firing is a certain event. Explicit expressions for $P(x \mid y, t)$ or $F(B \mid y, t)$, the probability density of firing interval, are not derivable. However, an explicit expression for the Laplace transform $f(B \mid y, t)$ of $F(B \mid y, t)$ can be derived. From Eqs. (3.2.13) and (3.2.15), $f(B \mid y, t)$ satisfies the differential equation and the boundary conditions

$$sf(B \mid y, s) = \left(m - \frac{y}{\tau}\right)\frac{\partial f}{\partial y} + \frac{\sigma^2}{2}\frac{\partial^2 f}{\partial y^2} \tag{22}$$

$$f(B \mid B, s) = 1, \qquad \lim_{y \to -\infty} f(B \mid y, s) = 0 \tag{23}$$

Making the transformation (17) these equations become

$$\frac{\sigma^2}{2\tau}\frac{\partial^2 f^*}{\partial {\xi_0}^2} - \xi_0 \frac{\partial f^*}{\partial \xi_0} - s\tau f^* = 0 \tag{24}$$

where

$$f(B \mid y, t) = f^*(B^* \mid \xi_0, t) \tag{25a}$$

B^* is defined by Eq. (19b) and

$$f^*(\xi_0 \mid B^*, s) = 1, \qquad \lim_{\xi_0 \to -\infty} f^*(B^* \mid \xi_0, s) = 0 \tag{25b}$$

A solution to the differential equation (24) (Whittaker and Watson, 1952) is

$$u(\xi_0, s) = \exp\left(\frac{\tau}{2\sigma^2}{\xi_0}^2\right) D_{-s\tau}((2\tau)^{1/2}\xi_0/\sigma) \tag{26}$$

where $D_{-s\tau}(x)$ is the parabolic cylindrical function (or Weber function) with the following integral representation

$$D_{-s\tau}(x) = \frac{\exp(-x^2/4)}{\Gamma(s\tau)} \int_0^\infty \exp\left(-x\eta - \frac{\eta^2}{2}\right) \eta^{s\tau-1}\, d\eta \tag{27}$$

Therefore, Eq. (26) becomes

$$u(\xi_0, s) = \frac{1}{\Gamma(s\tau)} \int_0^\infty \exp[-(2\tau)^{1/2}\xi_0\eta/\sigma - \eta^2/2]\, \eta^{s\tau-1}\, d\eta \tag{28}$$

Since $\lim_{\xi_0 \to -\infty} u(\xi_0, s) = 0$, the boundary conditions (25b) are satisfied if we choose

$$F^*(B^* \mid \xi_0, t) = \frac{u(\xi_0, s)}{u(B^*, s)} \tag{29}$$

as the solution of Eq. (24). Transforming back into the variables y and B, the final expression for the Laplace transform of first passage time density becomes

$$f(B \mid y, s) = u\left(\frac{y}{\tau} - m, s\right) \Big/ u\left(\frac{B}{\tau} - m, s\right) \tag{30}$$

where u is defined either by Eq. (26) or Eq. (28). The first representation is more convenient when tables of parabolic cylindrical functions are available (Abramowitz and Stegun, 1964, p. 700).

The form (30) of $f(B \mid y, s)$ is quite complicated for the calculation of moments of first passage time, since the variable s is in the index of the parabolic cylindrical function. It is more convenient to use directly the expressions given in Table 3.11. For the process under discussion

$$\pi(x) = \exp\left\{-2\int^x \frac{(m-\eta/\tau)}{\sigma^2}\, d\eta\right\} = \exp\left\{\frac{(x-m\tau)^2}{\sigma^2\tau}\right\} \tag{31}$$

Therefore, from column 1 of Table 3.11, the average firing time is

$$M_1(B \mid y) = \frac{(\pi\tau)^{1/2}}{\sigma} \int_y^B \exp\left(\frac{x-m\tau}{\sigma\tau^{1/2}}\right)^2 \phi\left(\frac{x-m\tau}{\sigma\tau^{1/2}}\right) dx \tag{32}$$

where $\phi(x)$ is defined by

$$\phi(x) = \frac{2}{\pi^{1/2}} \int_{-\infty}^x \exp(-\xi^2)\, d\xi \tag{33}$$

The expression (32) is simplified by introducing the dimensionless variable

$$v = (x-m\tau)/\sigma\tau^{1/2} \tag{34}$$

The resulting expression is

$$M_1(B|y) = 2\tau \int_{v_0}^{v_B} dv \exp(v^2) \int_{-\infty}^{v} \exp(-w^2)\, dw \tag{35}$$

where

$$v_B \equiv (B-m\tau)/\sigma\tau^{1/2}, \qquad v_0 = (y-m\tau)/\sigma\tau^{1/2} \tag{36}$$

By making the substitution $\eta = v+w$, the expression (35) is further simplified into

$$\frac{1}{\tau} M_1(B|y) = \int_0^{\infty} d\eta \exp(-\eta^2)[\exp(2\eta v_B) - \exp(2\eta v_0)] \Big/ \eta$$

$$= \sum_{n=1}^{\infty} 2^{n-1} \frac{\Gamma(n/2)}{\Gamma(n+1)} (v_B{}^n - v_0{}^n) \tag{37}$$

where the last equality is derived by expanding the bracketed term in the integrand and using standard integral formulas [Abramowitz and Stegun, 1964; Eqs. (7.4.4) and (7.4.5)]. Equation (37) exhibits the dependence of average firing time on two parameters v_B and v_0 [defined by Eq. (36)] which incorporate in them a combination of the properties of the input (m and σ) and of the neuron (B and τ).

The higher moments of firing time can be obtained using the formula given in Table 3.11 for the jth moment

$$M_j(B|y) = 2j \int_y^B d\eta\, \pi(\eta) \int_{-\infty}^{\eta} \frac{M_{j-1}(B|\xi)}{\sigma^2 \pi(\xi)}\, d\xi \tag{38a}$$

which in view of Eq. (31) becomes

$$M_j(B|y) = \frac{2j}{\sigma^2} \int_y^B d\eta \exp\left(\frac{\eta - m\tau}{\sigma\tau^{1/2}}\right)^2 \int_{-\infty}^{\eta} M_{j-1} \exp\left[-\left(\frac{\xi - m\tau}{\sigma\tau^{1/2}}\right)^2\right] d\xi \tag{38b}$$

This expression can be used to calculate moments recursively.

It should be noted that Eq. (30) can be numerically Laplace inverted to give $F(B|y,t)$. If λ_j are the poles of $f(B|y,s)$ [which form a discrete set (Titchmarsh, 1962) and are nonpositive (Keilson, 1964)], with the concomitant residues β_j, then

$$F(B|y,t) = \sum_{j=0}^{\infty} \beta_j \exp(\lambda_j t) \tag{39}$$

Keilson and Ross (1971) have given computational techniques to determine the poles and residues, and hence $F(B|y,t)$ can be readily found via high-speed computers.

The analytical results derived above are implemented after we discuss the case of variable threshold.

Variable threshold. We now remove the artificial restriction of a fixed threshold and examine the effect of refractoriness. As noted in Section 3.2, when the threshold varies in time, there is no general theory for calculating first passage time density, and hence no extensive literature on neuron firing exists when refractoriness has been included. We will closely follow the treatment given in a recent paper (Clay and Goel, 1973) where variable threshold is considered.

We assume that during the absolute refractory period τ_A, the soma membrane potential returns to a reset level y and that it remains at that level throughout the absolute refractory period. This hypothesis is physiologically reasonable, since the action potential may, in fact, partially propagate back over the soma and into the dendritic tree, thereby behaving as an active reset mechanism of duration τ_A (Eccles, 1964). Consequently, we need only add τ_A to the average interval between pulses. The neuron then gradually recovers firing capability during the next few milliseconds of the relative refractory period, when it can be made to fire again if it receives excitatory input much larger than when the cell is quiescent. We model this behavior by a simple threshold function $B(t)$, where $B(t)$ is a function which is initially very large or infinite, and then returns monotonically to the stationary level B. Examples of such simple functions are

$$B(t) = B + \beta e^{-t/\gamma} \qquad \text{with} \quad \beta/B \gg 1 \tag{40}$$

and

$$B(t) = Be^{+\alpha/t} \qquad \alpha > 0 \tag{41}$$

The latter was first suggested by Hagiwara (1954) on the basis of experiments on the sciatic nerve of the frog semitendinous muscle.

For the profile (40) with $\gamma = \tau$, an exact solution for the first passage time density $F(B \mid y, t)$ can be derived. For other values of γ and for the profile (41) an approximation technique must be used. This approximation technique has been used for the profile (40) with $\gamma = \tau$ to test the accuracy of the technique.

Let us first make the transformations

$$u = (x - m\tau)\, e^{t/\tau} \tag{42a}$$

$$w = e^{2t/\tau} \tag{42b}$$

on the FP equation (6a) and threshold function (40) (with $\gamma = \tau$). The resulting equations are

$$\frac{\partial P'}{\partial w} = \frac{\sigma^2 \tau}{4} \frac{\partial^2 P'}{\partial u^2} \tag{43}$$

$$B'(w) = \beta + (B - m\tau)\, w^{1/2} \tag{44}$$

where

$$P(x\,|\,y,t)\,dx = P'(u\,|\,u_0, w\,|\,w_0)\,du \tag{45a}$$

w_0 and u_0 are the initial values of w and u, respectively, i.e.,

$$u_0 = y - m\tau, \qquad w_0 = 1 \tag{45b}$$

Using Eqs. (42a) and (42b), Eq. (45a) becomes

$$P = P'w^{1/2} \tag{46}$$

For $B = m\tau$, the threshold (44) becomes

$$B'(w) = \beta$$

a constant threshold. Equation (43) is the FP equation for a Wiener process with $a(u) = 0$. Hence, from Table 3.4 [or from Eq. (10)]

$$P' = \{\pi\tau\sigma^2(w-w_0)\}^{-1/2}\left\{\exp\left[-\frac{(u-u_0)^2}{\sigma^2\tau(w-w_0)}\right] - \exp\left[-\frac{(u+u_0-2\beta)^2}{\sigma^2\tau(w-w_0)}\right]\right\} \tag{47}$$

Transforming back to the variables y and t, by using Eqs. (42) and (45), we get

$$P(x\,|\,y,t) = \frac{e^{t/\tau}}{[\pi\tau\sigma^2(e^{2t/\tau}-1)]^{1/2}} \times\left[\exp\left\{-\frac{(x-y)^2 e^{2t/\tau}}{\sigma^2\tau(e^{2t/\tau}-1)}\right\} - \exp\left\{\frac{[(x+y-2\tau)\,e^{t/\tau}-2\beta]^2}{\sigma^2\tau(e^{2t/\tau}-1)}\right\}\right] \tag{48}$$

The density of first passage time to the variable threshold $B(t) = m\tau + \beta e^{-t/\tau}$, derived from Eq. (48), in view of relation (3.2.24), is given by

$$F(B(t)\,|\,y,t) = \frac{2[(m\tau-y)\,e^{t/\tau}+\beta]\,e^{2t/\tau}}{[\pi\sigma^2\tau^3(e^{2t/\tau}-1)^3]^{1/2}}\exp\left\{-\frac{[\beta+(m\tau-y)\,e^{t/\tau}]^2}{\sigma^2\tau(e^{2t/\tau}-1)}\right\} \tag{49}$$

For $m \neq B/\tau$ but still $\gamma = \tau$, the Laplace transform of the first passage time density for the variable threshold (40) can be calculated by using one of the so-called Wald's identities, used frequently in probability theory. One such identity is (Daniels, 1969; Clay and Goel, 1973)

$$1 = \int_{w_0}^{\infty} \exp\{k(u_T-u_0) + k^2\sigma^2\tau(w_0-w)/4\}\,F(u_T\,|\,u_0; w\,|\,w_0)\,dw, \qquad k \geqslant 0 \tag{50}$$

where u_T is the threshold value of u, i.e., from Eqs. (40) and (42a)

$$u_T = \beta + (B-m\tau)\,w^{1/2} \tag{51}$$

Multiplying Eq. (50) by k^{s-1}, performing a simple algebraic manipulation, and integrating over k, we obtain

$$\int_0^\infty dk\, k^{s-1} \exp(ku_0 - k\beta - k^2\sigma^2\tau/4)$$

$$= \int_0^\infty dk\, k^{s-1} \int_1^\infty \exp\{k(B-m\tau)\, w^{1/2} - k^2\sigma^2\tau w/4\}\, F\, dw \equiv Q \quad (52)$$

If we let $\eta = kw^{1/2}$, the integral on the right-hand side of Eq. (52) becomes

$$Q = \int_0^\infty d\eta\, \eta^{s-1} \exp\{\eta(B-m\tau) - \sigma^2\eta^2\tau/4\} \int_1^\infty w^{-s/2} F\, dw$$

Since

$$\int_1^\infty w^{-s/2} F\, dw = f(u_{\mathrm{T}} \mid u_0, s) \quad (53)$$

where f is the Laplace transform of F in t space, Eq. (52) becomes

$$f(u_{\mathrm{T}} \mid u_0, s) = \frac{\int_0^\infty dk k^{s-1} \exp[k(u_0 - \beta) - k^2\sigma^2\tau/4]}{\int_0^\infty dk\, k^{s-1} \exp[k(B-m\tau) - k^2\sigma^2\tau/4]}$$

$$= \frac{D_{-s}(\xi)}{D_{-s}(\xi_0)} \exp[(\xi^2 - \xi_0{}^2)/4] \quad (54)$$

where

$$\xi = 2^{1/2}(\beta - u_0)/\sigma\tau^{1/2}, \qquad \xi_0 = 2^{1/2}(B - m\tau)/\sigma\tau^{1/2} \quad (55)$$

and $D_{-s}(x)$ is the parabolic cylindrical function used earlier in this section [Eq. (27)]. The average first passage time $M_1(B \mid y)$ can be readily calculated by differentiating Eq. (54) and using the relation

$$M_1(B \mid y) = -\left.\frac{\partial f}{\partial s}\right|_{s=0}$$

The resulting expression is

$$M_1(B \mid y) = \frac{\tau}{2} \sum_{k=0}^{\infty} 2^{k+1} \frac{\xi_0^{2k+2} - \xi^{2k+2}}{(2k+1)!!\,(k+1)} + 2\pi^{1/2}[\xi_0 F(\tfrac{1}{2}; \tfrac{3}{2}; \xi_0{}^2) - \xi F(\tfrac{1}{2}; \tfrac{3}{2}; \xi^2)] \quad (56)$$

where $F(a; b; z)$ is the confluent hypergeometric function (Abramowitz and Stegun, 1964, Chapter 13) and $(2k+1)!! = 1 \cdot 3 \cdot 5 \cdots (2k+1)$. The average frequency of firing of the neuron is then given by $[\tau_{\mathrm{A}} + M_1(B \mid y)]^{-1}$.

For other variable thresholds, analytical results for $\langle t \rangle$ have not been obtained without making further approximations. A reasonable way (Johannesma, 1968; Clay and Goel, 1973) to approach the problem is to assume that the threshold is constant for all time, but that all input processes are slowed down by a factor $j(t)$ related to the threshold functions (40) and (41) by

$$j(t) = B/B(t) \tag{57}$$

In other words, the time between input pulses is effectively lengthened for small t, so that the first passage time to threshold is increased. Mathematically this is the same as replacing the input frequencies f_e (f_i) by $j(t)f_e$ [$j(t)f_i$], and the rate of decay of the membrane potential by $\tau/j(t)$. The resulting FP equation is

$$\frac{\partial}{\partial y}\left(\frac{y}{\tau} - m\right)P + \frac{\sigma^2}{2}\frac{\partial^2 P}{\partial y^2} = \frac{1}{j(t)}\frac{\partial P}{\partial t} = \frac{\partial P}{\partial I} \tag{58a}$$

where

$$I(t) = \int_0^t j(u)\, du \tag{58b}$$

and

$$P(B \mid y, I(t)) = 0 \tag{58c}$$

That is, the solution to this problem is the same as for a fixed threshold, but with the transformations

$$t \to I(t) \tag{59a}$$

$$F(B(t) \mid y, t) \to F(B \mid y, I(t)) \tag{59b}$$

and

$$F(B \mid y, I(t))\, dI = F(B \mid y, t)\, j(t)\, dt \tag{59c}$$

Consequently, by using the expression (39) for $F(B \mid y, t)$, the expression for the first passage time density for the refractory threshold can be obtained. The details of the calculations of $F(B \mid y, t)$ are given by Clay and Goel (1973).

To implement the analysis just discussed to the calculation of the input–output curves for the model neuron, we need parameters. The difficulty is that there is very little good experimental information concerning the parameters of the model for cortical neurons. Most of the data on individual neuronal behavior come from experiments on motoneurons, and these results cannot be transferred in toto to other types of neurons because of known differences in behavior. However, intracellular recording techniques have yielded much information about reticulospinal neurons (Willis and Magni, 1964), which are much closer to cortical neurons in the neural hierarchy, and so we employ these results for the model. The experiments indicate that the absolute refractory period τ_A is 1.5 msec and the duration of the relative refractory period τ_R is about 5 msec. We choose τ_R to correspond to the time elapsed after $t = \tau_A$ at which the neuron returns to 90% effectiveness for generating another action potential, i.e., $j(\tau_R) = 0.9$. For the Hagiwara threshold, Eq. (41), this condition immediately determines α, the single parameter of the threshold. We have

$$j(\tau_R) = 0.9 = \exp(-\alpha/\tau_R) \tag{60}$$

which gives $\alpha \simeq 0.1\tau_R = 0.5$ msec.

For the exponential threshold, Eq. (40),

$$j(\tau_R) = 0.9 = \left[1 + \frac{\beta}{B}\exp(-\tau_R/\gamma)\right]^{-1} \tag{61}$$

Of course both of the parameters β and γ cannot be determined by Eq. (61), but we wish β/B to be large because the threshold is large at time $t = \tau_A$ and, since the input–output results are insensitive to the value of β as long as $\beta/B \gg 1$, we arbitrarily choose $\beta/B = 100$. Equation (61) then gives $\gamma \simeq 0.1\tau_R$. A graphical comparison of the two thresholds is given in Fig. 8.8.

The other two membrane parameters we need are τ and B. For mammalian neurons $\tau \simeq 5$ msec; i.e., $\tau = \tau_R$. The value of B, the height of the threshold relative to the effect on the soma potential of a single excitatory input pulse, is not known accurately for cortical neurons. In previous studies [cf. Stein

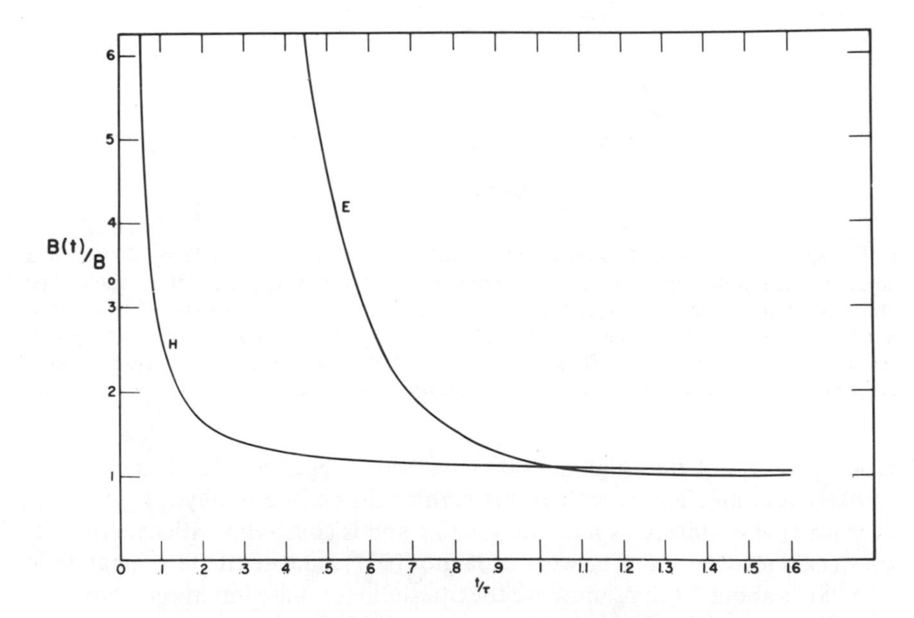

Fig. 8.8 Threshold functions $B(t)/B$ for the exponential case (E) for which $B(t)/B = 1+\beta/Be^{-t/\gamma}$ with $\gamma = 0.1\tau$ and for the Hagiwara case (H) for which $B(t)/B = e^{\alpha/t}$ with $\alpha = 0.1\tau$.

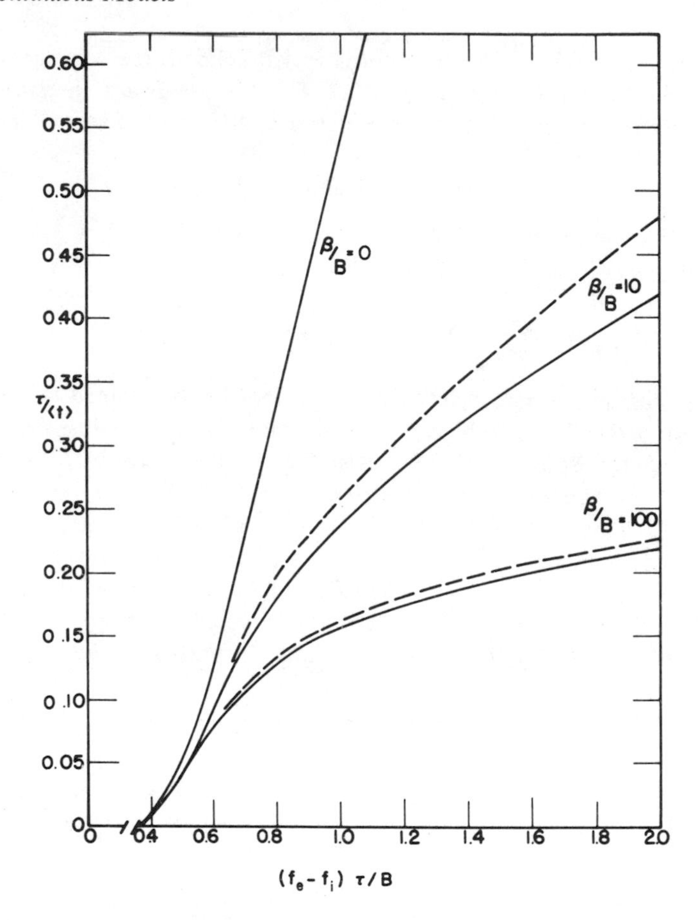

Fig. 8.9 Comparison of exact results (solid lines) with approximate results (dashed lines) for the input–output response of the model neuron for the exponential threshold. The threshold decay constant $\gamma = \tau$, the stationary level $B = 100$, the ratio of inhibitory input to excitatory input is 0.9, and the initial ($t = 0$) value of the threshold is chosen as B ($\beta/B = 0$, no refractoriness), $11B$ ($\beta/B = 10$), and $101B$ ($\beta/B = 100$). The approximate and exact results coincide for $\beta = 0$, and they are nearly identical for large values of β/B.

(1967)], values of B as high as 1000 have been suggested. However, it seems unlikely that an element with B this high could be of any physiological use. A value of $B \sim 100$ seems more reasonable and is consistent with experiments on spinal motoneurons. The work of Kuno (1971) indicates that the magnitude of a PSP is about 0.1 mV, and since the threshold for pulse initiation is typically 10 mV, a value of 100 is not an unreasonable choise for B.

We now have all of the membrane parameters necessary for evaluation of input–output behavior under stochastic stimulation. To check the validity of the approximation technique outlined in the preceding section, we take $j(t)$ to be $(1+(\beta/B)e^{-t/\tau})^{-1}$ corresponding to the threshold profile with the exact solution (54), and compare the two methods. The results are shown in Fig. 8.9 where we see that the two approaches give practically the same response curves, especially as β/B is increased. (In this graph we have taken τ_A to be zero to show directly how relative refractory behavior influences the response of the model neuron.) Consequently, the approximation method is very reliable. It can be used for other rapidly decaying thresholds such as the Hagiwara or exponential threshold with $\gamma < \tau$.

The characteristic response of elements with these thresholds is shown in Fig. 8.10 where the values of the parameters B, τ, τ_A, and τ_R given above have

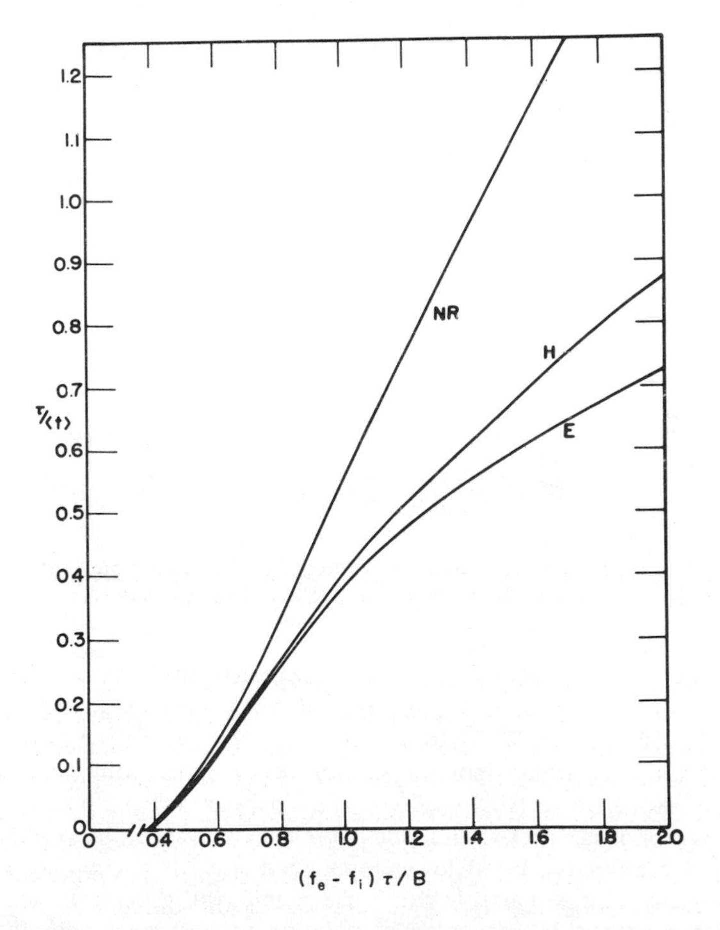

Fig. 8.10 Input–output response for the H and E thresholds with absolute fractoirness included. The NR curve corresponds to a nonrefractory neuron.

been used, and the initial or reset value of the membrane potential is taken to be zero. For comparison, we have also given the response curve for the constant threshold case (no refractory period). For varying threshold, below a certain input frequency, there is a cutoff in the neuron response, and for high input frequency there is a saturation in response. This result is not meant to be the response curve of any specific system, but it undoubtedly possesses some of the features of mammalian cortical neurons, such as a cutoff in response below a certain input frequency and a saturation in response for high input frequencies. Furthermore, the general shape of the curves agrees quite well with the experimental response curves of stochastically stimulated cat spinal motoneurons (Redman *et al.*, 1968), which are given in Fig. 8.11. In

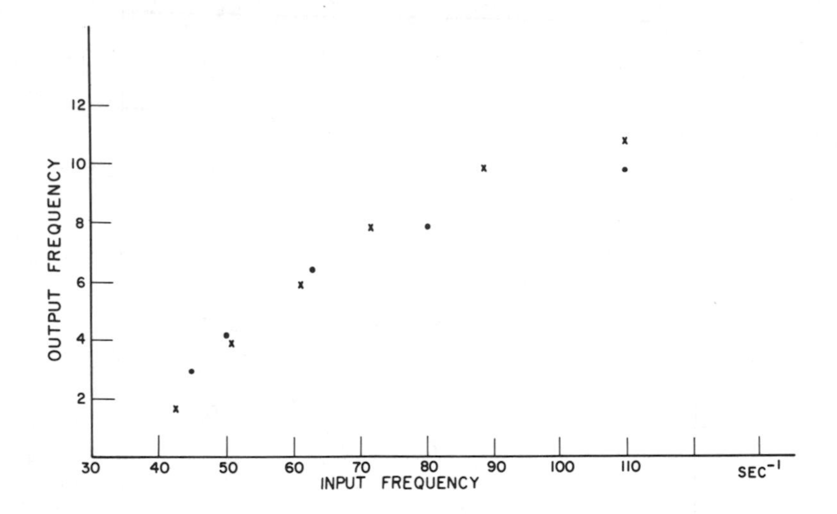

Fig. 8.11 Experimental input–output data taken from Redman *et al.* (1968), where the dots correspond to Poisson input, and the crosses to asynchronous input.

these experiments stochastic input was generated either via a weak γ-ray source or via a few periodic input trains, all of the same frequency but summated asynchronously. The latter is not a true stochastic process. However, for observation times less than the shortest period of the compound process, the input appears essentially random. The difference in temporal summation by the neuron is clearly shown in the figure. When asynchronous stimulation is used, the cell saturates at lower input than when it is stimulated purely stochastically. Consequently, there are significant differences in neural behavior produced by stochastic stimulation as compared with the more commonly used periodic stimulation techniques. The models presented in this chapter are primarily intended for the former type of experiments.

References

Abramowitz, M., and Stegun, I. A., eds. (1964). "Handbook of Mathematical Functions." Nat. Bur. Stand., Washington, D. C.

Capocelli, R. M., and Ricciardi, L. M. (1971). Diffusion approximation and first passage time problem for a model neuron. *Kybernetik* (*Berlin*) **8**, 214.

Clay, J. R. (1972). Models for Firing of a Neuron. Ph.D. Thesis, Univ. of Rochester, Rochester, New York.

Clay, J. R., and Goel, N. S. (1973). Diffusion models for firing of a neuron with a varying threshold. *J. Theore*). *Biol.* **39**, 633.

Daniels, H. E. (1969). The minimum of a stationary markov process superimposed on a U-shaped trend. *J. Appl. Probability* **6**, 399.

Eccles, J. C. (1964). "The Physiology of Synapses." Academic Press, New York.

Eccles, J. C. (1969). "The Inhibitory Pathways of the Central Nervous System." Banner House, Springfield, Illinois.

Goel, N. S., Richter-Dyn, N., and Clay, J. R. (1972). Discrete stochastic models for firing of a neuron. *J. Theoret. Biol.* **34**, 155.

Hagiwara, S. (1954). Analysis of interval fluctuation of the sensory nerve impulse. *Jap. J. Physiol.* **14**, 234.

Johannesma, P. I. M. (1968). Diffusion models for the stochastic activity of neurons, *in* 'Neural Networks" (E. R. Caianello, ed.) p. 116. Springer-Verlag, Berlin and New York.

Keilson J. (1964). A review of transient behaviour in regular diffusion and birth-death processes. *J. Appl. Probability* **1**, 247.

Keilson, J., and Ross, H. (1971). Passage time distributions of the Ornstein-Uhlenbeck process. Rep. 71-08. Center for System Sci. ,Univ. of Rochester, Rochester, New York.

Kuno, M. (1971). Quantum aspects of central and ganglionic synaptic transmission in vertebrates. *Physiol. Rev.* **51**, 647.

Mountcastle, V. B. (1967). The problem of sensing and the neural coding of sensory events, *in* "The Neurosciences" (G. C. Quarton *et al.*, eds.), p. 393. Rockefeller Univ., New York.

Redman, S. J., Lampard, D. G., and Annal, P. J. (1968). Monosynaptic stochastic stimulation of cat spinal motoneurons. II. Frequency transfer characteristics of tonically discharging motoneurons. *J. Neurophysiol.* **31**, 499.

Stein, R. B. (1967). Some models of neuronal variability. *Biophys. J.* **7**, 37.

Stevens, C. F. (1966). "Neurophysiology: A Primer." Wiley, New York.

ten-Hoopen, M. (1966). Probabilistic firing of neurons considered as a first passage problem. *Biophys. J.* **6**, 435.

Titchmarsh, E. C. (1962). "Eigen-function Expansions Associated with Second-Order Differential Equations." Oxford Univ. Press (Clarendon), London and New York.

Whittaker, E. T., and Watson, G. N. (1952). "A Course of Modern Analysis," 4th ed. Cambridge Univ. Press, London and New York.

Willis, W. D., and Magni, F. (1964). The properties of reticulo-spinal neurons. *Progr. Brain Res.* **12**, 53.

Additional References

Eysel, U. T. (1971). Computer simulation of the impulse pattern of muscle spindle afferents under static and dynamic conditions. *Kybernetik* (*Berlin*) **10**, 171.

Griffith, J. S. (1971). "Mathematical Neurobiology." Academic Press, New York.

Sugiyama, H., Moore, G. P., and Perkel, D. H. (1970). Solutions for a stochastic model of neuronal spike production. *Math. Biosci.* **8**, 323.

9

Chemical Kinetics

In this chapter we focus our attention on the stochastic modeling of biological systems at the lowest level of organization, namely, the molecular level, and in particular, on the kinetics of biochemical reactions. Since three comprehensive reviews exist on the stochastic approach to chemical kinetics (Montroll, 1967; McQuarrie, 1968; Teramoto *et al.*, 1971), and since in the second review the chemical kinetics of biological systems is discussed, we are brief in our discussion and analysis. We discuss the kinetics of three biochemical reactions: the conformational changes in the biopolymers, in particular DNA, the biosynthesis of biopolymers, and the enzyme reaction. Our discussion is meant to update and complement the discussion by McQuarrie (1968).

9.1 Conformational Changes of Biopolymers

It is now a well-known fact that the stability of the structure of macromolecules, in particular of biopolymers [deoxyribonucleic acid (DNA), ribonucleic acid (RNA), and proteins], is environment dependent [see Goel and Montroll (1968), Wartell and Montroll (1972), Poland and Scheraga (1970)]. For example, upon appropriate conditions, the well-known double-stranded helical structure of DNA (proposed by Watson and Crick) could be transformed into two separated strands in a random coil form. The enzymatic action of the proteins can also be drastically altered by appropriate change in the environment. This alteration is believed to be due to changes in the structure of the protein. The typical changes in the environment which cause structural transformation are changes in the ionic strength, temperature, acidity or basicity (pH change), concentration of certain organic solvents, or concentrations of other agents, e.g., urea, dyes, metallic ions. The structural transformation is referred to as melting, denaturation, helix-coil transition, and uncoiling. Both the equilibrium and kinetic aspects of structural transformations have been investigated theoretically and experimentally. In this section, we show how the kinetics of the process of denaturation and renaturation of a DNA molecule can be studied using the techniques of Chapter 2. According to the Watson–Crick model, a DNA molecule has the form of a double helix with the two strands coupled through bonds which connect bases (small molecules) on one strand to the ones on the other strand. There are four kinds of bases, adenine (A), thymine (T), cytosine (C), and guanine (G). A is always linked to T and G to C. The strength of the two links are different, the G–C link being stronger. For the purpose of illustrating the application of the methods of Chapter 2, we discuss DNA molecules with only one kind of bond. Such DNA molecules can be synthesized, and therefore our analysis can be compared with the experimental data.

Experimentally, denaturation of DNA molecules is usually followed by measuring the ultraviolet absorption of a dilute solution of DNA as a function of environmental parameters, e.g., temperature. When DNA is transformed from natural to the denatural form, the absorbance of light (260 mμ) increases by 40%. This increase reflects a change in the electron configuration of the bases (nucleotides) due to destruction of the double helical structure. If the additional reasonable assumption that the change in absorbance is proportional to the fraction of broken bonds is made, then the change in optical density gives the change in the number of unbroken bonds.

The chemical reactions involved in the process of denaturation and renaturation are

$$2A_0 \underset{\mu_1}{\overset{\lambda_0}{\rightleftharpoons}} A_1 \underset{\mu_2}{\overset{\lambda_1}{\rightleftharpoons}} A_2 \cdots \underset{\mu_N}{\overset{\lambda_{N-1}}{\rightleftharpoons}} A_N$$

where A_n stands for a DNA molecule with n intact bonds, and the various λ's and μ's are the reaction rate constants. Thus λ_0 is the reaction rate constant for the addition of one bond in two separated strands and μ_1 that for breaking the last intact bond of DNA. Let $P_{n,m}(t)$ be the probability that there are n

intact bonds at time t, given that there were m bonds at $t = 0$; then the probabilistic formulation of the reaction above is given by

$$dP_0/dt = -\lambda_0 P_0^2 + \mu_1 P_1 \tag{1a}$$

$$dP_1/dt = \lambda_0 P_0^2 - (\mu_1 + \lambda_1) P_1 + \mu_2 P_2 \tag{1b}$$

$$dP_n/dt = \lambda_{n-1} P_{n-1} - (\mu_n + \lambda_n) P_n + \mu_{n+1} P_{n+1}, \qquad n \neq 0, 1, N \tag{1c}$$

$$dP_N/dt = \lambda_{N-1} P_{N-1} - \mu_N P_N \tag{1d}$$

where, for simplicity in notation, we have suppressed the subscript m.

Equations (1) are exactly the same equations as (2.0.3) for the discrete birth and death process, with the exception of a quadratic term $\lambda_0 P_0^2$ in Eqs. (1a) and (1b) for $\dot{P}_0$ and $\dot{P}_1$. However, these nonlinearities can be ignored in those cases where $P_0(t) \approx 0$ physically, i.e., when only a negligible number of molecules have a small number of bonds. Such a situation is encountered in temperatures above or in the vicinity of the melting point—the temperature at which, on the average, half of the bonds are intact, and also during the early stages of the process of melting of DNA at any temperature. If we denote by λ and μ the probabilities per unit time of the closing of one broken bond and the breaking of one intact bond, respectively, then we have the following forms for λ_n and μ_n:

$$\lambda_n = \lambda(N-n), \qquad \mu_n = \mu n \tag{2}$$

For these forms the states $n = 0, N$ are reflecting states ($\lambda_N = 0, \mu_0 = 0$), in accordance with the physical situation, and λ, μ are temperature dependent.

The explicit expressions for $P_{n,m}(t)$, $\langle n \rangle$, and $\mathrm{var}(n)$ can be found in Table 2.1, row 8, with λ replaced by $-\lambda$ and v by λN. We do not discuss $P_{n,m}(t)$ here since there is no experimental method to measure it and the expression is too complicated to yield insight into the mechanism of the process. Instead we discuss $\langle n \rangle$ and $\mathrm{var}(n)$, which are given by

$$\langle n \rangle = \frac{\lambda}{\lambda+\mu} N + \left(m - \frac{\lambda}{\lambda+\mu} N \right) \exp[-(\lambda+\mu)t] \tag{3}$$

$$\begin{aligned} \mathrm{var}(n) = & -m(\lambda-\mu)(\lambda+\mu)^{-1} \exp[-(\lambda+\mu)t](1 - \exp[-(\lambda+\mu)t]) \\ & + N\lambda(\lambda+\mu)^{-2}(1 - \exp[-(\lambda+\mu)t])(\lambda \exp[-(\lambda+\mu)t] + \mu) \end{aligned} \tag{4}$$

where m is the initial number of intact bonds.

These equations describe the relaxation kinetics of denaturation and renaturation of DNA (Goel, 1968), i.e., the temporal behavior of a solution of DNA molecules when the solution is first allowed to come to equilibrium in some temperature (e.g., near the melting point), and then by a sudden increase or decrease of temperature the previous equilibrium (which determines m in these equations) is shifted. The average number of bonds relaxes then to a new equilibrium value, given by $N_0 = N\lambda/(\lambda+\mu)$, which is temperature dependent since λ and μ are. The relaxation to the new equilibrium point $\langle n\rangle(t) - N_0$ is exponential in time with the rate $\lambda+\mu$. The steady-state distribution is a binomial distribution in $\lambda(\lambda+\mu)$

$$P_n^* = \binom{N}{n}\left(\frac{\lambda}{\lambda+\mu}\right)^n\left(\frac{\mu}{\lambda+\mu}\right)^{N-n} \tag{5}$$

and for $N\lambda\mu(\lambda+\mu)^{-2} \geqslant 10$ is closely approximated by a normal distribution (see the discussion on p. 124).

In this process, the variance of the number of intact bonds increases from 0 at $t = 0$ to its equilibrium value

$$\text{var}(n)(t=\infty) \equiv V_\infty = N\lambda\mu(\lambda+\mu)^{-2} = \frac{\mu}{\lambda N} N_0^2 \tag{6a}$$

At equilibrium the coefficient of variation, given by

$$V_\infty^{1/2}/N_0 = (\mu/\lambda N)^{1/2} \tag{6b}$$

is small for N large enough, and the average well represents the process in this stage. To analyze the time-dependent behavior of the process, we rewrite the variance as

$$\begin{aligned}
\text{var}(n) &= \frac{1-\exp[-(\lambda+\mu)t]}{\lambda+\mu} \\
&\quad \times \{N_0(\lambda\exp[-(\lambda+\mu)t]+\mu) - m(\lambda-\mu)\exp[-(\lambda+\mu)t]\} \\
&= \frac{\mu}{\lambda}\frac{N_0^2}{N} + \frac{\lambda-\mu}{\lambda+\mu}(N_0-m)\exp[-(\lambda+\mu)t] \\
&\quad + \frac{m(\lambda-\mu)-N_0\lambda}{\lambda+\mu}\exp[-2(\lambda+\mu)t]
\end{aligned} \tag{7}$$

Now $\text{var}(n)(t) > V_\infty$ for some t if there exists $t_m > 0$ for which $d\,\text{var}(n)/dt = 0$, i.e.,

$$\exp[-(\lambda+\mu)t_m] = \frac{(\lambda-\mu)(N_0-m)}{2[N_0\lambda - m(\lambda-\mu)]} = \tfrac{1}{2}\left[1 - \frac{\mu N_0}{\lambda N_0 - (\lambda-\mu)m}\right] \tag{8}$$

Such a t_m exists if and only if

$$\left|\frac{\mu N_0}{\lambda N_0 - (\lambda-\mu)m}\right| < 1 \tag{9}$$

This condition is satisfied when

$$\lambda < \mu \quad \text{and} \quad N_0 < m \qquad \text{or} \qquad \lambda > \mu \quad \text{and} \quad N_0 > m \tag{10}$$

with $\lambda N_0 - m(\lambda-\mu) > 0$ in both cases. The reverse inequality cannot hold since then condition (9) implies $m > N$. The maximal value of $\text{var}(n)$ in the above two cases is at t_m,

$$\text{var}(n)_{\max} = \frac{\mu {N_0}^2}{\lambda N} + \frac{(\lambda-\mu)^2(N_0-m)^2}{4(\lambda+\mu)[N_0\lambda - m(\lambda-\mu)]} < \frac{\mu {N_0}^2}{\lambda N} + \frac{(\lambda-\mu)(N_0-m)}{2(\lambda+\mu)} \tag{11}$$

and in all other cases $\text{var}(n)$ is less than V_∞ for all $t > 0$. Noting that at the melting point $N_0 = N/2$ and therefore $\lambda = \mu$ while at higher (lower) temperatures $N_0 < N/2$ ($N_0 > N/2$) and therefore $\lambda < \mu$ ($\lambda > \mu$), we can conclude the following from the analysis above:

(a) For temperatures below (above) the melting point [$\lambda > \mu$ ($\lambda < \mu$)], the process of denaturation and renaturation of the DNA molecule is less (more) fluctuating around the average when the temperature of the DNA solution is increased to the given temperature from a lower temperature, i.e., $m > N_0$, than when it is cooled to this temperature from a higher temperature, i.e., $m < N_0$.

(b) For $\lambda > \mu$ and $m > N_0$ (relaxation after an increase in temperature below the melting point),

$$N_0 \leqslant \langle n \rangle(t) < m, \qquad \text{var}(n)(t) \leqslant V_\infty \tag{12}$$

and for all $t > 0$ the coefficient of variation is bounded by its equilibrium value $(\mu/\lambda N)^{1/2}$.

(c) For $\lambda < \mu$ and $m < N_0$ (relaxation after a decrease in temperature above the melting point),

$$m < \langle n \rangle(t) \leqslant N_0, \qquad \text{var}(n)(t) \leqslant V_\infty \tag{13}$$

and for all $t > 0$ the coefficient of variation is bounded by $(\mu/\lambda N)^{1/2} N_0/m$. Since our model is valid only for temperatures in which the probability to have a molecule with only few bonds is negligible, m cannot be much smaller than N_0 when $\lambda < \mu$. For $m = rN_0$, $\frac{1}{4} < r < 1$, the bound for the coefficient of variation becomes $r^{-1}(\mu/\lambda N)^{1/2}$, and again for moderate N the average closely describes the temporal behavior of the DNA molecules.

(d) For temperatures in the vicinity of the melting point $\lambda \approx \mu$, and for m of the order N_0 ($N_0/2 < m < 3N_0/2$), $\text{var}(n)$ in Eq. (11) is not significantly greater than V_∞, and the coefficient of variation in the two cases, given by Eq. (10), is well estimated by its value at $t = \infty$, given by Eq. (6b).

(e) For temperatures much lower than the melting point ($\lambda \gg \mu$), the maximal value of the variance of a process that starts after a decrease in temperature ($m < N_0$), up to first order in μ/λ, is

$$\begin{aligned} \operatorname{var}(n)_{\max} &= \frac{\mu}{\lambda(1+\mu/\lambda)^2} N + \frac{(1-\mu/\lambda)^2}{4(1+\mu/\lambda)} \frac{[N/(1+\mu/\lambda)-m]^2}{[N/(1+\mu/\lambda)-m(1-\mu/\lambda)]} \\ &\approx \frac{N-m}{4} + \left(\frac{\mu}{\lambda}\right)\frac{m}{2} < \frac{N}{4} + \left(\frac{\mu}{\lambda}\right)\left(\frac{m}{2}\right) \end{aligned} \tag{14}$$

The bound to the coefficient of variation when $m = rN$ is $N^{-1/2}[1+2r\mu/\lambda]/2r$ which is again of the order $N^{-1/2}$ for $r > \frac{1}{4}$.

As discussed earlier in this section, experimentally the state of the DNA molecules can be followed by following the optical density of the solution. Spatz and Baldwin (1965) observed that for dAT:dAT (synthetic DNA with A and T alternating in each of the two strands), the relative optical density (or in other words the average fraction of intact bonds) varies approximately exponentially with time with rates which increase as the temperature increases. Therefore $\lambda + \mu$ in our proposed model is an increasing function of temperature while λ is a decreasing function and μ is an increasing function of temperature. In the model discussed above each intact bond is given the same probability rate μ, to be broken and each broken bond is given the same probability rate, λ, to be closed. The resulting process describes a situation in which bonds are breaking and closing randomly along the chain. For the special case when renaturation and denaturation occur only in consecutive bonds (zipping and unzipping), either the first intact bond is to be broken, or the last broken bond is to be closed. Each of these two bonds is located between an intact and a broken bond, and we can assume that the probability of breaking or closing is independent of the number of already open bonds, i.e.,

$$\lambda_n = \nu, \qquad \mu_n = \rho, \qquad 1 < n < N-1 \tag{15}$$

The equation for $\langle n \rangle$, according to Eq. (2.0.8), is

$$d\langle n \rangle/dt = \langle \lambda_n \rangle - \langle \mu_n \rangle, \qquad \langle n \rangle(0) = m \tag{16}$$

When the probabilities of having $0, 1, N-1, N$ intact bonds are negligible (i.e., in the vicinity of the melting temperature) the terms in the averages $\langle \lambda_n \rangle$ and $\langle \mu_n \rangle$ corresponding to these states are negligible, and Eq. (16) can be approximated by

$$d\langle n \rangle/dt = (\nu-\rho) \sum_{n=2}^{N-2} P_{n,m}(t) = \nu - \rho, \qquad \langle n \rangle(0) = m \tag{17}$$

The solution to this equation is

$$\langle n \rangle = m + (\nu - \rho)t \tag{18}$$

while the variance, which satisfies Eq. (2.0.13), is given by

$$\operatorname{var}(n) = (\nu+\rho)\,t \tag{19}$$

These two equations are valid for times in which the chain is still far from either being almost open or almost closed. For short times the coefficient of variation is of the order of $(\nu+\rho)^{1/2}t^{1/2}/m$ and the linear behavior of the average is a good approximation. Thus at temperatures such that $\nu(T) > \rho(T)$ [$\nu(T) < \rho(T)$] bonds tend to be closed (opened), and the kinetics are linear in time. These characteristics, not observed experimentally, might indicate that the mechanism of zipping and unzipping is only secondary to the main process of random breaking and closing of bonds along the molecule's chain.

There are other special cases for which Eqs. (1) can be solved in spite of the nonlinear terms in (1a) and (1b). Let us add Eqs. (1a) and (1b) to obtain

$$d(P_0+P_1)/dt = -\lambda_1 P_1 + \mu_2 P_2 \tag{20}$$

which is a linear equation. We replace Eq. (1b) of the set (1) by Eq. (20) and propagate all the nonlinearities to Eq. (1a). If P_0 is known, the set of Eqs. (20), (1c), and (1d) is a set of linear master equations with an inhomogeneous term dP_0/dt. The solution of such equations is possible for special λ_n and μ_n (e.g., $\lambda_n = \nu$, $\mu_n = \rho$, Goel, 1965), but is beyond the scope of this monograph. The nonlinear equation for P_0 can be solved either when $\mu_1 = 0$ (i.e., once the first bond is formed between two single-stranded molecules, it does not break) so that $P_0 \sim (1+\lambda_0 t)^{-1}$, or when P_0 is constant (i.e., there exists an infinite reservoir of single-stranded DNA molecules). Under these assumptions, the kinetics is mostly of renaturation. Detailed solutions for the Laplace transform of P_n are given by Goel (1965), and the corresponding experimental data for the kinetics of renaturation of denatured DNA can be found in articles by Subirana (1966) and Subirana and Doty (1966).

9.2 Biosynthesis of Macromolecules

DNA, being one of the most important constituents of the cell, is replicated continuously with the continuous replications of cells in a living organism. The process of replication involves the partial denaturation of the DNA helix into two strands, each of the two strands acting as a template for biosynthesis of another strand, through polymerization of basic units (nucleotide) that are assumed to be present in the solution. The actual process of replication is complex and still not well understood [for review, see Watson (1970)]; it involves simultaneous (enzymatic) unwinding of the parent DNA and the formation of the daughter helices after new strands have been biosynthesized. A similar process presumably occurs in the transcription of DNA into messenger RNA which is eventually translated into a sequence of amino acids in a protein (Watson, 1970). In this section, we use the theory of birth and death processes to study the kinetics of template-mediated biopolymer synthesis.

For simplicity, let us assume that all the sites of a template are identical (i.e., parent DNA is homopolymeric). In the model we will be using, it is assumed that (1) each template molecule has K sites; (2) replication (and hence biosynthesis) occurs only in one direction from one end of the template, with a single growing center for each daughter; (3) the newly formed strand does not leave the template unless it has K units; and (4) once the site K is reached, the growing center is quickly released with the formation of daughter DNA helix completed.

Let $P_{n,m}(t)$ be the probability of a growing center being at site n at time t, given that it was at site m at time $t = 0$. If λ_n and μ_n are the probabilistic reaction rate constants for the growing center to move from site n to site $n+1$ and from site n to site $n-1$, respectively, then

$$dP_{1,m}/dt = -\lambda_1 P_{1,m} + \mu_2 P_{2,m} \tag{1a}$$

$$dP_{n,m}/dt = \lambda_{n-1} P_{n-1,m} - (\lambda_n + \mu_n) P_{n,m} + \mu_{n+1} P_{n+1,m} \tag{1b}$$

$$dP_{K,m}/dt = \lambda_{K-1} P_{K-1,m} \tag{1c}$$

In writing Eq. (1a) we have used the fact that if the growing center reaches site 1, due to the incoming monomers it will move toward site 2; and in writing Eq. (1c) we have used the fact that completion of replication occurs at site K. Equations (1) are the same equations as for birth and death processes and hence can be analyzed by the methods of Chapter 2, where site $n = 1$ is reflecting and site K is absorbing.

The forms of λ_n and μ_n are not a priori known. As a first approximation we take them to be independent of n, i.e.,

$$\lambda_n = \nu, \qquad \mu_n = \rho \tag{2}$$

This case is identical to the case we discuss in the next chapter in connection with photosynthesis. An analytical solution for $P_{n,m}(t)$ can be obtained (see Appendix C) but is quite complicated. However, expressions for the average time for the process to reach a site N for the first time and its variance can be calculated by using the expressions given in Table 2.3. Let us consider the case when the growing center is at site 1, i.e.,

$$P_{n,m}(0) = \delta_{n,1} \tag{3}$$

The average time in this case is given by (see column 1, row 3 of Table 2.3)

$$M_{N,1} = \sum_{i=1}^{N-1} \sum_{n=0}^{i} \lambda_n^{-1} \Pi_{n+1,i} \tag{4}$$

where

$$\Pi_{i,j} = \frac{\mu_i \mu_{i+1} \cdots \mu_j}{\lambda_i \lambda_{i+1} \cdots \lambda_j}, \qquad i \leqslant j, \qquad \Pi_{i,i-1} = 1$$
$$= \left(\frac{\rho}{v}\right)^{j-i+1}, \qquad i \leqslant j \tag{5}$$

Substituting Eq. (5) into Eq. (4) and carrying out the summation, we obtain

$$M_{N,1} = \frac{1}{v(1-\alpha)}[N-1+\alpha(1-\alpha^{N-1})], \qquad \alpha \equiv \rho/v \tag{6}$$

Substitution of $N = K$ in this expression gives the average time for the growing center to reach the site K, i.e., for the completion of the formation of a daughter DNA.

Under conditions which favor the duplication of DNA, ρ is very small and $\alpha \ll 1$. Equation (6) then becomes

$$\langle t \rangle(N) \equiv M_{N,1} \simeq \frac{N-1+\alpha}{v(1-\alpha)} \tag{7}$$

i.e., the growth kinetics is linear in N with rate constant $v(1-\alpha)$. For the irreversible case ($\alpha = 0$), the rate constant for the linear kinetics is v. The expression for $P_{n,1}(t)$ in this case can be obtained from Table 2.1, row 1, and is given by

$$P_{n,1}(t) = \frac{e^{-vt}(vt)^{n-1}}{(n-1)!}, \qquad n < K \tag{8}$$

which is the Poisson distribution. From Eq. (8) the probability density function of the time to completion of the replication is given by

$$F_{k,1}(t) = \lambda_{K-1} P_{K-1,1}(t) = \frac{e^{-vt}(vt)^{K-1}}{t(K-2)!} \tag{9}$$

Let us now write down an expression for v in terms of experimentally measurable quantities. The addition of one bond requires three events: (1) an empty site immediately adjacent to the growing center, (2) absorption of a monomer at this site, and (3) the incorporation of this monomer in the growing chain. If the rate of unwinding leading to event 1 is larger than the rate of incorporation (event 3), an empty site adjacent to the growing center will always be available. We can therefore write

$$v = pk_t \tag{10}$$

where p is the probability that the site in front of the growing center contains an absorbed monomer, and k_t is the rate of incorporation. The probability p will depend on the concentration of monomers c and the adsorption energy E.

Maniloff (1969), on the basis of simple thermodynamic analysis of absorption–desorption of monomer equilibrium, shows that

$$p = Ace^{-E/kT} \tag{11}$$

where A is some constant parameter, k the Boltzmann constant, and T the temperature in absolute degrees. Thus, p and hence v change linearly with the concentration of monomers and exponentially with the monomer adsorption energy.

From the dependence on v of average time for a certain degree of polymerization [Eq. (7)], we conclude that the biosynthesis is not very sensitive to the concentration of monomers, unless the concentration is changed by a few orders of magnitude. Therefore changes in monomer pool cannot serve as a sharp "on–off" mechanism for regulating biosynthesis. On the other hand, the sensitivity of biosynthesis to the adsorption energy suggests a regulatory mechanism through changes in the adsorption energy (e.g., binding of some molecule or macromolecule to the template).

We summarize this section by pointing out various generalizations of the analysis of kinetics of biosynthesis. Zimmerman and Simha (1965, 1966) allow multicenter growth and consider the case when the process of unwinding of DNA is not much faster than the process of incorporation of monomer. McDonald *et al.* (1968) consider the reversible case.

9.3 Enzyme Kinetics

The simplest enzyme is believed to follow the one-substrate–one-intermediate–one-product mechanism

$$\mathrm{E} + \mathrm{S} \underset{k2}{\overset{k1}{\rightleftharpoons}} \mathrm{X} \underset{k4}{\overset{k3}{\rightleftharpoons}} \mathrm{E} + \mathrm{P} \tag{1}$$

where E, S, X (= ES), and P represent the enzyme, substrate, intermediate, and product species, respectively. The intermediate species is the complex between the enzyme and substrate species. This mechanism is known as the Michaelis–Menten mechanism after L. Michaelis and M. L. Menten who in 1913 suggested the mechanism and analyzed it deterministically. According to this mechanism, the time dependence of the concentration of various species is given by

$$dE/dt = -k_1 ES + (k_2+k_3) X - k_4 EP \tag{2a}$$

$$dS/dt = -k_1 ES \tag{2b}$$

$$dX/dt = k_1 ES - (k_2+k_3) X + k_4 EP \tag{2c}$$

$$dP/dt = k_3 X - k_4 EP \tag{2d}$$

where the symbol defining a species is italicized to represent the concentration of that species, and k_1, k_2, k_3, and k_4 are the rate constants for the various steps as indicated in Eq. (1). Equations (2a) and (2c), and Eqs. (2b), (2c), and (2d) imply that

$$E + X = \text{constant} = E_0 \tag{3a}$$

$$S + X + P = \text{constant} = S_0 \tag{3b}$$

where E_0 and S_0 denote the initial concentrations of enzyme and substrate. These equations are the conservation equations. Equation (3a) implies the conservation of enzyme; i.e., the initial concentration of enzyme should be equal to the concentration of free enzyme plus that of enzyme complex at any instant. Equation (3b) implies the conservation of substrate; i.e., the initial concentration of substrate should be equal to the sum of concentrations of free substrate, substrate–enzyme complex, and product at any instant. Equations (2) are not amenable to an analytical solution. However, two assumptions are usually made which lead to a simple analytical solution. One assumption is that $k_4 = 0$ and the other is a so-called steady-state assumption. According to this assumption, $dX/dt = 0$; i.e., the concentration of the enzyme–substrate complex is constant. This latter assumption is not valid at the initial stages of the reaction, but after a certain initial period, it is perhaps valid since whatever additional complex is formed will most likely dissociate either into E and S or into E and P, the latter being more likely. According to this assumption, from Eq. (2c)

$$X = k_1 ES/(k_2+k_3) \tag{4a}$$

which, using Eq. (3a), becomes

$$X = \frac{E_0 S}{K_m + S} \tag{4b}$$

where

$$K_m = (k_2+k_3)/k_1 \tag{4c}$$

Substituting Eq. (4b) into Eq. (2d) we get

$$\frac{dP}{dt} = \frac{k_3 E_0 S}{K_m + S} \equiv V \tag{5}$$

This equation describes the rate of production of the product (also known as the velocity of the reaction or relative activity of the enzyme) and is known as the Michaelis–Menten equation. K_m is known as the Michaelis constant. Equation (5) is schematically plotted in Fig. 9.1 with V taking half of its maximum value at $S = K_m$. Most of the individual enzymes follow the kinetics described by Eq. (5). It should be noted that although the behavior of an enzyme

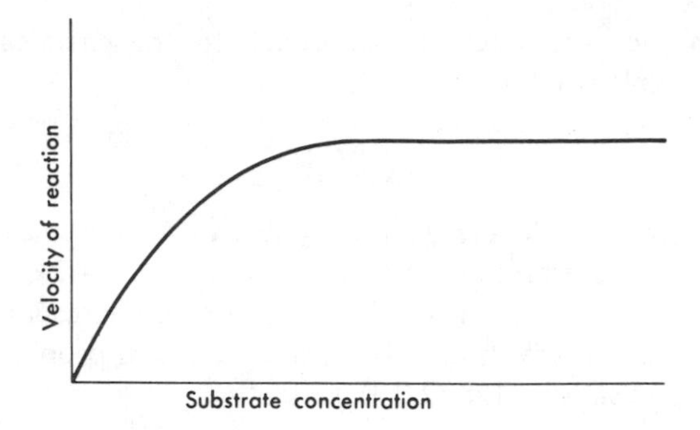

Fig. 9.1 Michaelis–Menten kinetics.

may be like that shown in Fig. 9.1, its mechanism may be more complex than the Michaelis–Menten mechanism (Fisher and Hoagland, 1968).

In natural biochemical systems, the enzymes are proteins that are biopolymers. The enzymatic activity of these proteins depend on the configuration of the protein, which may be very sensitive to the environment. The environment (e.g., pH, ionic strength) changes continuously, causing changes in enzyme activity. In most cases, a chemical reaction is a complex reaction involving several simple reactions and enzymes. In such cases, it is possible that, due to the fluctuations in the environment, the change in the activity of each of the enzymes may be such that the rate of product formation deviates significantly from the mean. Because of the exceptionally precise, accurate, and efficient nature of the biological systems, these deviations may be quite damaging to the functioning of the biological system. It is therefore important to study enzyme kinetics by using a stochastic approach.

In the discrete stochastic approach [see McQuarrie (1968), and Montroll (1967) for review] the numbers of reactant and product molecules are treated as discrete random variables, and the reaction is defined by the probability functions of the states of the reaction system. In the Michaelis–Menten enzyme kinetics, there are four random variables E, S, X, and P, but due to the conservation equations (3), only two of them are independent. Therefore, the process is effectively a bivariate process. In a realistic complex chemical reaction, the number of random variables is much larger and the process is multivariate. The investigation of the biological systems which lead to bivariate and multivariate processes is beyond the scope of this monograph. In spite of this, because of the importance of the problem, we briefly describe the stochastic approach to give the readers an idea of the types of equations to be solved. We also discuss a model system which describes the Michaelis–Menten kinetics, when a suitable approximation is made, and which can be treated as a univariate process.

Let us introduce the stochastic approach to the chemical kinetics by considering a simple reaction

$$A \xrightarrow{k} B \tag{6}$$

Let the random variable $X(t)$ be the number of A molecules in the system at time t. This variable takes the values $n = m, m-1, m-2, \ldots, 1, 0$ where m is the initial number of A molecules. The evolution of the reaction is described by a pure death process with $\mu_n = kn$ as the transition probability rate. The probability $P_{n,m}(t)$ satisfies Eq. (2.0.3), i.e.,

$$dP_{n,m}(t)/dt = k(n+1)\,P_{n+1,m}(t) - knP_n(t) \tag{7}$$

The mean value $\langle n \rangle$ of the variable $X(t)$ and the variance $\mathrm{var}(n)$ satisfy Eqs. (2.0.8) and (2.0.13), i.e.,

$$d\langle n \rangle/dt = -k\langle n \rangle \tag{8a}$$

$$d\,\mathrm{var}(n)/dt = -2k\,\mathrm{var}(n) + k\langle n \rangle \tag{8b}$$

so that

$$\langle n \rangle = me^{-kt} \tag{9a}$$

$$\mathrm{var}(n) = me^{-kt}(1-e^{-kt}) \tag{9b}$$

Note that the mean value in the stochastic model is the deterministic result, which, incidentally, is true only for unimolecular reactions. Further, for later use, we note that after a short time $t = \Delta t$, the change in the mean and variance, in terms of the transition probability rates, $\lambda_m = 0$, $\mu_m = km$, are

$$\Delta\langle n \rangle = -mk\,\Delta t = (\lambda_m - \mu_m)\,\Delta t \tag{10a}$$

$$\Delta\,\mathrm{var}(n) = mk\,\Delta t = (\lambda_m + \mu_m)\,\Delta t \tag{10b}$$

For a bimolecular reaction, e.g.,

$$A + B \xrightarrow{k} C \tag{11}$$

the approach is basically the same except that the transition probability rate for the number of molecules of species C to increase from n to $n+1$ is k times the product of the number of molecules of A and B.

Having introduced the basic stochastic approach, let us go back to the Michaelis–Menten mechanism (1) for the enzyme reaction. For this mechanism, if e, s, x, and p denote the (integer) values taken by the four random variables representing the number of enzyme, substrate, intermediate, and product molecules at time t, then the probability density $P(e,s,x,p;t)$

satisfies the equation

$$\frac{dP(e,s,x,p;t)}{dt} = k_1(e+1)(s+1)P(e+1,s+1,x-1,p;t) + k_2(x+1)P(e-1,s-1,x+1,p;t) + k_3(x+1)P(e-1,s,x+1,p-1;t) + k_4(e+1)(p+1)P(e+1,s,x-1,p+1;t) - (k_1 es + k_2 x + k_3 x + k_4 ep)P(e,s,x,p;t) \tag{12}$$

Using the conservation equations (3), this equation reduces to

$$\frac{dP(e,s;t)}{dt} = k_1(e+1)(s+1)P(e+1,s+1;t) + k_2(E_0-e+1)P(e-1,s-1;t) + k_3(E_0-e+1)P(e-1,s;t) + k_4(e+1)(S_0-E_0-s+e+1)P(e+1,s;t) - [k_1 es + (k_2+k_3)(E_0-e) + k_4 e(S_0-E_0-s+e)]P(e,s,t) \tag{13}$$

with $P(e,s;t)$ replacing $P(e,s,x,p;t)$ and

$$P(e,s;0) = \begin{cases} 1, & e = E_0, \quad s = S_0 \\ 0, & \text{otherwise} \end{cases} \tag{14}$$

A solution for $P(e,s;t)$ has not yet been obtained. Bartholomay (1962a, b), by assuming $\langle(e)(s)\rangle = \langle e\rangle\langle s\rangle$, showed that the stochastic mean agrees with the deterministic mean. Jachimowski *et al.* (1964) showed that these results hold even if the covariance of the random variables is assumed to be zero only during the early stages of the reaction, provided $S_0 \gg E_0$. Darvey and Staff (1967) looked more carefully into the difference between the deterministic solution and the stochastic mean and showed that

$$\sigma_x(\infty)/\varepsilon_x > 2(S_0/E_0)^{1/2} \tag{15}$$

where

$$\varepsilon_x = \langle x\rangle_\infty - X_\infty, \qquad \varepsilon_x \geqslant 0 \tag{16}$$

is the difference between the stochastic mean and the deterministic solution in the steady state, and $\sigma_x(\infty)$ is the variance of the random variable describing the number of molecules of the enzyme–substrate complex in the steady state. Thus, in the steady state the stochastic mean is always greater than or equal to the deterministic solution, and when $S_0 \gg E_0$ (which is true in most *in vitro* experiments), the stochastic mean is adequately expressed by the deterministic solution.

We now describe a model system that has some characteristics of the simple enzyme kinetics. In this model system (Hawkins and Rice, 1971), the enzyme system is connected to two reservoirs of S and P through semipermeable membrane, and the physical processes which bring equilibrium between the reservoirs and the system are so fast that equilibrium is always established, no matter what value X (the intermediate complex) assumes. In other words, the reactions of Eq. (1) are so slow that they cannot disturb the equilibrium between system and reservoirs. This condition is satisfied if the chemical potentials characterizing the reservoirs are independent of the concentration of X (Hawkins and Rice, 1971). Within this approximation, we may take S and P, respectively, equal to S^* and P^*, which are independent of time and are determined by the properties of the reservoirs, and describe the system by a single random variable E or X. [The sum of E and X is a constant, by Eq. (3a).] Let us choose the variable E and, for consistency with our notation of Chapter 2, denote by n the number of molecules of E. The process is described by the probability density $P_n(t) \equiv P(e, s; t)$ which satisfies an equation similar to Eq. (13) (except that now s is kept fixed), i.e.,

$$\begin{aligned} dP_n(t)/dt &= k_1(n+1)\,S^* P_{n+1}(t) + (k_2+k_3)[E_0 - (n-1)]\,P_{n-1}(t) \\ &\quad + k_4(n+1)\,P^* P_{n+1}(t) \\ &\quad - [k_1 nS^* + (k_2+k_3)(E_0-n) + k_4 nP^*]\,P_n(t) \end{aligned} \tag{17}$$

i.e.,

$$dP_n(t)/dt = \lambda_{n-1} P_{n-1}(t) - (\lambda_n+\mu_n)\,P_n(t) + \mu_{n+1} P_{n+1}(t) \tag{18}$$

where

$$\mu_n = \mu n, \qquad \mu = (k_1 S^* + k_4 P^*) \tag{19a}$$

$$\lambda_n = \lambda(E_0-n), \qquad \lambda = (k_2+k_3) \tag{19b}$$

and we have used the conservation equation (3a). This process is therefore a birth and death process with linear transition probabilities and with the states $n = 0$ and $n = E_0$ acting as reflecting states. For this process we can carry out the analysis of Chapter 2, and since by Eq. (19b) $\lambda_{E_0} = 0$ and $\mu_0 = 0$, the mean value and variance of n satisfy Eqs. (2.0.8) and (2.0.13). Now, λ_n and μ_n for the present process are of the same form as for renaturation and denaturation of DNA [see Eq. (9.1.2)]. Thus $\langle n \rangle$ and $\operatorname{var}(n)$ are given by Eqs. (9.1.3) and (9.1.4) with λ and μ given by Eqs. (19), and both m and N replaced by E_0. Let us consider the expression for $\langle n \rangle$, i.e.,

$$\begin{aligned} \langle n \rangle &= E_0 \exp[-(\lambda+\mu)\,t] + \frac{\lambda E_0}{\lambda+\mu}\{1 - \exp[-(\lambda+\mu)\,t]\} \\ &= \frac{E_0 \lambda}{\lambda+\mu}\left\{1 + \frac{\mu}{\lambda}\exp[-(\lambda+\mu)\,t]\right\} \\ &= \frac{E_0 K_m}{K_m+S^*}\left\{1 + \frac{S^*}{K_m}\exp[-k_1(S^*+K_m)\,t]\right\} \end{aligned} \tag{20}$$

where for comparison with the Michaelis--Menten treatment, we have set $k_4 = 0$ and K_m as the Michaelis constant. The average number of the enzyme–substrate complex molecules is

$$\langle X \rangle = E_0 - \langle n \rangle = \frac{E_0 \mu}{\lambda + \mu}\{1 - \exp[-(\lambda+\mu)t]\}$$

$$= \frac{E_0 S^*}{K_m + S^*}\{1 - \exp[-k_1(S^* + K_m)t]\} \tag{21}$$

Thus the concentration of the enzyme–substrate complex reaches the steady-state value [Eq. (4b)] used in the Michaelis–Menten treatment, exponentially, with the relaxation time $[k_1(S^* + K_m)]^{-1}$. One can likewise discuss the time dependence of variance using Eq. (9.1.4). Let us just discuss the steady-state value of the variance, which is

$$\mathrm{var}(n)(t \to \infty) = E_0 \lambda\mu(\lambda+\mu)^{-2} = E_0 K_m S^*(K_m + S^*)^{-2}$$

$$= \mathrm{var}(X)(t \to \infty) = (K_m/E_0 S^*)\langle X \rangle^2 \qquad (t \to \infty) \tag{22a}$$

so that the coefficient of variation is

$$[(\mathrm{var}(X))^{1/2}/\langle X \rangle]_\infty = (K_m/E_0 S^*)^{1/2} \tag{22b}$$

compared to the zero variance of the deterministic model of Michaelis–Menten.

The steady-state probability distribution of n, from Table 2.7, is given by the binomial distribution

$$P_n^* = \frac{(\lambda/\mu)^n \binom{E_0}{n}}{(1+\lambda/\mu)^{E_0}} = \frac{\binom{E_0}{n}(K_m/S^*)^n}{(1+K_m/S^*)^{E_0}} = \binom{E_0}{n}\left(\frac{K_m}{S^*+K_m}\right)^n \left(\frac{S^*}{S^*+K_m}\right)^{E_0-n} \tag{23}$$

Hawkins and Rice (1971) obtained the continuous analog of Eq. (23) by assuming enzyme concentration to change continuously. They used the Fokker–Planck equation instead of the master equation (18), with coefficients determined by Eqs. (10) and (19):

$$a(E) = (k_2+k_3)E_0 - (k_1 S^* + k_2 + k_3)E = \lambda E_0 - (\lambda+\mu)E \tag{24a}$$

$$b(E) = (k_2+k_3)E_0 + (k_1 S^* - k_2 - k_3)E = \lambda E_0 + (\mu - \lambda)E \tag{24b}$$

The rate at which P is being produced is determined by Eq. (2d) with $k_4 = 0$; i.e., the average rate of production is k_3 times $\langle X \rangle$. Equation (21) gives the time dependence of the rate of production of P (which is equal to the rate at which the reservoir is taking away P from the system), in contrast to the Michaelis–Menten treatment where this rate is assumed to be time independent. Equation (22b) times $k_3{}^2$ gives the variance in this rate.

For completeness we point out that Heyde and Heyde (1969) obtained the expression for the time-dependent joint probability distribution of E and P during the initial phase of the reaction (so that $S \simeq S_0$) with $k_4 = 0$. They compared their results for the reaction catalyzed by triose phosphate isomerase (Heyde and Heyde, 1971) and concluded that for this one-substrate–one-product–enzyme reaction, the stochastic fluctuations in concentrations of E, X, S, and P are negligible. By doing approximate analysis of the case when S is not taken equal to S_0, they showed that the stochastic fluctuations will in general be negligible except at the beginning of the reaction.

References

Bartholomay, A. F. (1962a). Enzymatic reaction rate theory: A stochastic approach. *Ann. N.Y. Acad. Sci.* **96**, 897.

Bartholomay, A. F. (1962b). A stochastic approach to statistical kinetics with application to enzyme kinetics. *Biochemistry* **1**, 223.

Darvey, I. G., and Staff, P. J. (1967). The application of the theory of Markov processes to the reversible one substrate—one intermediate—one product enzymic mechanism. *J. Theoret. Biol.* **14**, 157.

Fisher, J. R., and Hoagland, V. D., Jr. (1968). A systematic approach to enzyme kinetics of multisubstrate enzyme systems. *Advances in Med. Biol. Phys.* **12**, 163.

Goel, N. S. (1965). Denaturation and Renaturation of the DNA Molecule. Ph. D Thesis, Univ. of Maryland, College Park.

Goel, N. S. (1968). Relaxation kinetics of denaturation of DNA. *Biopolymers* **6**, 55.

Goel, N. S., and Montroll, E. W. (1968). Denaturation and renaturation of DNA. II. Possible use of synthetic periodic copolymers to establish model and parameters. *Biopolymers* **6**, 731.

Hawkins, R., and Rice, S. A. (1971). Study of concentration fluctuations in model systems. *J. Theoret. Biol.* **30**, 579.

Heyde, C. C., and Hyde, E. (1969). A stochastic approach to a one substrate, one product enzyme reaction in the initial velocity phase. *J. Theoret. Biol.* **25**, 159.

Heyde, C. C., and Heyde, E. (1971). Stochastic fluctuations in a one substrate one product enzyme system; Are they ever relevant?. *J. Theoret. Biol.* **30**, 395.

Jachimowski, C. J., McQuarrie, D. A., and Russell, M. E. (1964). A stochastic approach to enzyme-substrate reactions. *Biochemistry* **3**, 1732.

MacDonald, F. T., Gibbs, J. H., and Pipkin, A. C. (1968). Kinetics of biopolymerization on nucleic acid templates. *Biopolymers* **6**, 1.

McQuarrie, D. A. (1968). "Stochastic Approach to Chemical Kinetics." Methuen, London.

Maniloff, J. (1969). Theoretical considerations of biopolymer synthesis. *J. Theoret. Biol.* **23**, 441.

Montroll, E. W. (1967). Stochastic processes and chemical kinetics, *in* "Energetics in Metallurgical Phenomenon" (W. M. Mueller, ed.), Vol. 3, p. 123. Gordon & Breach, New York.

Pipkin, A. C. (1966). Kinetics of synthesis and/or conformational changes of biological macromolecules. *Biopolymers* **4**, 3.

Poland, D., and Scheraga, H. A. (1970). Theory of Helix-Coil Transitions in Biopolymers. Academic Press, New York.

Spatz, H. C., and Baldwin, R. L. (1965). Study of the folding of the dAT copolymer by kinetics measurements of melting. *J. Mol. Biol.* **11**, 213.

Subirana, J. A. (1966). Kinetics of renaturation of denatured DNA. II. Products of the reaction. *Biopolymers* **4**, 171.

Subirana, J. A., and Doty, P. (1966). Kinetics of renaturation of denatured DNA. I. Spectrophotometric results. *Biopolymers* **4**, 171.

Teramoto, E., Shiegesada, N., Nakajima, H., and Sato, K. (1971). Stochastic theory of reaction kinetics. *Advances in Biophys.* **2**, 155.

Wartell, R. M., and Montroll, E. W. (1972). Equilibrium denaturation of natural and of periodic synthetic DNA molecules. *Advances in Chem. Phys.* **22**, 129.

Watson, J. D. (1970). "Molecular Biology of the Gene." Benjamin, New York.

Zimmerman, J. M., and Simha, R. (1965). The kinetics of multicenter growth along a template. *J. Theoret. Biol.* **9**, 156.

Zimmerman, J. M., and Simha, R. (1966). The kinetics of cooperative unwinding and template replication of biological macromolecules. *J. Theoret. Biol.* **13**, 106.

10

Photosynthesis

In its broadest sense, photosynthesis is the conversion of light energy into chemical energy by living organisms. Some (photosynthetic) bacteria, algae, and higher green plants store this chemical energy in the form of carbohydrates and sugars. All of these organisms contain one or another form of chlorophyll within their cellular structures (e.g., chloroplast). Many (30–300) of these molecules cooperate as a "photosynthetic unit" (PSU) to "process" the energy of many photons efficiently. In the generally accepted model of a PSU, 30–300 of the chlorophyll molecules are arranged in such a way that photoexcitations (known as excitons), originating from absorption of photons anywhere in the network of chlorophyll molecules, travel through the network and are then absorbed by a special molecule T ("trap"). This trapping (absorption by the trap) accompanied by absorption of the exciton energy by T is followed, usually, by a biochemical reaction leading to the formation of carbohydrates and sugars and emission of O_2. Occasionally the absorbed energy may be transformed into fluorescence or heat. The movement of the exciton through the network of chlorophyll molecules may be assumed to be a random process with trapping molecules as the absorbing states. The network most likely is three dimensional, and the random process is a trivariate process with the three coordinates, specifying the location of the exciton, as the three random variables. Because of the limitations of this monograph, we only consider the movement of excitons through a one-dimensional network. In particular we calculate an expression for the average time for trapping of an exciton which is confined to a one-dimensional lattice of chlorophyll molecules, and remark only briefly on two- and three-dimensional lattices. It is hoped that the analysis for the one-dimensional lattice will give the reader an idea of the nature and importance of stochastic modelling in understanding, in part, the process of photosynthesis.

The motion of excitons through a network of N identical molecules (assuming no traps) can be described by the master equation

$$dP_{n,m}/dt = \sum_{k=1}^{N} (F_{n,k} P_{k,m} - F_{k,n} P_{n,m}) \tag{1}$$

In Eq. (1), $P_{n,m}(t)$ is the probability that the exciton is localized on the nth molecule at time t given that initially ($t = 0$) it was at the mth molecule; $F_{i,k} > 0$ is the rate constant (transition probability rate) for transfer of the exciton from the kth to the ith molecule. If the Tth molecule is the trap, Eq. (1) is to be modified to

$$dP_{n,m}/dt = \sum_{k=1, k \neq T}^{N} (F_{n,k} P_{k,m} - F_{k,n} P_{n,m}) - F_{T,n} P_{n,m} \tag{2}$$

where $F_{T,i}$ is the probability rate for exciton transfer from the ith molecule to T. In writing Eq. (2), we assume that the exciton "decays" only due to trapping. The decay due to other processes, e.g., radiation damping, radiationless conversion to intramolecular vibrational mode, can be taken into account in a manner described later in this chapter.

If we make a reasonable assumption that transition occurs only between nearest neighbors, then for a one-dimensional lattice, Eq. (1) becomes

$$dP_{n,m}/dt = \lambda_{n-1} P_{n-1,m} - (\mu_n + \lambda_n) P_{n,m} + \mu_{n+1} P_{n+1,m} \tag{3}$$

where

$$\lambda_n \equiv F_{n+1,n}, \qquad \mu_n \equiv F_{n-1,n} \tag{4}$$

Equation (3) is identical to Eq. (2.0.3) and hence can be analyzed likewise.

We consider two basically different one-dimensional lattices (PSU): one (type I) which has a trap on one end, and the other (type II) which has traps at both ends. A lattice that has its only trap in the middle can be considered as two lattices, each of type I, and a lattice with more than one trap can be considered as a combination of lattices of types I and II. This combination of lattices is possible since an exciton cannot travel from one side of a trap to the other.

For a lattice (PSU) of type I, let us number the lattice points as $0, 1, 2, \ldots, N$ with the point N as the trapping point. Since N is the trap and the exciton does not leave the lattice, N is the absorbing state and 0 is the reflecting state ($\mu_0 = 0, \lambda_0 > 0$). Since the process is restricted between a reflecting and an absorbing state, we can analyze the process using the methods of Section 2.2 and, in particular, of Section 2.2A. As we have seen in Section 2.2, an analytical solution of Eq. (3) is only possible for very special λ_n and μ_n (see Table 2.2). However, an analytical expression for other quantities which give insight into the process can be obtained for arbitrary λ_n and μ_n (see Section 2.2A). One of these quantities that can be experimentally measured is the average time for the exciton to reach the trap. When the photon is initially absorbed by the mth molecule, this expression is given in column 1, row 3 of Table 2.3, with $l = 0$, i.e., by

$$M_{N,m} = \sum_{i=m}^{N-1} \sum_{n=0}^{i} \lambda_n^{-1} \Pi_{n+1,i} \tag{5}$$

where

$$\Pi_{i,j} = \frac{\mu_i \mu_{i+1} \cdots \mu_j}{\lambda_i \lambda_{i+1} \cdots \lambda_j}, \qquad i \leqslant j, \qquad \Pi_{i,i-1} = 1 \tag{6}$$

If we assume that each chlorophyll molecule has the same probability of first absorbing a photon, then the average time for trapping of an exciton is

$$\langle t \rangle = \frac{1}{N} \sum_{m=0}^{N-1} M_{N,m} = \frac{1}{N} \sum_{i=0}^{N-1} (i+1) \sum_{n=0}^{i} \lambda_n^{-1} \Pi_{n+1,i} \tag{7}$$

One can also calculate other quantities describing the behavior of the process, e.g., higher moments of absorption time, probability of reaching the reflecting state without getting absorbed, etc. We will not do so because these quantities are not measurable experimentally, at least not with any reasonable ease.

We now consider a special set of transition probabilities, namely, $\lambda_i = v$, $\mu_i = \rho$, with $\alpha = v/\rho$. For this case an analytical expression for $P_{n,m}(t)$ can also be obtained (see Appendix C), which is quite complicated. The expression (7) for the average time for absorption becomes

$$\langle t \rangle = \frac{1}{N} \sum_{i=0}^{N-1} \frac{(i+1)}{v\alpha^i} \sum_{n=0}^{i} \alpha^n = \frac{1}{Nv} \sum_{i=0}^{N-1} \frac{(i+1)(i-\alpha^{i+1})}{\alpha^i(1-\alpha)} \tag{8}$$

which can be summed to give

$$\langle t \rangle_\alpha = \frac{1}{2Nv\alpha^{N-1}(1-\alpha)^3} [2N - 2(N+1)\alpha + 2\alpha^{N+1} - N(1+N)(\alpha-1)^2\alpha^N] \tag{9}$$

Here $\alpha \neq 1$ corresponds to the case when transitions in one direction are more probable than in the other. It is very unlikely that, in general, a natural PSU admits asymmetric transitions. However, such asymmetric transitions in a natural PSU can result by a special kind of thermal vibration of the lattice, in which the average exciton transfer time is equal to one-half the period of the vibration (Elliott *et al.*, 1969) or by an external field (electric) which might increase the transition probabilities in one direction relative to the other. For $\alpha = 1$, i.e., when the probability rates of transition in both directions are the same, $\Pi_{i,j} = 1$, and from Eq. (7) [or from Eq. (9)], the average time for absorption is

$$\langle t \rangle_{\alpha=1} = \frac{(N+1)(N+\frac{1}{2})}{3\nu} \tag{10}$$

For lattice (PSU) of type II, with traps on both ends (or, equivalently, if the chlorophyll molecules are arranged on a ring), 0 and N are absorbing states. Thus the process can be analyzed by using the results of Sections 2.2 and 2.2B. For example, starting from position m, the probability that the exciton will be eventually absorbed at the trap N (rather than at 0) is given by the expression in Table 2.5, column 2, row 1, with $l = 0$ and $u = N$, i.e., by

$$R_{N,m} = \sum_{i=0}^{m=1} \Pi_{1,i} \Big/ \sum_{i=0}^{N-1} \Pi_{1,i} \tag{11}$$

where $\Pi_{i,j}$ is given by Eq. (6). For $\lambda_i = \nu$, $\mu_i = \rho$, this reduces to

$$R_{N,m} = \frac{\alpha^{N-m} - \alpha^N}{1 - \alpha^N}, \qquad \alpha = \nu/\rho \tag{12a}$$

For $\alpha \to 1$,

$$R_{N,m} \xrightarrow[\alpha \to 1]{} m/N \tag{12b}$$

Thus, for equal transition probability rates of going toward each of the two traps (at N and at 0), the probability of the exciton to be trapped at N depends linearly on its initial distance from the trap at 0. The other quantity of main interest is the average time for the exciton to get trapped (either at 0 or N) starting at the position m. This is given in Table 2.5 by the expression for M_m (with $l = 0$, $u = N$),

$$M_m = \sum_{i=m}^{N-1} \sum_{n=1}^{i} \lambda_n^{-1} \Pi_{n+1,i} - R_{0,m} \sum_{i=1}^{N-1} \sum_{n=1}^{i} \lambda_n^{-1} \Pi_{n+1,i} \tag{13a}$$

where

$$R_{0,m} = 1 - R_{N,m} \tag{13b}$$

If we assume that each chlorophyll molecule has the same probability of first absorbing a photon, then the average time for absorption is

$$\langle t\rangle = \frac{1}{N-1}\sum_{m=1}^{N-1} M_m \tag{14}$$

For the special case of $\lambda_i = \nu$, $\mu_i = \rho$, $\nu/\rho = \alpha$ ($\neq 1$), there is an analytical expression for $P_{n,m}(t)$ (see Table 2.2, row 4) and M_m [Eq. (9)] becomes

$$M_m = \frac{\alpha}{\nu(\alpha-1)}\left\{N\frac{(1/\alpha)^m-1}{(1/\alpha)^N-1} - m\right\} \tag{15}$$

Substituting this equation into Eq. (14) we obtain

$$\langle t\rangle = \frac{1}{2\nu}\frac{N}{N-1}\frac{\alpha}{\alpha-1}\left\{\frac{1+\alpha}{1-\alpha} + N\frac{\alpha^N+1}{\alpha^N-1}\right\} \tag{16}$$

With the notation

$$\delta \equiv (\alpha-1)/(\alpha+1), \qquad \alpha = (1+\delta)/(1-\delta) \tag{17}$$

and after simple algebraic manipulation, Eq. (16) becomes

$$\langle t\rangle = \frac{1}{2\nu}\frac{1+\delta}{2\delta}\frac{N^2}{N-1}\left\{\coth(N\tanh^{-1}\delta) - \frac{1}{N\delta}\right\} \tag{18}$$

For $\alpha = 1$, Eqs. (15) and (16) reduce to

$$M_m(\alpha = 1) = m(N-m)/2\nu \tag{19a}$$

$$\langle t\rangle(\alpha = 1) = N(N+1)/12\nu \tag{19b}$$

Equations (15), (18), and (19) can be written in terms of the average number of transitions $\langle K\rangle_m$ and $\langle K\rangle$ before exciton is trapped, if we note that the probability of no transition within time t is equal to $\exp(-(\nu+\rho)t)$, so that the average time in which there is no transition is

$$\lim_{T\to\infty}\frac{1}{T}\int_0^T t\exp(-(\nu+\rho)t)\,dt = (\nu+\rho)^{-1}$$

In other words, the average time between two transitions is $(\nu+\rho)^{-1}$. Therefore, the average number of transitions before the exciton is trapped is

$$\langle K\rangle_m = \langle t\rangle_m/(\nu+\rho), \qquad \langle K\rangle = \langle t\rangle/(\nu+\rho) \tag{20}$$

From Eqs. (18) and (19b),

$$\langle K\rangle = \frac{1}{2N\delta}\frac{N^3}{N-1}\left\{\coth(N\tanh^{-1}\delta) - \frac{1}{N\delta}\right\} \tag{21}$$

$$\langle K\rangle(\alpha = 1) = \frac{N(N+1)}{6} \tag{22}$$

Equation (22) was derived by Montroll (1969a) and Elliott *et al.* (1969), and Eq. (21) by Elliott *et al.* (1969) by the method of random walk in which both time and state space are taken to be discrete.

The results derived above are true when there are only two absorbing states at both ends of the state space. However, the problem of more than two absorbing states at $n = N_1, N_2, \ldots, N_r$ can be reduced to the problem of two absorbing states at both ends. For, if at $t = 0$ the exciton is between absorbing states N_j and N_{j+1}, the number of states between these two absorbing states is $N_{j+1} - N_j \equiv S_j$, and hence the average time for absorption is the same as for the problem with two absorbing states at the ends, 0 and S_j. When the absorbing states are distributed periodically with period S, then $\langle t \rangle$ is given by Eq. (18) with N replaced by S.

To analyze exciton trapping for two- and three-dimensional photosynthetic units, we have to consider bivariate and trivariate processes. This is beyond the scope of the present monograph. For completeness we note that two- and three-dimensional PSU have been analyzed by Montroll (1969b) using methods of random walk (both time and state space discrete) for a variety of lattices with equal probabilities of transitions to the nearest neighbor lattice points. The average number of steps n taken by the random walker before being trapped are as follows (Montroll, 1969b).

a. *For two-dimensional lattices.*

$$\langle K \rangle = [c_1 N \log N + c_2 N + c_3 + c_4 N^{-1} + O(N^{-1})] N(N-1)^{-1} \tag{23}$$

where c_1, c_2, c_3, and c_4 are constants which depend on the lattice. For a square lattice,

$$c_1 = 1/\pi, \qquad c_2 = 0.195056166, \qquad c_3 = -0.11696481,$$
$$c_4 = -0.05145650 \tag{24a}$$

for a triangular lattice,

$$c_1 = 3^{1/2}/2\pi, \qquad c_2 = 0.235214021, \qquad c_3 = -0.251407596 \tag{24b}$$

and for a hexagonal lattice,

$$c_1 = 3^{3/2}/2\pi, \qquad c_2 = 0.06620698, \qquad c_3 = -0.2542227888 \tag{24c}$$

b. *For simple cubic lattice.*

$$\langle K \rangle = 1.5164N + O(N) \tag{25}$$

Numerical calculations of trap-limited lifetimes have been made by Pearlstein (1966, 1967) and Robinson (1967) for two- and three-dimensional lattices. Some analytical work has been carried out by ten Bosch and Ruijgrok (1963), Knox (1968), and Sanders *et al.* (1971).

Before we present the experimental data and their analysis in terms of migration and trapping of excitons, let us point out another analysis of the process of migration and trapping. In this analysis (Pearlstein, 1966) the excitation transfer is considered as a diffusion process. This is especially true when the trap concentration is low. In this analysis, $P_{n,m}(t)$ is taken equal to $P(\mathbf{r}_n, \mathbf{r}_m, t)$ where $\mathbf{r}_i$ is the position vector of the ith molecule in the network. If $P(\mathbf{r}_i, t)$ is now considered to be a continuous function of $\mathbf{r}_i$ (for convenience we have suppressed $\mathbf{r}_m$, the position vector of the initial location of the exciton) rather than a function on discrete states, the right-hand side of Eq. (1) may be expanded in a Taylor series to give

$$\frac{\partial}{\partial t} P(\mathbf{r}_i, t) = \sum_{k=1}^{N} F_{i,k}[(\mathbf{r}_k - \mathbf{r}_i) \cdot \nabla P(\mathbf{r}_i, t) + \tfrac{1}{2}(\mathbf{r}_k - \mathbf{r}_i)^2 \nabla^2 P(\mathbf{r}_i, t) + \cdots] \tag{26}$$

where ∇ and ∇^2 are the gradient and Laplacian operators, respectively. For symmetric transition probabilities and regular lattices, the first term in the brackets vanishes identically, and to lowest nonvanishing order in the expansion, Eq. (26) becomes the diffusion equation

$$\frac{\partial P}{\partial t} = \frac{\Lambda}{2} \nabla^2 P \tag{27a}$$

where

$$\Lambda = h^2 \sum_k F_{i,k} s_{i,k}^2 \tag{27b}$$

In Eq. (27b) h is the lattice constant and $s_{i,k}$ is the magnitude (in units of h) of the lattice vector to the kth site from the ith site. For only nearest neighbor transitions, Eq. (27b) becomes

$$\Lambda = (q/2D)\,\nu h^2 \tag{28}$$

where

$$F_{i,k} = \begin{cases} \nu, & i,k \quad \text{nearest neighbors} \\ 0, & \text{otherwise} \end{cases}$$

q is the coordination number of the lattice, and D its dimensionality. For typical situations, $\nu \simeq 10^{11}$ sec^{-1}, $h = 10^{-7}$ cm, $(q/2D) \simeq 1$, and $\Lambda = 10^{-3}$ cm^2/sec.

For a one-dimensional lattice, Eq. (27a) is the FP equation with $a(x) = 0$, $b(x) = \Lambda$, and hence can be analyzed as in Chapter 3. We will not carry out the detailed analysis because, as we noted earlier in this chapter, the one-dimensional lattice does not realistically represent the PSU unit. Pearlstein (1966) has carried out the detailed analysis for one-, two-, and three-dimensional lattices.

We now discuss the type of experimental data which can be analyzed in order to understand the nature of the PSU, the process of movement of excitons and subsequent trapping, and in particular, whether the PSU behaves like a two-dimensional or three-dimensional lattice for the movement of exciton. From Eqs. (23) and (25) the average number of steps (and hence trapping time) taken by an exciton before it is trapped is larger for the three-dimensional than for the two-dimensional lattice. Thus trapping time provides the clue for the dimensionality of the lattice.

Consider a lattice with random locations of traps. When the lattice is exposed to light, excited sites are produced. The rate at which these sites are produced will depend on the intensity of light (I). For light that is not too intense we can assume this rate to be linearly proportional to I, i.e.,

$$\frac{d}{dt}[\mathrm{B}^*] = kI \tag{29}$$

where $[\mathrm{B}^*]$ is the concentration of excited states and k is the rate constant. As noted earlier in this chapter, the absorbed energy at an excited site can be transformed either into fluorescence and heat, or into biochemical energy by eventually getting trapped in a trapping site. If we assume that the last process depletes B^* by an effective rate $k_3[\mathrm{P}]$, where $[\mathrm{P}]$ is the concentration of traps, then

$$d[\mathrm{B}^*]/dt = kI - (k_1+k_2)[\mathrm{B}^*] - k_3[\mathrm{P}][\mathrm{B}^*] \tag{30}$$

where k_1 and k_2 are the rate constants of the depletion of B^* by the conversion of absorbed energy into fluorescence and heat, respectively. In the steady state $d[\mathrm{B}^*]/dt = 0$ and the quantity $\phi_{\mathrm{F}} = k_1[\mathrm{B}^*]/kI$ becomes

$$\phi_{\mathrm{F}} = \frac{k_1}{k_1+k_2+k_3[\mathrm{P}]} \tag{31}$$

ϕ_{F} is known as fluorescence yield and is a measure of the absorbed energy being given out as fluorescent light. In the absence of traps (or when k_3 is effectively zero), the fluorescent yield is

$$\phi_{\mathrm{F}}{}^0 = k_1/(k_1+k_2) \tag{32}$$

or

$$\phi_{\mathrm{F}}{}^0/\phi_{\mathrm{F}} = 1 + k_3\tau[\mathrm{P}] \tag{33}$$

where

$$\tau \equiv (k_1+k_2)^{-1} \tag{34}$$

is the lifetime of an exciton in the absence of trapping. k_3 is related to the process of movement of exciton and trapping. By measuring $1/\phi_{\mathrm{F}}$ while varying $[\mathrm{P}]$ (which is detected by small changes in absorption strength of the

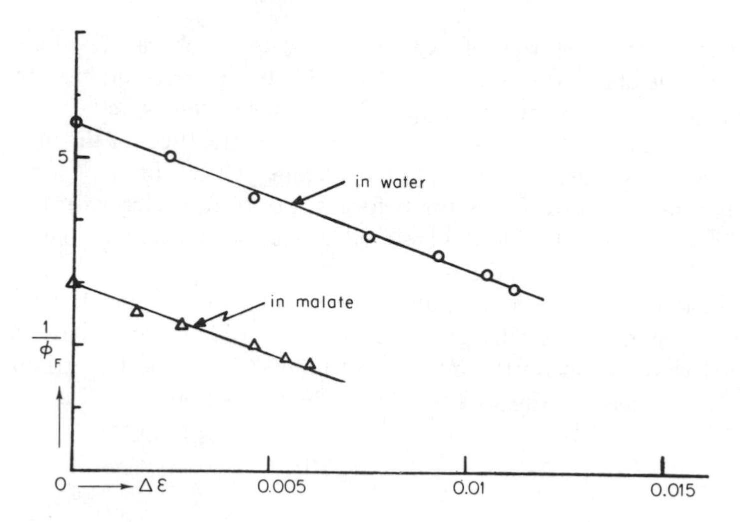

Fig. 10.1 Variation in $1/\phi_F$ (arbitrary units) with $\Delta\varepsilon$ for *Rhodospirilum rubrum* (Vredenberg and Duysens, 1963).

PSU), k_3 can be measured. In Fig. 10.1 are shown the results of Vredenberg and Duysens (1963) on the bacterium *Rhodospirillum rubrum*. $\Delta\varepsilon$ is proportional to $[P]_0 - [P]$, where $[P]_0$ is the initial concentration of traps. The reciprocal fluorescence yield is reduced by a factor of 2 for $\Delta\varepsilon \sim 0.0125$.

We may now derive the relation between k_3 and the process of movement of exciton and trapping. k_3 can be written as

$$k_3 = v_{B*}\sigma_p \tag{35}$$

where v_{B*} is the characteristic speed with which an exciton travels, and σ_p is the capture cross section at the trap and characterizes the probability of trapping once the exciton is at the trap. If r is the rate of transfer between two molecules, v_{B*} can be approximately written as

$$v_{B*} = rh \tag{36a}$$

where h is the average distance between chlorophyll (bacteriochlorophyll) molecules (the lattice constant). A better expression for v_{B*} is obtained by noting that when q neighbors (the coordination number of the lattice) compete for the excitation, the initial excitation can be lost q times as fast as when there is only one neighbor. The resulting expression is

$$v_{B*} = qrh \tag{36b}$$

For one-dimensional lattice, $q = 2$ and the rate of hopping between two molecules is $[(\nu+\rho)/2]^{-1}$ so that

$$v_{B*} = h/(\nu+\rho) \tag{36c}$$

σ_{B*} can be approximated by a geometrical cross section, i.e., by

$$\sigma_p = h^2 \tag{37}$$

Substituting Eqs. (36b) and (37) into Eq. (35) we get

$$k_3 = qr/[B], \qquad [B] \equiv 1/h^3 \tag{38}$$

and Eq. (33) becomes

$$\phi_F{}^0/\phi_F = 1 + qr\tau[P]/[B] \tag{39}$$

where [B] is the concentration of sites in the lattice, $qr\tau$ is the number of steps the exciton takes during its lifetime, and [P]/[B] is the chance of hitting a trap and being trapped.

By the experimental data of Fig. 10.1, the coefficient of [P]/[B], $qr\tau$, can be calculated. In principle, $qr\tau$ can be correlated with the form of the lattice by results of stochastic models such as those given in Eqs. (21), (23), and (25). The derived dimensionality of the lattice is not very reliable because of the uncertainty in the computation of τ and r. (For green plants, there is additional complication due to two distinct types of traps.) The most widely held view is that PSU is a three-dimensional unit with layered structure and interlayer coupling weaker than the intralayer coupling.

References

Elliott, R. A., Lakatos, K., and Knox, R. S. (1969). The effect of lattice vibrations on trap limited exciton lifetimes. *J. Statist. Phys.* **1**, 253.

Knox, R. S. (1968). On the theory of trapping of excitation in the photosynthetic unit. *J. Theoret. Biol.* **21**, 244.

Montroll, E. W. (1969a). Random walks on lattices containing traps. *J. Phys. Soc. Japan Suppl.* **26**, 6.

Montroll, E. W. (1969b). Random walks on lattices. III. Calculation of first-passage times with application to exciton trapping on photosynthetic units. *J. Mathematical Phys.* **10**, 753.

Pearlstein, R. M. (1966). Migration and Trapping of Excitation Quanta in Photosynthetic Units. Ph. D. Thesis, Univ. of Maryland, College Park.

Pearlstein, R. M. (1967). Migration and trapping of excitation quanta in photosynthetic units. *Brookhaven Nat. Lab. Symp.* **19**, 8.

Robinson, G. W. (1967). Excitation transfer and trapping in photosynthesis. *Brookhaven Nat. Lab. Symp.* **19**, 16.

Sanders, J. W., Ruijgrok, Th. W., and ten Bosch, J. J. (1971). Some remarks on the theory of trapping of excitons in the photosynthetic unit. *J. Mathematical Phys.* **12**, 534.

ten Bosch, J. J., and Ruijgrok, Th. W. (1963). A remark on the energy transfer in biological systems. *J. Theoret. Biol.* **4**, 225.

Vredenberg, W. J., and Duysens, L. N. M. (1963). Transfer of energy from bacteriochlorophyll to a reaction centre during bacterial photosynthesis. *Nature* (*London*) **197**, 355.

11

Epilogue

In this monograph we have reviewed two approaches to the stochastic modeling of complex biological systems. In these approaches, we limited ourselves to systems which are either inherently or approximately describable by a single random variable for which the difference-differential or partial differential equation satisfied by its probability function or density is a linear and homogeneous equation. Further, this random variable is either unrestricted or is restricted by boundaries which are either perfectly absorbing or perfectly reflecting. Relaxing these limits would have been inconsistent with the introductory nature of the monograph. In this last chapter, for the sake of completeness and for the more inquisitive readers, we briefly describe various generalizations of the processes and techniques discussed and used in this monograph and introduce some basically different techniques. These generalizations and the different techniques should allow one to get a better insight and more accurate results based on fewer assumptions for the biological systems discussed herein. In addition, it makes other interesting complex biological systems amenable to stochastic modeling. We also describe some of these systems.

The processes and techniques of Chapters 2 and 3 can be generalized into the following

(a) Restricted univariate diffusion process where the boundaries are partially reflecting and partially absorbing, i.e., the boundaries are accessible and the process returns to the interval between the boundaries after staying in the boundary for some time. The time for which the process stays at a boundary must be exponentially distributed for the process to remain Markovian. Such processes are known as *elementary return processes* and are discussed by Feller (1952, 1954) and Bharucha-Reid (1960). In addition to the functions $a(x)$ and $b(x)$, the process is defined by time-independent probabilities of the process jumping back from each boundary to points between the boundaries and to the other boundary. The resulting generalized Fokker–Planck equation includes terms which describe the probability "masses" at the boundaries.

(b) Birth and death and diffusion processes where the number of random variables is more than one. Such a process is known as a *multivariate process.* Throughout the monograph we commented on the need for such processes and the techniques to study them. We could have understood the growing population better if we had techniques for studying even bivariate birth and death processes. Instead of learning only the behavior of the total number of individuals, we could have learned the total number of "births" (or "deaths") by using another variable which keeps track of the number of transitions in one direction. The techniques used for analyzing the multivariate processes are basically the extensions of the techniques of Chapters 2 and 3. A multivariate birth and death process, with k random variables $x_1, x_2, \ldots, x_k$, is characterized by the probability function $P_{i_1, i_2, \ldots, i_k}$ of these variables being in the states $i_1, i_2, \ldots, i_k$ where $i_1, \ldots, i_k$ are integers. The transitions are allowed to occur only between neighboring points, i.e., a change in the state of only one of the variables by one integer. The master equations relate $dP_{i_1, i_2, \ldots, i_k}/dt$ to $P_{i_1, i_2, \ldots, i_k}, P_{i_1 \pm 1, i_2, \ldots, i_k}, P_{i_1, i_2 \pm 1, \ldots, i_k}, \ldots, P_{i_1, i_2, \ldots, i_k \pm 1}$ with coefficients which are transition probabilities per unit time between neighboring states. For unrestricted processes, the method of generating function can be used to calculate the probability density, $P_{i_1, \ldots, i_k}$. The generating function is defined by

$$G(z_1, z_2, \ldots, z_k; t) = \sum P_{i_1, \ldots, i_k} z_1^{i_1} z_2^{i_2} \cdots z_k^{i_k}$$

and satisfies a linear partial differential equation, first order in time, and involving k variables $z_1, z_2, \ldots, z_k$. The discussion of generating functions and other aspects of these processes may be found in the works of Keilson (1964) and Bailey (1964).

The multivariate diffusion process similar to its univariate counterpart is described by forward and backward diffusion equations which are linear partial differential equations involving first-order time derivative and first- and second-order derivatives with respect to the space random variables with coefficients which are the rates of change of the mean and the covariances of the random variables. The forward diffusion equation, as for the univariate case, is solved by using one of many standard methods for solving linear partial differential equations, e.g., separation of variables and eigenfunction expansions. For example, for the k-variate Ornstein–Uhlenbeck process, the forward diffusion equation is

$$\frac{\partial P}{\partial t} = -\sum_i \lambda_i \frac{\partial}{\partial x_i}(x_i P) + \tfrac{1}{2}\sum_{ij} \sigma_{ij} \frac{\partial^2 P}{\partial x_i\, \partial x_j} \tag{1}$$

where λ_i and σ_{ij} are some constants. The fundamental solution of this equation is given in Wang and Uhlenbeck (1945). The distribution is a k-dimensional Gaussian distribution with the average values

$$\langle x_i \rangle = y_i \exp(\lambda_i t) \tag{2}$$

and the variance

$$\mu_{ij} \equiv \langle (x_i - \langle x_i \rangle)(x_j - \langle x_j \rangle) \rangle = -\frac{\sigma_{ij}}{\lambda_i + \lambda_j}[1 - \exp(\lambda_i + \lambda_j)\, t] \tag{3}$$

where y_i are the initial values of the x_i. Further discussion of the multivariate diffusion process is presented by Krishnaiah (1965, 1969).

(c) Birth and death or diffusion process inhomogeneous in time, where the parameters λ_n and μ_n, or equivalently $a(x)$ and $b(x)$, which characterize the process, become time dependent. The evolution of the process depends not only on $t - t_0$ (t_0 being the initial time) but on both t and t_0. Discussion of these *inhomogeneous processes* can be found in the book by Bailey (1964). A technique for analyzing them involves generating function and Laplace transform dependent on the form of $\lambda_n(t)$ and $\mu_n(t)$. Some special inhomogeneous processes can be analyzed by an analysis which is a simple extension of that given in Chapters 2 and 3. If, as in Sections 5.2 and 8.2, for the inhomogeneous process

$$\lambda_n(t) = i(t)\, \lambda_n, \qquad \mu_n(t) = i(t)\, \mu_n$$

or equivalently,

$$a(x,t) = i(t)\, a(x), \qquad b(x,t) = i(t)\, b(x)$$

where $i(t)$ is some function of time, then by the transformation

$$t \to I(t), \qquad I(t) = \int_0^t i(\tau)\, d\tau$$

the inhomogeneous process is transformed into one which is homogeneous in time. If as $t \to \infty$, $\lambda_n(t)$ and $\mu_n(t)$ [or $a(x,t)$ and $b(x,t)$] tend to a limit, then $\lim_{t\to\infty} P$ exists and is the solution of the corresponding diffusion equation for the homogeneous process, characterized by the limiting values of the parameters λ_n and μ_n [or $a(x,t)$ and $b(x,t)$].

(d) Process with both continuous and discrete transitions in the state space. The probability density for such a process satisfies an integrodifferential equation when the size of the jumps is distributed according to some continuous distribution. This integrodifferential equation turns into a difference-differential equation when there is only a discrete set of allowed jump sizes. As an example consider the process of firing of a neuron where the potential of the membrane changes continuously and deterministically (due to decay) as well as discretely and randomly, due to inputs which arrive according to some distribution. The integrodifferential equation can be derived by noting that

$$P(x,t+\Delta t) = P(x+x\,\Delta t/\tau,t)(1-\alpha\,\Delta t) + \alpha\,\Delta t \int P(x-z,t)\,h(z)\,dz \quad (4)$$

where $P(x,t) \equiv P(x\,|\,y,t)$, τ^{-1} is the rate of decay, α the probability of arrival of input per unit time, and $h(z)\,dz$ the probability that the input size is in the interval $(z, z+dz)$. Taking the limit $\Delta t \to 0$, we get

$$\frac{\partial P(x,t)}{\partial t} = \frac{x}{\tau} P(x,t) - \alpha P(x,t) + \alpha \int P(x-z,t)\,h(z)\,dz \quad (5)$$

In case $h(z)$ is nonzero only for a discrete set of z values, the integral in the right-hand side becomes a sum.

The discussion of the above-mentioned processes can be found in the literature on *queuing* theory [see, e.g., Takács (1962)].

(e) Process in discrete time and state space. Such a process is termed a Markov chain if it is Markovian, and is characterized by a set of transition probabilities $p_{i,j}$ of transition from state i to state j in unit time, which in general may depend on time. When these probabilities are independent of time, the process is known as a stationary Markov chain. If

$$(p_0(n), p_1(n), \ldots, p_s(n))$$

are the distributions of the process in the states $0, 1, 2, \ldots, s$ at time n (also called the nth generation), the probability distribution at time $(n+1)$ is given by

$$p_j(n+1) = \sum p_{i,j}\, p_i(n) \quad (6)$$

There are two special cases of stationary Markov chains which have been studied in detail (Freedman, 1971).

1. *Branching process.* If x_n is the random variable describing the process

at time n, then the Markov chain is called a branching process if

(a) the process starts at state 1, $x_0 = 1$, and

(b) if $x_n = j$, then x_{n+1} is distributed as the sum of j independent variables, each with the same distribution $q(i)$, $i = 0, 1, 2, \ldots, s$, with $\sum_{i=0}^{s} q(i) = 1$.

This implies that

$$p_i(n+1) = \sum p_j(n) \left[\sum q(i_1) q(i_2) \cdots q(i_j)\right] \tag{7}$$

where the inner sum runs over all the i_j such that

$$i_1 + i_2 + \cdots + i_j = i \tag{8}$$

while the outer sum runs over all possible values of x_n. As an example, for a growing population of identical individuals when time can be counted in units of generations, $p_j(n)$ is the probability of population having j individuals in the nth generation, and $q(i)$ is the probability of each individual becoming i individuals in the next generation. [For simple birth and death processes, with $\lambda_n = \lambda n$, $\mu_n = \mu n$, $q(2) = \lambda$, $q(0) = \mu$, and $q(1) = 1-(\lambda+\mu)$.] A discussion of branching processes may be found in the work of Mode (1971) and Karlin (1966).

2. *Random walks.* When $p_{ij} = 0$ for $|i-j| > 1$, i.e., transitions are allowed only between nearest neighbors, the Markov chain is analogous to birth and death processes and is called a random walk. We have already mentioned this approach briefly in connection with photosynthesis. The approach has been used extensively in many physical and chemical systems, e.g., configuration statistics of polymers, polymers in solution, polymer absorption; in solid-state physics, especially lattice vibrations, luminescence, diffusion in solids, Ising model of ferromagnetism; in chemical kinetics, and so on. A recent review may be found in the book by Barber and Ninham (1970). The approach has limitations in that analytical expressions for various quantities giving insight into the process can only be obtained when the transition probabilities from one state to the neighboring states are independent of the state. When transitions to all neighboring states are equally probable, expressions for the interesting quantities exist also for bivariate and trivariate processes. Additionally, for this case, a continuous time description of the lattice walk is possible; i.e., one can allow each step to occur at random time governed by a preassigned distribution (Montroll and Weiss, 1965). If this distribution is different from the exponential distribution (which characterizes the birth and death processes), the process is no longer Markovian and the future behavior of the random variable depends not only on its present state but also on its past behavior [see (g)]. Such a process can be approximated by a process with several stages, such that for each state the time interval is exponentially distributed with a different parameter for each stage (Jensen, 1954).

(f) Process with time intervals between occurrence of two consecutive events independently and identically distributed (as in a Poisson process or birth and death process with $\lambda_n = \nu$, $\mu_n = \rho$). Such a process is known as a simple renewal process and has been studied quite extensively in the literature in applied probability [see, e.g., Cox (1962)].

The theory of simple renewal processes can be very useful in further understanding the firing of a neuron. If we take the first passage time distribution of firing of a neuron as the distribution between two firing events, the process of the sequence of firings is a renewal process and quantities such as the distribution of the number of firings in a given time interval, and the distribution of the time at which a certain, say rth, firing occurs, can be calculated [see Srinivasan and Rajamannar (1970a, b, c) for some aspects of firing of a neuron using the theory of renewal processes]. The renewal processes can be generalized (Cox, 1962) to include intervals of different types, characterized by different probability distributions, the sequence of which can also be stochastic. The theory of these generalized renewal processes can be applied to improve the understanding of extinction and reestablishment of colonies, recurrence of an epidemic, and firing of a neuron with a refractory period.

(g) Non-Markovian process. As noted in (e) above, for such a process the future behavior of the random variable depends not only on its present state but also on its past history. Such dependence may arise due to one of many reasons. We have already mentioned one such reason in (e), namely, nonexponential distribution of time intervals between two consecutive events. Another reason is a nonvanishing time delay between the cause and the effect. There are many methods of analyzing a non-Markovian process, the basis of most common ones being the following:

(i) finding a random variable which describes some aspects of the process whose behavior is Markovian,

(ii) adding supplementary random variables which represent different possible states in the past, so that the multivariate process is a Markovian one (Cox, 1955), and

(iii) instead of considering the process for all $t > 0$, choosing a selected set of discrete points in time, such that the resulting process is a Markov process. Such a process is known as imbedded Markov process (Kendall, 1953).

(h) System subject to a nonwhite noise. Such a system can be studied by generalization of the techniques of Chapter 3. The generalization is done by constructing a system (described by a dynamical equation) such that its output for white noise input is the known nonwhite noise. The nonwhite noise to the complex system is then replaced by this system and white noise. If the complex system is exactly or approximately describable by a single variable in the absence of noise, the above approach will involve the analysis of a bivariate process, the second variable being the variable describing the constructed system. Keilson and Mermin (1959) give the construction of the system for the so-called shot noise.

Having briefly discussed generalizations of the processes and techniques of Chapters 2 and 3, let us now briefly describe two techniques which are quite different from the ones discussed in this monograph.

a. *Statistical mechanics.* This technique, which was introduced by J. Willard Gibbs during the beginning years of this century in an effort to explain some inconsistencies in thermodynamics, has been very successfully used to analyze a variety of complex physical systems with a large number of identical "components" (e.g., atoms, molecules, particles) and with almost complete ignorance (with the exception of a few measurable microscopic quantities). Four basic assumptions are made about the system to be analyzed:

(i) There exists a constant of motion, i.e., a constant which is time invariant.

(ii) All possible copies of the system compatible with whatever characteristics of the system are given (e.g., a constant of motion) constitute an ensemble in which each copy is weighed equally.

(iii) In phase space, in which the state of each member of the large ensemble is represented by a point, the density of phase points is conserved, i.e., Liouville's theorem holds.

(iv) The time averages of physical quantities over a single system equal the corresponding averages over the ensemble (ergodic theorem).

The last assumption is the most difficult to justify and has only been rigorously justified for the case of perfect gas, consisting of idealized perfectly elastic, spherical molecules. Nevertheless, this technique has been successfully used to explain a wide variety of physical phenomena, e.g., conductivity of metals, blackbody radiations, thermal expansion of solids, superfluidity, etc. For these physical systems, Hamiltonian (energy) is a constant of motion, the phase space variables are space and momentum coordinates of various particles, and Liouville's theorem holds. There are a variety of complex biological systems for which the variables of the phase space are position and momentum coordinates. For these systems statistical mechanics has been successfully applied. One of the authors (Goel, 1972) has recently reviewed applications of statistical mechanics to understand the cooperative processes in biological systems. These processes include conformational transitions in biopolymers, small-molecule–large-molecule reactions (enzyme action, biosynthesis, etc.), conformational changes in membrane and nerve excitation, aggregation of cells into tissues, and polarization phenomena in society. For those systems for which the phase space variables are not position and momentum coordinates, one must justify the use of statistical mechanics before the results can be trusted. Kerner (1957) first applied statistical mechanics to such a system, namely, the population of interacting biological species. Since then it has been applied to an assembly of interacting biochemical oscillators, an assembly of neurons (nervous system), and an assembly of growing cells [see Goel *et al.* (1971) for review and references]. For the applicability of statistical mechanics to such systems, the dynamical equations must be special, i.e., it should be possible to write them in a quasi-Hamiltonian form so that assumptions (i) and (iii) described above can be satisfied. In some of the biological systems mentioned above, the more realistic dynamical equations have been "tortured" by making drastic assumptions so as to satisfy the conditions mentioned above. In all these cases, the authors presumably felt that the effects of deviation of the realistic situation from the ideal situation will cancel out and the statistical mechanical analysis does give useful and correct insight into the system. The justification for their feelings is their better judgment. One of the present authors has carefully studied the application of statistical mechanics to the population of interacting biological species (Goel *et al.*, 1971) and found that any slight deviation from the ideal situation violates the justification of the use of statistical mechanics. For example, if for the population of interacting species one incorporates the realistic characteristics of the system of a finite time delay between the amount of food consumed and the population growth due to this food, or of dependence of biomass transfer efficiency from a prey to a predator on the prey and the predator species, one no longer has a constant of motion.

b. *Information theory and Bayesian statistics.* This technique is based on an alternate formulation of statistical mechanics due to Jaynes (1957a, b) which is useful from a practical point of view. He derives the distribution function by maximizing the "ignorance" (negative of information content) such that the distribution function is consistent with the known macroscopic averages, if any, about the system. When the macroscopic average is energy, which is a constant of motion, the distribution function is exactly the same as that from statistical mechanics. If, at a later time, one has more information about the system (more known averages), one can use this information, using Bayesian statistics, to sharpen the distribution function. This procedure has been very effectively applied in engineering problems and decision making (Tribus, 1969) and its use in Volterra–Lotka ecological dynamics has been suggested by Hammann and Bianchi (1970). In this approach, the ergodic theorem is ignored and it is assumed that if the distribution obtained is a sharp one, the calculated averages represent the averages of the system.

In conclusion, we point out some areas of biology which we have not discussed and in which stochastic modeling is very useful. We have not included these areas either because of the limited insight obtained when the techniques covered in this monograph are used, or because compacting the extensive literature in a chapter would not have done justice to the area.

One biologically interesting speculation, based on the analysis of uni-, bi-, and trivariate diffusion processes, has been put forth by Adam and Delbrück (1968). They argue that organisms handle some of the problems of timing and efficiency, when small numbers of molecules and their diffusion are involved, by reducing the dimensionality in which diffusion takes place from three-dimensional space to two-dimensional surface diffusion. They calculate the average time for the process to diffuse from one point to another, as a function of dimensionality of diffusion, by solving one-, two-, and three-dimensional equations and show that it increases with increase in dimension. Thus in chemical reactions involving diffusion, for greater rates of reactions, chemical molecules will prefer reaching their destination area (where the chemical reaction will take place) not directly by free diffusion in three-dimensional space, but by subdividing the diffusion process into successive stages of lower spatial dimensionality. Adam and Delbrück apply their analysis in detail to the chemoreception of a particular odorant, sex attractant bombykol, exuded by the female and perceived by the male over great distances, of the species *Bombyx mori* (silkworm moth).

Another important and emerging area is of morphogenesis, which may be approximately defined as the sum of processes whereby the form of an organism is developed The development of numerous small regions of tissue appears to be functionally related and to behave in a coordinated fashion to produce new structures and regions as time progresses. A classical explanation of how the relative positions of different regions are specified is that it is done by the distribution of certain chemical or chemicals which diffuse. There are many examples of cells and organisms displaying movement determined by the distribution of chemicals. [For review, see Trinkaus (1969), Goodwin and Cohen (1969), Keller and Segel (1971a, b).] Crick of Watson–Crick fame is currently leading a group which is investigating this explanation (Crick, 1970).

A comparatively older but important area is the compartmental analysis using tracers. The basic idea behind the method is to use a certain chemical in trace amount and follow its flow (kinetics) through the complex system which is composed of "compartments" connected to each other. If the flow of tracer follows the path of going from one compartment to the other without feedbacks, the concentration of the tracer in the nth compartment is quite accurately given by the deterministic equation

$$dx_n/dt = k_{n-1}x_{n-1} - (k_n + k_n')x_n + k'_{n+1}x_{n+1} \tag{9}$$

where k_n is the rate constant for transfer of tracer from the nth to the $(n+1)$th compartment, and k_n' from the nth to the $(n-1)$th compartment. In the above equation dx_n/dt has been assumed to be linearly proportional to x_n, which is reasonable because the tracer is in minute amount. By analyzing Eq. (9) several properties about the complex system can be determined, e.g., the number of compartments. There exists a large amount of literature on this subject and we refer the reader to the book by Rescigno and Segre (1966) and the recent papers by Sheppard (1971) and Thakur *et al.* (1972) where references to the previous work can be found.

References

Adam, G., and Delbrück, M. (1968). Reduction of dimensionality in biological diffusion processes, *in* "Structural Chemistry and Molecular Biology" (A. Rich and N. Davidson, eds.), p. 198. Freeman, San Francisco, California.

Bailey, N. T. J. (1964). "The Elements of Stochastic Processes with Applications to the Natural Sciences." Wiley, New York.

Barber, M. N., and Ninham, B. W. (1970). "Random and Restricted Walks; Theory and Applications." Gordon & Breach, New York.

Bharucha-Reid, A. T. (1960). "Elements of the Theory of Markov Processes and their Applications." McGraw-Hill, New York.

Cox, D. R. (1955). The analysis of non-Markovian stochastic processes by the inclusion of supplementary variables. *Proc. Cambridge Philos. Soc.* **51**, 313.

Cox, D. R. (1962). "Renewal Theory." Methuen, London.

Crick, F. (1970). Diffusion in embryogenesis. *Nature (London)* **225**, 420.

Feller, W. (1952). The parabolic differential equations and the associated semi-groups of transformations. *Ann. of Math.* **55**, 468.

Feller, W. (1954). Diffusion processes in one dimension. *Trans. Amer. Math. Soc.* **77**, 1.

Freedman, D. (1971). "Markov Chains." Holden-Day, San Francisco, California.

Goel, N. S. (1972). Cooperative processes in biological systems. *Progr. Theoret. Biol.* **2**, 213.

Goel, N. S., Maitra, S. C., and Montroll, E. W. (1971). "On the Volterra and other nonlinear models of interacting populations. *Rev. Modern Phys.* **43**, 231; "On the Volterra and Other Nonlinear Models of Interacting Populations." Academic Press, New York.

Goodwin, B., and Cohen, M. H. (1969). A phase-shift model for the spatial and temporal organization of developing systems. *J. Theoret. Biol.* **25**, 49.

Hamann, J. R., and Bianchi, L. M. (1970). Stochastic population mechanics in the relational systems formalism: Volterra-Lotka ecological dynamics. *J. Theoret. Biol.* **28**, 175.

Jaynes, E. T. (1957a). Information theory and statistical mechanics. *Phys. Rev.* **106**, 620.

Jaynes, E. T. (1957b). Information theory and statistical mechanics. II. *Phys. Rev.* **108**, 171.

Jensen, A. (1954). "A Distribution Model Applicable to Economics." Munksgaard, Copenhagen.

Karlin, S. (1966). "A First Course in Stochastic Processes." Academic Press, New York.

Keilson, J. (1964). A review of transient behavior in regular diffusion and birth-death processes. *J. Appl. Probability* **1**, 247.

Keilson, J., and Mermin, N. D. (1959). The second-order distribution of integrated shot noise. *IRE Trans. Information Theory* **IT-5** (2), 75.

Keller, E. F., and Segel, L. A. (1971a). Model for chemotaxis. *J. Theoret. Biol.* **30**, 225.

Keller, E. F., and Segel, L. A. (1971b). Traveling bands of chemotactic bacteria: A theoretical analysis. *J. Theoret. Biol.* **30**, 235.

Kendall, D. G. (1953). Stochastic processes occurring in the theory of queues and their analysis by the method of imbedded Markov chains. *Ann. Math. Statist.* **24**, 338.
Kerner, E. H. (1957). A statistical mechanics of interacting biological species. *Bull. Math. Biophys.* **19**, 121.
Krishnaiah, P. R., ed. (1965). "Multivariate Analysis I." Academic Press, New York.
Krishnaiah, P. R., ed. (1969). "Multivariate Analysis II." Academic Press, New York.
Mode, C. J. (1971). "Multiple Branching Processes; Theory and Applications." Amer. Elsevier, New York.
Montroll, E. W., and Weiss, G. (1965). Random walks on lattices II. *J. Mathematical Phys.* **6**, 167.
Rescigno, A., and Segre, G. (1966). "Drug and Tracer Kinetics." Ginn (Blaisdell), Boston Massachusetts.
Sheppard, C. W. (1971). Stochastic models for tracer experiments with the circulation III. The lumped catenary system. *J. Theoret. Biol.* **33**, 491.
Srinivasan, S. K., and Rajamannar, G. (1970a). Renewal point processes and neuronal spike trains. *Math. Biosci.* **6**, 331.
Srinivasan, S. K., and Rajamannar, G. (1970b). Counter models and dependent renewal point processes related to neuronal firing. *Math. Biosci.* **7**, 27.
Srinivasan, S. K., and Rajamannar, G. (1970c). Addendum to counter models and dependent renewal point processes related to neuronal firing. *Math. Biosci.* **9**, 29.
Takács, L. (1962). "Introductions to the Theory of Queues." Oxford Univ. Press, London and New York.
Thakur, A. K., Rescigno, A., and Schafer, D. E. (1972). On the stochastic theory of compartments: I. A single compartment system. *Bull. Math. Biophys.* **34**, 53.
Tribus, M. (1969). "Rational Descriptions, Decisions, and Designs." Pergamon, Oxford.
Trinkaus, J. P. (1969). "Cells into Organs." Prentice Hall, Englewood Clifls, New Jersey.
Wang, M. C., and Uhlenbeck, G. E. (1945). On the theory of the Brownian motion II. *Rev. Modern Phys.* **17**, 323.

Additional References

Chandresekhar, S. (1943). Stochastic problems in physics and astronomy. *Rev. Modern Phys.* **15**, 1.
Çinlar, E. (1969). Markov renewal theory. *Adv. Appl. Probability* **1**, 123.
Çinlar, E. (1972). Superposition of point processes, *in* "Stochastic Point Processes" (P. A. W. Lewis, ed.), p. 549. Wiley, New York.
Cox, D. R., and Lewis, P. A. W. (1972). Multivariate point processes. *Proc. Symp. Math. Statist. Probability, 6th, Berkeley, California*, Vol. 3, p. 401.
Hoel, P. G., Port, S. C., and Stone, C. J. (1972). "Introduction to Stochastic Processes." Houghton, Boston, Massachusetts.
Lewis, P. A. W., ed. (1972). "Stochastic Point Processes." Wiley, New York.
Srinivasan, S. K., and Rangan, A. (1970). Stochastic Model for the Quantum Theory in Vision. *Math. Biosci.* **9**, 31.
Wax, N., ed. (1954). "Noise and Stochastic Processes." Dover, New York.
Wolpert, L. (1969). Positional information and the spatial pattern of cellular differentiation. *J. Theoret. Biol.* **25**, 1.

Appendixes

Appendix A

Calculation of $\langle n \rangle$ ***and*** var(n)

In this appendix we solve the differential equations (2.0.8) and (2.0.13) satisfied by $\langle n \rangle$ and var(n) subject to the initial conditions (2.0.14b). As noted in Chapter 2, when λ_n and μ_n are at most linear in n, the solution of Eqs. (2.0.8) and (2.0.13) is known in a closed form. We derive this solution first.

Let $\lambda > 0$, $\mu > 0$, and let

$$\lambda_n = \lambda n + \nu, \qquad \mu_n = \mu n + \rho, \qquad n \geqslant -\rho/\mu \geqslant -\nu/\lambda \tag{1}$$

We have taken $\rho\lambda < \mu\nu$, and restricted the process to states with $n \geqslant -\rho/\mu$ to ensure that the transition probability rates λ_n and μ_n are nonnegative for all t. Substituting Eq. (1) into Eqs. (2.0.8) and (2.0.13), we get

$$d\langle n \rangle/dt = (\lambda - \mu)\langle n \rangle + (\nu - \rho) \tag{2}$$

$$d\,\text{var}(n)/dt = 2(\lambda - \mu)\,\text{var}(n) + (\lambda + \mu)\langle n \rangle + (\rho + \nu) \tag{3}$$

The solution of Eq. (2) for the initial condition (2.0.14b) is

$$\langle n \rangle = m\sigma + \frac{(\nu-\rho)\gamma}{\lambda}\frac{\sigma-1}{\gamma-1} \tag{4}$$

where

$$\sigma \equiv e^{(\lambda-\mu)t} \tag{5a}$$

$$\gamma \equiv \lambda/\mu \tag{5b}$$

Substituting Eq. (4) into Eq. (3) and integrating, we obtain

$$\operatorname{var}(n) = \frac{m\sigma(\sigma-1)(\gamma+1)}{(\gamma-1)} + \frac{(\nu-\rho)\gamma(\sigma-1)(\gamma\sigma-1)}{\lambda(\gamma-1)^2} \tag{6}$$

In Table 2.1 we have given the forms of Eqs. (4) and (6) when one or more of the parameters λ, μ, ν, ρ vanish.

Let us now discuss the case when λ_n and μ_n depend nonlinearly on n. As noted in Section 2.0, for this case the differential equation for $\langle n \rangle$ involves $\langle n^2 \rangle$,† the differential equation for which, in turn, involves $\langle n^3 \rangle$, and so on. This set of coupled equations can be solved approximately. Some of the widely used approximation methods are as follows:

(a) The open coupled set of equations is decoupled by approximately expressing higher moments in terms of lower moments. If one is interested in the moments $\langle n^i \rangle$, $i \leqslant k$, k a preassigned integer, then the approximation involves expressing all the moments higher than the kth moment, appearing in the first k equations, in terms of the first k moments. This can be done in two ways: One is to express the higher moments as products of the lower moments, e.g., $\langle n^l \rangle = \langle n^k \rangle \langle n^{l-k} \rangle$, $l > k$. The other way (Wang, 1971) is to assume that $\{\langle n^l \rangle\}^{1/l} \equiv \xi_l$ is a smooth function of l and then to express ξ_l, $l > k$, in terms of $\xi_1, \ldots, \xi_k$ by polynomial extrapolation. For example, for $k = 1$, the extrapolation of ξ_2 in terms of ξ_1 is

$$\xi_2 = \xi_1 \qquad \text{or} \qquad \{\langle n^2 \rangle\}^{1/2} = \langle n \rangle \tag{7a}$$

and for $k = 2$

$$\xi_3 = 2\xi_2 - \xi_1 \qquad \text{or} \qquad \{\langle n^3 \rangle\}^{1/3} = 2\{\langle n^2 \rangle\}^{1/2} - \langle n \rangle \tag{7b}$$

In general,

$$\xi_l = \sum_{i=1}^{k} (-1)^{k-i} \binom{l-i-1}{k-i} \binom{l-1}{i-1} \xi_i \tag{7c}$$

The resulting set of k nonlinear ordinary differential equations obtained in either way can be numerically integrated by one of many known methods,

† There is an exception. If $\lambda_n = \theta n^2 + \lambda n + \nu$, $\mu_n = \theta n^2 + \mu n + \rho$, then $\langle n \rangle$ satisfies the differential equation [cf. Eq. (2.0.8)] $d\langle n \rangle/dt = (\lambda - \mu)\langle n \rangle + \nu - \rho$.

e.g., the Runge–Kutta method (Lapidus and Seinfeld, 1971). The resulting expressions for the moments can be improved by increasing k, i.e., by truncating the hierarchy of moment equations at a higher order, at the expense of making computations more difficult.

(b) There are two other methods which have been used to calculate $\langle n \rangle$ and $\langle n^2 \rangle$ in connection with stochastic theory of chemical kinetics of bimolecular reactions where λ_n and μ_n are quadratic in n (McQuarrie, 1968). In one of the methods, a simple functional dependence of $\langle n^2 \rangle / \langle n \rangle^2$ on t is guessed, e.g., $\langle n^2 \rangle = \langle n \rangle^2 e^{pt}$. This dependence, together with Eq. (2.0.8), is then used to solve $\langle n \rangle$ in terms of p, t, and the parameters in the expression for λ_n and μ_n. The parameter p is determined by demanding that $d\langle n^2 \rangle / dt|_{t=0}$, as calculated by Eq. (2.0.11), is the same as that calculated by using the guessed dependence mentioned above. In the other method, a bimolecular process is treated as a unimolecular process, but with time-dependent rate constants obtained from the deterministic case. In our notation, this procedure is equivalent to solving the equation

$$dn/dt = \lambda_n - \mu_n$$

for n, and then using this solution $n(t)$ to express λ_n and μ_n as linear functions of n with time-dependent coefficients. As noted above, for the resulting linear λn, μn, the equations for moments can be solved exactly. For further details and evaluation of these approximations, we refer interested readers to the review article by McQuarrie (1968).

References

Lapidus, L., and Seinfeld, J. H. (1971). "Numerical Solution of Ordinary Differential Equations." Academic Press, New York.

McQuarrie, D. A. (1968). "Stochastic Approach to Chemical Kinetics." Methuen, London.

Wang, Y. K. (1971). Noise Effects and Transient Behavior for a Laser. Ph. D. Thesis, Yale Univ., New Haven, Connecticut.

Appendix B

Differential Equation for the Generating Function for Birth and Death Processes

In this appendix we derive the differential equation satisfied by the generating function of a birth and death process

$$G(t, z) = \sum_{n=-\infty}^{\infty} P_{n,m}(t) z^n \tag{1}$$

for a given set of transition probabilities, λ_n and μ_n, which are polynomials in n.

Let $N(t)$ denote the random variable describing the state of the process at time t. Let us define $\Delta N(t)$ by

$$\Delta N(t) = N(t+\Delta t) - N(t) \tag{2}$$

Let $G(t,z)$ denote the probability generating function for $N(t)$, $G(t+\Delta t, z)$ for $N(t+\Delta t)$, and $\Delta G(t,z)$ for $\Delta N(t)$. Let $\underset{t}{E}$ denote the expectation value of any function of N at time t [averaging over all possible values of $N(t)$] and let $\underset{t+\Delta t}{E}$ denote the expectation at time $t+\Delta t$ [averaging over all possible values of $N(t+\Delta t)$]. Similarly, let $\underset{\Delta t|t}{E}$ denote the conditional expectation at the end of interval Δt [averaging over all possible values of $\Delta N(t)$], given the value of $N(t)$ at time t. With these definitions, from Eqs. (1) and (2)

$$G(t+\Delta t, z) = \underset{t+\Delta t}{E}\{z^{N(t+\Delta t)}\} = \underset{t+\Delta t}{E}\{z^{N(t)+\Delta N(t)}\}$$

$$= \underset{t}{E}\left\{\underset{\Delta t|t}{E}[z^{N(t)+\Delta N(t)}]\right\} = \underset{t}{E}\left\{z^{N(t)}\underset{\Delta t|t}{E}[z^{\Delta N(t)}]\right\}$$

Therefore

$$\frac{\partial G(t,z)}{\partial t} = \lim_{\Delta t\to 0}\frac{G(t+\Delta t,z)-G(t,z)}{\Delta t} = \lim_{\Delta t\to 0}\underset{t}{E}\left\{z^{N(t)}\underset{\Delta t|t}{E}\left(\frac{z^{\Delta N(t)}-1}{\Delta t}\right)\right\} \tag{3}$$

Since $\Delta N(t)$ takes the values -1, 0, 1 for Δt sufficiently small, with probabilities $\mu_{N(t)}\,\Delta t$, $1-\mu_{N(t)}\,\Delta t-\lambda_{N(t)}\,\Delta t$, and $\lambda_{N(t)}\,\Delta t$, respectively,

$$\lim_{\Delta t\to 0}\underset{\Delta t|t}{E}\frac{z^{\Delta N(t)}-1}{\Delta t} = \lambda_{N(t)}z + \mu_{N(t)}z^{-1} - \lambda_{N(t)} - \mu_{N(t)}$$

and Eq. (3) becomes

$$\frac{\partial G(t,z)}{\partial t} = \underset{t}{E}\left\{\left[\lambda_{N(t)}(z-1) + \mu_{N(t)}\left(\frac{1}{z}-1\right)\right]z^{N(t)}\right\} \tag{4}$$

Denoting by $f_1(n)$ the function λ_n and by $f_{-1}(n)$ the function μ_n, and since $z\,\partial z^n/\partial z = nz^n$, we get for $\lambda n, \mu n$ polynomials in n

$$\lambda_{N(t)}z^{N(t)} = f_1(N)z^N = f_1\left(z\frac{\partial}{\partial z}\right)z^N \tag{5}$$

$$\mu_{N(t)}z^{N(t)} = f_{-1}(N)z^N = f_{-1}\left(z\frac{\partial}{\partial z}\right)z^N$$

Inserting Eq. (5) into Eq. (4) we get

$$\frac{\partial G(t,z)}{\partial t} = \underset{t}{E}\left\{\left[(z-1)f_1\left(z\frac{\partial}{\partial z}\right) + \left(\frac{1}{z}-1\right)f_{-1}\left(z\frac{\partial}{\partial z}\right)\right]z^{N(t)}\right\} \tag{6}$$

When the order of differentiation and expectation can be interchanged, Eq. (6) becomes

$$\frac{\partial G(t,z)}{\partial t} = \left[(z-1)f_1\left(z\frac{\partial}{\partial z}\right) + \left(\frac{1}{z}-1\right)f_{-1}\left(z\frac{\partial}{\partial z}\right)\right] G(t,z) \tag{7}$$

which is the required partial differential equation for the generating function for given transition probabilities λ_n [$\equiv f_1(n)$] and μ_n [$\equiv f_{-1}(n)$].

Appendix C

Calculation of $P_{n,m}(t)$ for a Process with Constant Transition Probabilities and a Reflecting State

In this appendix we illustrate the method described in Section 2.2 for the calculation of $P_{n,m}(t)$ for a process with one or two boundary states. The process we consider has the state 0 as a reflecting state (so that the process always stays in the states $n \geqslant 0$), and the transition probability rates

$$\lambda_n = \nu, \qquad \mu_n = \rho \tag{1}$$

According to Eq. (2.1.3), the differential equation satisfied by the generating function is

$$\partial G(t,z)/\partial t = (z-1)(\nu-\rho/z)\,G(t,z) \tag{2}$$

with the initial condition

$$G(0,z) = z^m + \sum_{n=1}^{\infty} d_n z^{-n} \tag{3}$$

which is derived from the initial conditions on $P_{n,m}(t)$:

$$P_{n,m}(0) = \delta_{n,m}, \qquad n \geqslant 0; \qquad P_{-n,m}(0) = d_n, \qquad n > 0 \tag{4}$$

The general solution of Eq. (2) is

$$\begin{aligned} G(t,z) &= \psi(z)\exp[t(z-1)(\nu-\rho/z)] \\ &= \psi(z)\exp[-t(\nu+\rho)] \sum_{n=-\infty}^{\infty} I_n(2t(\nu\rho)^{1/2})(z(\nu/\rho)^{1/2})^n \end{aligned} \tag{5}$$

where $I_n(x)$ is the modified Bessel function of order n and $\psi(z)$ is an arbitrary function. In order to satisfy the initial condition (3), we choose

$$\psi(z) = z^m + \sum_{n=1}^{\infty} d_n z^n \tag{6}$$

To determine the set d_n, $n > 0$, we use the boundary condition (2.2.4a) for the reflecting state $n = 0$, i.e.,

$$dP_{0,m}/dt = \mu_1 P_{1,m} - \lambda_0 P_{0,m} \tag{7a}$$

Since $P_{n,m}(t)$ [which is the coefficient of z^n in the expansion of G given by Eq. (5)] satisfies the forward equation (2.0.3) for $-\infty < n < \infty$, and in particular for $n = 0$, the above condition is equivalent to the condition

$$\nu P_{-1,m}(t) - \rho P_{0,m}(t) = 0 \tag{7b}$$

Expanding Eq. (5) in powers of z, we obtain expressions for $P_{-1,m}(t)$ and $P_{0,m}(t)$ as the coefficients of z^{-1} and z^0, respectively, and then inserting the expressions into Eq. (7b), we obtain

$$0 = \rho \exp[-(\rho+\nu)t]\left\{-I_m(t)\alpha^{-m/2} - \sum_{n=1}^{\infty} d_n I_n \alpha^{n/2} + I_{m+1}\alpha^{-(m-1)/2} + \sum_{n=1}^{\infty} d_n I_{n-1}\alpha^{(n+1)/2}\right\} \tag{8}$$

where

$$\alpha = \nu/\rho, \qquad I_n = I_n(2t(\nu\rho)^{1/2}) \tag{9}$$

and $I_n(x)$ is the modified Bessel function of order n. Since Eq. (8) must hold for all t, the quantities d_n are determined uniquely by equating to zero the coefficient of each I_n, taking into account that $I_n = I_{-n}$. Thus the equations which determine d_n (derived by putting the coefficients of I_m, I_{m+1}, I_n, $n \neq m$, $m+1$, $n \geqslant 1$, and I_0, respectively, equal to zero) are

$$-\alpha^{-m/2} - d_m \alpha^{m/2} + d_{m+1}\alpha^{(m+2)/2} = 0 \tag{10a}$$

$$-d_{m+1}\alpha^{(m+1)/2} + \alpha^{-(m-1)/2} + d_{m+2}\alpha^{(m+3)/2} = 0 \tag{10b}$$

$$-d_n + d_{n+1}\alpha = 0 \tag{10c}$$

$$d_1 = 0 \tag{10d}$$

The final form of $G(t, z)$, with d_n satisfying Eqs. (10), becomes

$$G(t,z) = \exp[-t(\nu+\rho)]\left\{z^m + \alpha^{-m-1}z^{-m-1} + (1-\alpha)\sum_{j=m+2}^{\infty}(\alpha z)^{-j}\right\} \times \sum_{n=-\infty}^{\infty} \alpha^{n/2} z^n I_n(2t(\nu\rho)^{1/2}) \tag{11}$$

By definition (2.1.1) of the generating function, $P_{n,m}(t)$ for $n \geqslant 0$ is given by the coefficient of z^n in Eq. (11), i.e.,

$$P_{n,m}(t) = \alpha^{(n-m)/2}\exp[-(\nu+\rho)t]\left\{I_{n-m}(2t(\nu\rho)^{1/2}) + \alpha^{-1/2}I_{n+m+1}(2t(\nu\rho)^{1/2}) + (1-\alpha)\sum_{j=2}^{\infty}\alpha^{-j/2}I_{n+m+j}(2t(\nu\rho)^{1/2})\right\} \tag{12}$$

Appendix D

Analysis of a Process Confined between an Absorbing and a Reflecting State

In this appendix we derive the expressions for the probability density $F_{u,m}(t)$ of the time $T_{u,m}$ for absorption at the state u, and its moments $\overset{j}{M}_{u,m}$, when the random process is confined between an absorbing state u and a reflecting state l.

Our starting point is the backward equation (2.2.9) satisfied by $F_{u,m}(t)$, the boundary conditions (2.2.10) and (2.2.12), and the initial conditions (2.2.14). Taking the Laplace transform of the backward equation (2.2.9), we get the recurrence relation

$$sf_{u,m}(s) - F_{u,m}(0) = \lambda_m f_{u,m+1}(s) - (\lambda_m + \mu_m) f_{u,m}(s) + \mu_m f_{u,m-1}(s), \qquad l < m < u \tag{1}$$

satisfied by the Laplace transform $f_{u,m}(s)$ of $F_{u,m}(t)$ defined by Eq. (2.2.15). In deriving Eq. (1) we have used the relation

$$\int_0^\infty e^{-st}\left(\frac{d}{dt}F_{u,m}\right) dt = s\int_0^\infty e^{-st}F_{u,m}\, dt - F_{u,m}(0) \tag{2}$$

Similarly, the boundary conditions (2.2.10) and (2.2.12) on $F_{u,m}$ are transformed into the boundary conditions

$$sf_{u,l} - F_{u,l}(0) = \lambda_l f_{u,l+1} - \lambda_l f_{u,l} \tag{3}$$

$$f_{u,u}(s) = 0 \tag{4}$$

on $f_{u,m}(s)$. Substituting the initial conditions (2.2.14) into Eqs. (1) and (3), we get

$$\lambda_m f_{u,m+1}(s) = (s+\lambda_m+\mu_m) f_{u,m}(s) - \mu_m f_{u,m-1}(s), \qquad l < m < u-1 \tag{5a}$$

$$\lambda_{u-1} = (s+\lambda_{u-1}+\mu_{u-1}) f_{u,u-1}(s) - \mu_{u-1} f_{u,u-2}(s) \tag{5b}$$

$$\lambda_l f_{u,l+1}(s) = (s+\lambda_l) f_{u,l}(s) \tag{5c}$$

To solve these difference equations we define the quantities $Q_n(s)$ by the relation

$$Q_n(s) = f_{u,n}(s)/f_{u,l}(s) \tag{6}$$

so that Eqs. (5a) and (5c) become

$$\lambda_m Q_{m+1} = (s+\lambda_m+\mu_m) Q_m - \mu_m Q_{m-1}, \qquad l < m < u-1 \tag{7a}$$

$$Q_l = 1, \qquad Q_{l+1} = (s+\lambda_l)/\lambda_l \tag{7b}$$

and Eq. (5b) becomes

$$\lambda_{u-1}/f_{u,l}(s) = (s+\lambda_{u-1}+\mu_{u-1})\,Q_{u-1} - \mu_{u-1}\,Q_{u-2} \tag{8}$$

Equations (7) define $Q_n(s)$ recursively as a polynomial in s of degree $n-l$ for all $l \leqslant n \leqslant u-1$. For a given set of λ_n, μ_n, the expressions for $Q_n(s)$ can be used to calculate $f_{u,n}(s)$ by noting that from Eq. (7a), with $m = u-1$, and Eq. (8)

$$f_{u,l}(s) = 1/Q_u(s) \tag{9}$$

so that from Eq. (6)

$$f_{u,n}(s) = Q_n(s)/Q_u(s), \qquad l \leqslant n \leqslant u-1 \tag{10}$$

This expression for $f_{u,m}(s)$, which is the ratio of two polynomials, can be inverted to obtain $F_{u,m}(t)$ uniquely by using one of the many standard methods and extensive tables for inverting the Laplace transforms. However, for the calculation of various moments of $T_{u,m}$, defined by Eq. (2.2.19), a direct and simpler method is available which we now use.

Calculation of moments of $T_{u,m}$. Differentiating Eqs. (5) j times, letting $s = 0$, and using the defining equation (2.2.19) for the jth moment, $\overset{j}{M}_{u,m}$, we get the following recurrence relations for the moments

$$(\lambda_m+\mu_m)\overset{j}{M}_{u,m} - \lambda_m \overset{j}{M}_{u,m+1} - \mu_m \overset{j}{M}_{u,m-1} = j\,\overset{j-1}{M}_{u,m}, \qquad l < m < u-1 \tag{11a}$$

$$(\lambda_{u-1}+\mu_{u-1})\overset{j}{M}_{u,u-1} - \mu_{u-1}\overset{j}{M}_{u,u-2} = j\,\overset{j-1}{M}_{u,u-1} \tag{11b}$$

$$\lambda_l \overset{j}{M}_{u,l} - \lambda_l \overset{j}{M}_{u,l+1} = j\,\overset{j-1}{M}_{u,l} \tag{11c}$$

We can replace the boundary condition (11b) by the condition

$$\overset{j}{M}_{u,u} = 0 \tag{12}$$

if we let Eq. (11a) be valid for $m = u-1$ also. To solve Eq. (11a) recursively, we note that from Eq. (2.2.19),

$$\overset{0}{M}_{u,m} = f_{u,m}(0) = \int_0^\infty F_{u,m}(t)\,dt = 1, \qquad l \leqslant m \leqslant u-1 \tag{13}$$

The last equality follows from the observation that absorption is a certain event. Putting $j = 1$ in Eq. (11a), we obtain difference equations for the first moment $\overset{1}{M}_{u,m} \equiv M_{u,m}$. The solution of these equations can then be used to obtain $\overset{2}{M}_{u,m}$ by using Eq. (11a) with $j = 2$. This process can be continued j times to get an expression for $\overset{j}{M}_{u,m}$.

We now follow this procedure to derive the expression for $\overset{j}{M}_{u,m}$ explicitly. First we consider the difference equation

$$(\lambda_m + \mu_m) g_m - \lambda_m g_{m+1} - \mu_m g_{m-1} = 0, \qquad l < m < u \tag{14}$$

Defining

$$r_m = g_m - g_{m+1} \tag{15}$$

Eq. (14) becomes

$$\mu_m r_{m-1} = \lambda_m r_m \tag{16}$$

If we define $r_l = 1$, Eq. (16) admits the solution

$$r_m = \frac{\mu_{l+1}\mu_{l+2}\cdots\mu_m}{\lambda_{l+1}\lambda_{l+2}\cdots\lambda_m}, \qquad r_l = 1 \tag{17}$$

Once r_m is known, g_{m-1} is known, since from Eq. (15)

$$g_m = g_{m+1} + r_m = g_u + \sum_{i=m}^{u-1} r_i \tag{18a}$$

Equation (14) also admits $g_m =$ constant as a solution. Therefore, in Eq. (18a), we can choose $g_u = 0$ (g_u is a constant added to all g_m) so that

$$g_m = \sum_{i=m}^{u-1} r_i \tag{18b}$$

To determine the solution of Eq. (11a), we note that the homogeneous part of this equation is of the same form as Eq. (14), and so we take

$$\overset{j}{M}_{u,m} = \sum_{i=m}^{u-1} a_i r_i \tag{19}$$

as a trial solution of the inhomogeneous equation (11a). Substituting this trial solution into Eq. (11a) we get

$$\lambda_m r_m a_m - \mu_m a_{m-1} r_{m-1} = j \overset{j-1}{M}_{u,m} \tag{20}$$

which, using Eq. (16), becomes

$$a_m = a_{m-1} + (\lambda_m r_m)^{-1} j \overset{j-1}{M}_{u,m} = a_{l-1} + j \sum_{n=l}^{m} (\lambda_n r_n)^{-1} \overset{j-1}{M}_{u,n} \tag{21}$$

Subsituting Eq. (21) into Eq. (19), we get

$$\overset{j}{M}_{u,m} = a_{l-1} \sum_{i=m}^{u-1} r_i + j \sum_{i=m}^{u-1} r_i \sum_{n=l}^{i} (\lambda_n r_n)^{-1} \overset{j-1}{M}_{u,n} \tag{22}$$

Equation (22) is consistent with the boundary condition (12) and if, in addition, a_{l-1} is chosen to be equal to zero, then condition (11c) is satisfied since

$$\lambda_l\left(\overset{j}{M}_{u,l}-\overset{j}{M}_{u,l+1}\right) = \lambda_l j r_l(\lambda_l r_l)^{-1}\overset{j-1}{M}_{u,l} = j\overset{j-1}{M}_{u,l}$$

Thus from Eqs. (17) and (22), the required expression for the jth moment is

$$\overset{j}{M}_{u,m} = j\sum_{i=m}^{u-1}\sum_{n=l}^{i}\lambda_n^{-1}\Pi_{n+1,i}\overset{j-1}{M}_{u,n} = j\sum_{n=l}^{u-1}\lambda_n^{-1}\overset{j-1}{M}_{u,n}\sum_{i=\max(m,n)}^{u-1}\Pi_{n+1,i} \tag{23}$$

where

$$\Pi_{i,j} = \frac{\mu_i\mu_{i+1}\cdots\mu_j}{\lambda_i\lambda_{i+1}\cdots\lambda_j}, \qquad i \leqslant j, \qquad \Pi_{i,i-1} = 1 \tag{24}$$

$\Pi_{i,j}$ is the ratio of the transition probability rates from state j to state $i-1$ to that from state i to state $j+1$.

Usually one is interested mainly in the first moment $M_{u,m}$ and variance $V_{u,m}$ of the absorption time, rather than the higher moments. Therefore we will now use Eq. (23) to derive explicit expressions for $M_{u,m}$ and $V_{u,m}$.

Letting $j = 1$ in Eq. (23) and using Eq. (13), we get

$$M_{u,m} = \sum_{i=m}^{u-1}\sum_{n=l}^{i}\lambda_n^{-1}\Pi_{n+1,i} \tag{25}$$

To calculate variance we note that since $\{T_{i+1,i}\}$ are independent random variables,

$$T_{u,m} = \sum_{i=m}^{u-1} T_{i+1,i} \tag{26}$$

Therefore

$$M_{u,m} = \sum_{i=m}^{u-1} M_{i+1,i} \tag{27}$$

$$V_{u,m} = \overset{2}{M}_{u,m} - (M_{u,m})^2 = \sum_{i=m}^{u-1} V_{i+1,i} \tag{28}$$

If we regard state $(i+1)$ as an absorbing state and consider a process which starts from the ith state, from Eq. (23) we obtain

$$\overset{j}{M}_{i+1,i} = j\sum_{n=l}^{i}\lambda_n^{-1}\Pi_{n+1,i}\overset{j-1}{M}_{i+1,n} \tag{29}$$

so that

$$V_{i+1,i} = \overset{2}{M}_{i+1,i} - (M_{i+1,i})^2 = 2\sum_{n=l}^{i}\lambda_n^{-1}\Pi_{n+1,i}\sum_{r=n}^{i}M_{r+1,r} - (M_{i+1,i})^2 \tag{30}$$

Substituting this equation into Eq. (28) and changing the order of summation we obtain

$$V_{u,m} = \sum_{i=m}^{u-1} \left\{ 2 \sum_{n=l}^{i-1} \Pi_{n+1,i} (M_{n+1,n})^2 + (M_{i+1,i})^2 \right\} \tag{31}$$

Appendix E

Analysis of a Process Confined between Two Absorbing States

In this appendix we derive the expressions for the probability density $F_{u,m}(t)$ $[F_{l,m}(t)]$ of the time $T_{u,m}$ $(T_{l,m})$ for absorption at the state u (l), and its moments $\overset{j}{M}_{u,m}$ $\left(\overset{j}{M}_{l,m}\right)$, when the random process is confined between two absorbing states l and u. The corresponding quantities for time to absorption T_m (either at the state l or at state u) can be derived by using Eqs. (2.2.25).

Since as noted in Section 2.2B, $F_{u,m}(t)$ for $l+1 < m < u-1$ satisfies Eq. (2.2.9), its Laplace transform will satisfy Eq. (D.1) for $l+1 < m < u-1$, the Laplace transform of Eq. (2.2.9). Using the initial condition (2.2.27a), Eq. (D.1) becomes

$$\lambda_m f_{u,m+1}(s) = (s+\lambda_m+\mu_m) f_{u,m}(s) - \mu_m f_{u,m-1}(s), \qquad l+1 < m < u-1 \tag{1}$$

The boundary conditions (2.2.26) imply the boundary conditions

$$f_{u,l}(s) = 0, \quad f_{u,u}(s) = 0 \tag{2}$$

which, when substituted together with the initial conditions (2.2.27) into Eq. (D.1) for $m = u-1$ and $m = l+1$, give

$$\lambda_{u-1} = (s+\lambda_{u-1}+\mu_{u-1}) f_{u,u-1} - \mu_{u-1} f_{u,u-2} \tag{3}$$

$$\lambda_{l+1} f_{u,l+2} = (s+\lambda_{l+1}+\mu_{l+1}) f_{u,l+1} \tag{4}$$

Introducing the functions

$$q_n(s) = f_{u,n}(s)/f_{u,l+1}(s) \tag{5}$$

$$q_{l+1}(s) = 1 \tag{6a}$$

into Eqs. (4) and (1), we obtain

$$q_{l+2}(s) = (s+\mu_{l+1}+\lambda_{l+1})/\lambda_{l+1} \tag{6b}$$

$$\lambda_m q_{m+1} = (s+\lambda_m+\mu_m) q_m - \mu_m q_{m-1}, \qquad l+1 < m < u-1 \tag{7}$$

These recurrence relations imply that $q_n(s)$ is a polynomial in s of degree

$n-l-1$. For a given set of λ_n, μ_n, the form of $q_n(s)$, as derived by solving Eqs. (6) and (7), can be used to calculate $f_{u,n}(s)$ since from Eq. (3)

$$\lambda_{u-1}/f_{u,l+1} = (s+\lambda_{u-1}+\mu_{u-1})\,q_{u-1} - \mu_{u-1}\,q_{u-2} \tag{8}$$

and according to Eq. (7), with $m = u-1$,

$$f_{u,l+1} = 1/q_u(s) \tag{9}$$

so that from Eq. (5)

$$f_{u,n}(s) = q_n(s)/q_u(s), \qquad l+1 \leqslant n \leqslant u-1 \tag{10}$$

Thus $f_{u,n}$ is given by a ratio of two polynomials and in principle can be Laplace inverted to give $F_{u,n}(t)$.

The recurrence relations (1) with boundary conditions (2) for $f_{u,m}$ can be used to obtain $R_{u,m}$, the probability of ever reaching the state u, i.e., the probability of the process being absorbed at state u and not at state l. By definition this probability is given by

$$R_{u,m} = R_{u,m}(l) = \int_0^\infty F_{u,m}(t)\,dt = f_{u,m}(0) \tag{11}$$

Therefore from Eqs. (1), (3), and (4), $R_{u,m}$ satisfies the recurrence relations

$$\lambda_m R_{u,m+1} = (\lambda_m+\mu_m)\,R_{u,m} - \mu_m R_{u,m-1}, \qquad l+1 < m < u-1 \tag{12a}$$

$$\lambda_{u-1} = (\lambda_{u-1}+\mu_{u-1})\,R_{u,u-1} - \mu_{u-1}\,R_{u,u-2} \tag{12b}$$

$$\lambda_{l+1}\,R_{u,l+2} = (\lambda_{l+1}+\mu_{l+1})\,R_{u,l+1} \tag{12c}$$

Equation (12a) is similar to Eq. (D.14) and hence can be treated likewise. The general solution of Eq. (12a) is

$$R_{u,m} = a\sum_{i=m}^{u-1} r_i + b, \qquad l+1 \leqslant m \leqslant u-1 \tag{13}$$

where a and b are constants to be determined by the boundary conditions (12b) and (12c). From Eq. (12b), $b = 1$, and from Eq. (12c),

$$a = -\left[1+\sum_{l-l+1}^{u-1} r_i\right]^{-1} = -\left[\sum_{i-l}^{u-1} \Pi_{l+1,i}\right]^{-1} \tag{14}$$

where the last result follows from Eqs. (D.17) and (D.24). Therefore, finally,

$$R_{u,m} = 1 - \sum_{i=m}^{u-1} \Pi_{l+1,i} \Big/ \sum_{i=l}^{u-1} \Pi_{l+1,i} = \sum_{i=l}^{m-1} \Pi_{l+1,i} \Big/ \sum_{i=l}^{u-1} \Pi_{l+1,i} \tag{15}$$

Since the absorption is a certain event, the probability of absorption at state l (and not at state u) is

$$R_{l,m} = 1 - R_{u,m} = \sum_{i=m}^{u-1} \Pi_{l+1,i} \Big/ \sum_{i=l}^{u-1} \Pi_{l+1,i} \tag{16}$$

The recurrence relations (1), (3), and (4) for $f_{u,m}$, together with the defining equation (2.2.19) for the jth moment of $T_{u,m}$, give the recurrence relations for $\overset{j}{M}_{u,m}$. These relations, obtained by differentiating Eqs. (1), (3), and (4) j times and setting $s = 0$, are

$$(\lambda_m + \mu_m)\overset{j}{M}_{u,m} - \lambda_m \overset{j}{M}_{u,m+1} - \mu_m \overset{j}{M}_{u,m-1} = j\overset{j-1}{M}_{u,m}, \qquad l+1 < m < u-1 \tag{17a}$$

$$(\lambda_{u-1} + \mu_{u-1})\overset{j}{M}_{u,u-1} - \mu_{u-1}\overset{j}{M}_{u,u-2} = j\overset{j-1}{M}_{u,u-1} \tag{17b}$$

$$(\lambda_{l+1} + \mu_{l+1})\overset{j}{M}_{u,l+1} - \lambda_{l+1}\overset{j}{M}_{u,l+2} = j\overset{j-1}{M}_{u,l+1} \tag{17c}$$

where, from Eqs. (2.2.19) and (11),

$$\overset{0}{M}_{u,m} = f_{u,m}(0) = R_{u,m}, \qquad l+1 \leqslant m \leqslant u-1 \tag{18}$$

Since Eq. (17a) is identical to Eq. (D.11a) for $l+1 < m < u-1$, the solution (D.23) of Eq. (D.11a), with l replaced by $l+1$, is a particular solution of Eq. (17a). This solution also satisfies Eq. (17b), since (17b) is identical to Eq. (D.11b), satisfied by the solution (D.23). In order to satisfy the boundary condition (17c), we add to this particular solution a general solution of the homogeneous part of Eq. (17a) [which is identical to Eq. (D.14)]. Therefore we take

$$\overset{j}{M}_{u,m} = j\sum_{i=m}^{u-1}\sum_{n=l+1}^{i} \lambda_n^{-1}\Pi_{n+1,i}\overset{j-1}{M}_{u,n} + a\sum_{i=m}^{u-1}\Pi_{l+1,i} \tag{19}$$

as a trial solution of Eq. (17) and determine a by the condition (17c). Substituting Eq. (19) into Eq. (17c) and solving for a, we get

$$\overset{j}{M}_{u,m} = j\sum_{i=m}^{u-1}\sum_{n=l+1}^{i} \lambda_n^{-1}\Pi_{n+1,i}\overset{j-1}{M}_{u,n} - jR_{l,m}\sum_{i=l+1}^{u-1}\sum_{n=l+1}^{i} \lambda_n^{-1}\Pi_{n+1,i}\overset{j-1}{M}_{u,n} \tag{20}$$

the required expression for the jth moment. In particular, the first moment ($j = 1$) is given, in view of Eq. (11), by

$$M_{u,m} = \sum_{i=m}^{u-1}\sum_{n=l+1}^{i} \lambda_n^{-1}\Pi_{n+1,i}R_{u,n} - R_{l,m}\sum_{i=l+1}^{u-1}\sum_{n=l+1}^{i} \lambda_n^{-1}\Pi_{n+1,i}R_{u,n} \tag{21}$$

By similar arguments, the moments of $T_{l,m}$, $M_{l,m}$ satisfy the same recurrence relations as (17). Therefore, the expression for $M_{l,m}$ is found to be the same as Eq. (20) with $\overset{j-1}{M}_{u,n}$ in the right-hand side of Eq. (20) replaced by $\overset{j-1}{M}_{l,n}$ and $R_{u,n}$ in the right-hand side of Eq. (21) replaced by $R_{l,n}$.

Appendix F

Explicit Calculation of $f_{u,m}(s)$ for a Process with Linear Transition Probability Rates, Confined between Two Absorbing States

In this appendix we illustrate the method for calculating $f_{u,m}(s)$, discussed in Section 2.2.B and Appendix E, for a process confined between two absorbing states $n = 0$ and $n = u$, with transition probability rates

$$\lambda_n = \lambda n, \qquad \mu_n = \mu n, \qquad \gamma \equiv \lambda/\mu, \qquad 0 \leqslant n \leqslant u-1, \quad \lambda_u = \mu_u = 0 \quad (1)$$

From Eq. (2.2.28), $f_{u,m}(s)$ is given by

$$f_{u,m}(s) = q_m(s)/q_u(s), \qquad l+1 \leqslant m \leqslant u-1 \quad (2)$$

where, from Eqs. (2.2.17) and (2.2.29), the q's satisfy the recurrence relations

$$q_1 = 1 \quad (3a)$$

$$q_2 = 1 + \gamma^{-1} + s/\lambda \quad (3b)$$

and

$$q_{m+1}(s) = \left(1 + \gamma^{-1} + \frac{s}{m\lambda}\right) q_m(s) - \gamma^{-1} q_{m-1}(s) \quad (3c)$$

One method of solving these recurrence relations is to introduce a generating function

$$\phi(z) = \sum_{n=1}^{\infty} q_n[s(\lambda-\mu)]\, z^{n-1} \quad (4)$$

In view of Eqs. (3), $\phi(z)$ satisfies the differential equation

$$\frac{\phi'(z)}{\phi(z)} = \frac{1+\gamma+s(\gamma-1)-2z}{(1-z)(\gamma-z)} \quad (5a)$$

with

$$\phi(z=0) = q_1(s) = 1 \quad (5b)$$

The solution to this equation is

$$\phi(z) = (1-z)^{-s-1}(1-\gamma^{-1}z)^{s-1} \quad (6)$$

Expanding $\phi(z)$ in powers of z, we obtain

$$q_n((\lambda-\mu)s) = \sum_{k=0}^{n-1} \frac{(s+1)_{n-k-1}}{(n-k-1)!} \frac{(s-k)_k}{k!} (-\gamma)^{-k} \quad (7)$$

where

$$(a)_0 = 1, \qquad (a)_k = a(a+1)(a+2)\cdots(a+k-1) \quad (8)$$

From Eq. (7) we confirm the result of Section 2.2B that $q_n((\lambda-\mu)s)$ is a polynomial in s of degree $n-1$. Substituting Eq. (7) into Eq. (2), we get for $f_{u,m}(s)$ the expression

$$f_{u,m}(s) = \left[\sum_{k=0}^{m-1} \frac{(s\beta+1)_{m-k-1}(s\beta-k)_k}{(m-k-1)!k!}(-\gamma)^{-k}\right] \times \left[\sum_{k=0}^{u-1} \frac{(s\beta+1)_{u-k-1}}{(u-k-1)!}\frac{(s\beta-k)_k}{k!}(-\gamma)^{-k}\right]^{-1}, \quad l+1 \leqslant m \leqslant u-1 \tag{9}$$

where

$$\beta^{-1} = \lambda - \mu \tag{10}$$

Appendix G

Calculation of $P(x\,|\,y,t)$ *for the OU and Wiener Processes*

In this appendix we illustrate the method presented in Section 3.0 for the calculation of $P(x\,|\,y,t)$, for four cases:

(a) the unrestricted OU process,
(b) the OU process restricted to $x \geqslant 0$ by a reflecting boundary at $x = 0$,
(c) the Weiner process confined by one reflecting and one absorbing boundary, and
(d) the Weiner process confined by two absorbing boundaries.

(a) The OU process is characterized by $a(x) = -rx$, $b(x) = \sigma^2$. The FP equation for this process is

$$\frac{\partial P}{\partial t} = r\frac{\partial}{\partial x}(xP) + \tfrac{1}{2}\sigma^2\frac{\partial^2 P}{\partial x^2} \tag{1}$$

and can be transformed by the transformation (3.0.32)

$$z = x/\sigma, \qquad z_0 = y/\sigma, \qquad g(z\,|\,z_0,t) = \sigma p(x\,|\,y,t) \tag{2}$$

into an equation of the form (3.0.27):

$$\frac{\partial g}{\partial t} = \frac{\partial(rzg)}{\partial z} + \frac{1}{2}\frac{\partial^2 g}{\partial z^2} \tag{3}$$

To solve Eq. (3) we consider the eigenvalue equation (3.0.37) which, in view of Eqs. (3.0.38) and (3), is of the form

$$d^2\psi/dz^2 + (E+r-r^2z^2)\psi = 0 \tag{4}$$

For the unrestricted process $P(\pm\infty\,|\,y,t) = 0$ and according to Eq. (2), the eigenfunctions of Eq. (4) satisfy the boundary conditions

$$\psi(\pm\infty) = 0 \tag{5}$$

These conditions are in agreement with the nature of the boundaries $A = -\infty$, $B = +\infty$, which are natural boundaries by the classification of Section 3.1. The eigenvalues and normalized eigenfunctions of the eigenvalue problem (4), (5) are (Abramowitz and Stegun, 1964, p. 789)

$$E_n = 2nr \tag{6}$$

$$\psi_n(z) = H_n(r^{1/2}z)\exp(-rz^2/2)/[2^n n!(\pi/r)^{1/2}]^{1/2} \tag{7}$$

where $H_n(z)$ are the Hermite polynomials defined by

$$H_n(z) = (-1)^n \exp(z^2)(d/dz)^n \exp(-z^2) \tag{8}$$

so that

$$H_0 = 1, \qquad H_1 = 2z, \qquad H_2 = 4z^2, \qquad H_3 = 8z^2 - 12z, \quad \ldots \tag{9}$$

Using Eqs. (3.0.41) and (3.0.36a), which relates $g(z\,|\,z_0,t)$ with the eigenvalues (6) and eigenfunctions (7), we obtain

$$g(z\,|\,z_0,t) = \exp(-rz^2)(r/\pi)^{1/2}\sum_{n=0}^{\infty}\frac{H_n(r^{1/2}z_0)\,H_n(r^{1/2}z)}{2^n n!}\exp\{-nrt\} \tag{10}$$

and from Eq. (2)

$$P(x\,|\,y,t) = \exp(-\theta^2x^2)(\theta/\pi^{1/2})\sum_{n=0}^{\infty}\frac{H_n(\theta x)\,H_n(\theta y)}{2^n n!}\exp(-nrt), \qquad \theta \equiv r^{1/2}/\sigma \tag{11}$$

Using Mehler's formula for the sum (Titchmarsh, 1937),

$$\sum_{n=0}^{\infty}\frac{H_n(x)\,H_n(y)}{2^n n!\,\pi^{1/2}}\alpha^n = \frac{1}{\pi^{1/2}(1-\alpha^2)^{1/2}}\exp\left\{x^2 - \frac{(x-y\alpha)^2}{1-\alpha^2}\right\} \tag{12}$$

we can rewrite Eq. (11) in the form

$$P(x\,|\,y,t) = \frac{1}{(2\pi)^{1/2}V(t)}\exp\{-[x-m(t)]^2/2V^2(t)\} \tag{13a}$$

where

$$m(t) = y\exp(-rt), \qquad V^2(t) = (\sigma^2/2r)[1-\exp(-2rt)] \tag{13b}$$

(b) When the OU process is restricted by a reflecting boundary at $x = 0$, the eigenvalue problem to be solved consists of Eq. (4) and the boundary conditions

$$\psi(+\infty) = 0 \tag{14a}$$

$$\frac{d\psi}{dz}(0) = 0 \tag{14b}$$

where condition (14b) is derived from condition (3.2.8) and the transformation (2). The functions given by Eq. (7) satisfy condition (14a), and those with an even index are also symmetric, $\psi_n(-z) = \psi_n(z)$, and therefore satisfy condition (14b). The solution in this case is given in terms of a sum similar to the one in Eq. (11), but where the index n takes only even values,

$$P(x \mid y, t) = \exp(-\theta^2 x^2)(\theta/\pi^{1/2}) \sum_{n=0}^{\infty} \frac{H_{2n}(\theta x)\, H_{2n}(\theta y)}{4^n (2n)!} \exp(-2nrt),$$
$$\theta = r^{1/2}/\sigma \quad (15)$$

Thus, $P(x \mid y, t)$ of Eq. (15) is the even part of the function $P(x \mid y, t)$ of Eq. (11), denoted in the following by $P_a(x \mid y, t)$, i.e.,

$$P(x \mid y, t) = [P_a(x \mid y, t) + P_a(-x \mid y, t)]/2 \tag{16}$$

and is given explicitly by

$$P(x \mid y, t) = \frac{1}{(8\pi)^{1/2} V(t)} [\exp\{-[x - m(t)]^2/2V^2(t)\}$$
$$+ \exp\{-[x + m(t)]^2/2V^2(t)\}] \quad (17)$$

(c) The Weiner process is characterized by $a(x) = r$, $b(x) = \sigma^2$. The FP equation for this process is

$$\frac{\partial P}{\partial t} = -r \frac{\partial P}{\partial x} + \tfrac{1}{2}\sigma^2 \frac{\partial^2 P}{\partial x^2} \tag{18}$$

and can be transformed by the transformation (2) into

$$\frac{\partial g}{\partial t} = -\frac{r}{\sigma} \frac{\partial g}{\partial z} + \frac{1}{2} \frac{\partial^2 g}{\partial z^2}, \qquad z = \frac{x}{\sigma} \tag{19}$$

In this case, the eigenvalue equation of the form (3.0.37) is

$$\frac{d^2\psi}{dz^2} + \left(E - \frac{r^2}{\sigma^2}\right)\psi = 0 \tag{20}$$

The boundary conditions imposed on ψ, when the boundary A is reflecting and the boundary B is absorbing, are [see Eqs. (3.2.7) and (3.2.8)]

$$\frac{d\psi}{dz}\left(\frac{A}{\sigma}\right) = 0, \qquad \psi\left(\frac{B}{\sigma}\right) = 0 \tag{21}$$

Equation (20) has two independent solutions:

$$\psi_1(z) = \sin \lambda z, \qquad \psi_2(z) = \cos \lambda z, \qquad \lambda = [E-(r/\sigma)^2]^{1/2} \tag{22}$$

To satisfy the boundary conditions (21), we take a linear combination of these two solutions

$$\psi(z) = \alpha \sin \lambda z + \beta \cos \lambda z \tag{23}$$

such that

$$\alpha \cos(\lambda A/\sigma) - \beta \sin(\lambda A/\sigma) = 0 \tag{24a}$$

$$\alpha \sin(\lambda B/\sigma) + \beta \cos(\lambda B/\sigma) = 0 \tag{24b}$$

A pair α, β which satisfies Eqs. (24) exists if the determinant of this system of equations is zero, i.e., if

$$\cos(\lambda A/\sigma)\cos(\lambda B/\sigma) + \sin(\lambda A/\sigma)\sin(\lambda B/\sigma) = 0 \tag{25a}$$

or, equivalently, if

$$\cos[\lambda(B-A)/\sigma] = 0 \tag{25b}$$

Therefore the eigenvalues of the problem are determined by the condition

$$\frac{\lambda}{\sigma}(B-A) = \frac{\pi}{2} + n\pi \tag{26a}$$

and are given by

$$E_n = \left(\frac{r}{\sigma}\right)^2 + \left(\frac{2n+1}{2L}\pi\sigma\right)^2, \qquad L = B - A \tag{26b}$$

The corresponding eigenfunctions are given, according to Eqs. (23) and (24), by

$$\psi_n(z) = \beta_n\left[\cos\left\{\frac{(2n+1)\pi}{2L}\sigma z\right\} + \tan\left\{\frac{(2n+1)\pi}{2L}A\right\}\sin\left\{\frac{(2n+1)\pi}{2L}\sigma z\right\}\right] \tag{27}$$

where the β_n are chosen to normalize the $\psi_n(z)$. To simplify the discussion, we take $A = 0$. The eigenfunctions in this case are

$$\psi_n(z) = \left(\frac{2\sigma}{B}\right)^{1/2} \cos\frac{(2n+1)\pi}{2B}\sigma z, \qquad n \geqslant 0 \tag{28}$$

and by the relations (3.0.41) and (3.0.36a)

$$g(z\,|\,z_0,t) = \frac{2\sigma}{B}\exp\left[\frac{r}{\sigma}(z-z_0)\right]\sum_{n=0}^{\infty}\cos\left[\frac{(2n+1)\pi}{2B}\sigma z\right] \times \cos\left[\frac{(2n+1)\pi}{2B}\sigma z_0\right]\exp\left(-\frac{E_n t}{2}\right) \tag{29}$$

Transforming back to $P(x\,|\,y,t)$ by using Eq. (2), we get

$$P(x\,|\,y,t) = \frac{2}{B}\exp\left[\frac{r}{\sigma^2}\left(x-y-\frac{rt}{2}\right)\right]\sum_{n=0}^{\infty}\cos\left[\frac{(2n+1)\pi}{2B}x\right] \times \cos\left[\frac{(2n+1)\pi}{2B}y\right]\exp\left\{-\left(\frac{2n+1}{2B}\pi\sigma\right)^2\frac{t}{2}\right\} \tag{30}$$

(d) When the Weiner process is confined between two absorbing boundaries, the eigenfunction (23) must vanish on the boundaries. Therefore, the coefficients α and β are chosen to meet the conditions

$$\alpha\sin(\lambda A/\sigma) + \beta\cos(\lambda A/\sigma) = 0 \tag{31a}$$

$$\alpha\sin(\lambda B/\sigma) + \beta\cos(\lambda B/\sigma) = 0 \tag{31b}$$

There is a nontrivial solution to Eqs. (31) if

$$\sin(\lambda A/\sigma)\cos(\lambda B/\sigma) - \cos(\lambda A/\sigma)\sin(\lambda A/\sigma) = 0 \tag{32a}$$

or if

$$\sin[\lambda(B-A)/\sigma] = 0 \tag{32b}$$

The eigenvalues defined by (32b) are

$$(\lambda/\sigma)L = n\pi \qquad \text{or} \qquad E_n = (r/\sigma)^2 + (\sigma n\pi/L)^2, \qquad L = B - A \tag{33}$$

From Eqs. (31a) and (23),

$$\psi_n(t) \sim \{\cos(\lambda A/\sigma)\sin\lambda z - \sin(\lambda A/\sigma)\cos(\lambda A/\sigma)\} = \sin\lambda(z-A/\sigma)$$

and after normalization, we obtain

$$\psi_n(z) = [2\sigma/L]^{1/2}\sin\lambda(z-A/\sigma) \tag{34}$$

Using relations (3.0.41), (3.0.36a), and (2), we can write $P(x\,|\,y,t)$ as

$$P(x\,|\,y,t) = \left(\frac{2}{L}\right)\exp\left\{\frac{r}{\sigma^2}(x-y-rt/2)\right\}\sum_{n=1}^{\infty}\sin\frac{n\pi}{L}(x-A) \times \sin\frac{n\pi}{L}(y-A)\exp\left\{-\frac{\sigma^2 n^2\pi^2}{2L^2}t\right\} \tag{35}$$

Appendix H

Derivation of Moments of First Passage Time for a Process in a Continuous State Space

In this appendix we derive the explicit expressions for the various moments of $T(B|y)$, the first pasasge time for the random variable to take the value B, given that initially it had the value y.

As was shown in Sections 3.2A and 3.2B, the jth moment $M_j(B|y)$ satisfies the differential equation (3.2.18)

$$\tfrac{1}{2}b(y)\frac{\partial^2 M_j(B|y)}{\partial y^2} + a(y)\frac{\partial M_j(B|y)}{\partial y} + jM_{j-1}(B|y) = 0 \tag{1}$$

and the boundary conditions [Eqs. (3.2.20) and (3.2.45)]

$$M_j(B|B) = 0 \tag{2}$$

$$M_j(B|A) = 0 \qquad A\text{-absorbing} \tag{3a}$$

$$\frac{\partial}{\partial y}M_j(B|y)\bigg|_{y=A} = 0 \qquad A\text{-reflecting} \tag{3b}$$

To find the solution of Eq. (1) under the various boundary conditions, we first consider the homogeneous equation derived from Eq. (1) by putting $M_{j-1}(B|y) = 0$,

$$\tfrac{1}{2}b(y)h''(y) + a(y)h'(y) = 0, \qquad h'(y) \equiv dh/dy \tag{4}$$

The solution to this equation is

$$h(y) = \int^y \pi(\eta)\,d\eta \tag{5a}$$

where

$$\pi(\eta) = \exp\left\{-2\int^\eta d\xi\; a(\xi)\Big/b(\xi)\right\} \tag{5b}$$

Let us choose

$$M_j(B|y) = \int^y \pi(\eta)r(\eta)\,d\eta \tag{6}$$

as the trial solution, and let us substitute it into Eq. (1) to get an equation for $r(y)$,

$$\tfrac{1}{2}b(y)\pi(y)r'(y) = -jM_{j-1}(B|y) \tag{7}$$

which has the solution

$$r(y) = -2j\int^y d\eta\; M_{j-1}(B|\eta)\Big/b(\eta)\pi(\eta) \tag{8}$$

Equation (6) together with Eq. (8) defines the particular solution of Eq. (1) to which we add any solution of the homogeneous equation (4) to obtain finally the solution of Eq. (1) as

$$M_j(B|y) = 2j \int_y d\eta\, \pi(\eta) \int^{\eta} d\xi\, M_{j-1}(B|\xi) \Big/ b(\xi)\pi(\xi) + C \int_y \pi(\eta)\, d\eta \qquad (9)$$

The unspecified limits of integration and the constant C are determined by the boundary conditions (2) and (3).

To satisfy the boundary condition (2) at B, we take B as the upper limit of the integrals in Eq. (9), which then becomes

$$M_j(B|y) = 2j \int_y^B d\eta\, \pi(\eta) \int^{\eta} d\xi\, M_{j-1}(B|\xi) \Big/ b(\xi)\pi(\xi) + C \int_y^B \pi(\eta)\, d\eta \qquad (10)$$

When A is a reflecting boundary, the boundary condition (3b) can be satisfied by taking $C = 0$ and the unspecified lower limit in Eq. (10) as A. The final expression for $M_j(B|y)$ for a process confined between a reflecting boundary A and an absorbing boundary B becomes

$$M_j(B|y) = 2j \int_y^B d\eta\, \pi(\eta) \int_A^{\eta} d\xi\, M_{j-1}(B|\xi) \Big/ b(\xi)\pi(\xi) \qquad (11)$$

When A is an absorbing boundary, the boundary condition (3a) can be satisfied by taking

$$C = -2j \int_A^B d\eta\, \pi(\eta) \int^{\eta} \frac{M_{j-1}(B|\xi)}{b(\xi)\pi(\xi)}\, d\xi \Big/ \int_A^B \pi(\eta)\, d\eta \qquad (12)$$

The unspecified lower limits in Eqs. (10) and (12) can be taken as A, and the resulting expression for $M_j(B|y)$ is

$$M_j(B|y) = 2j \Bigg[\int_y^B d\eta\, \pi(\eta) \int_A^{\eta} \frac{M_{j-1}(B|\xi)}{b(\xi)\pi(\xi)}\, d\xi - \frac{\int_y^B \pi(\eta)\, d\eta}{\int_A^B \pi(\eta)\, d\eta} \int_A^B d\eta\, \pi(\eta) \int_A^{\eta} \frac{M_{j-1}(B|\xi)}{b(\xi)\pi(\xi)}\, d\xi \Bigg] \qquad (13)$$

Recalling the expression for $R(A|y)$ when A is an absorbing boundary [Eq. (3.2.44)], we can rewrite Eq. (13) in the form (3.2.46).

Index